Modern Polarographic Methods in Analytical Chemistry

MONOGRAPHS IN ELECTROANALYTICAL CHEMISTRY AND ELECTROCHEMISTRY

Consulting Editor

Allen J. Bard
Department of Chemistry
University of Texas
Austin, Texas

Electrochemistry at Solid Electrodes, *Ralph N. Adams*

Electrochemical Reactions in Nonaqueous Systems, *Charles K. Mann, Karen Barnes*

Electrochemistry of Metals and Semiconductors: The Application of Solid State Science to Electrochemical Phenomena, *Ashok K. Vijh*

Modern Polarographic Methods in Analytical Chemistry, *A. M. Bond*

Other Volumes in Preparation

Modern Polarographic Methods in Analytical Chemistry

A. M. BOND

Division of Chemical and Physical Sciences
Deakin University
Waurn Ponds, Victoria
Australia

CRC Press
Taylor & Francis Group
Boca Raton London New York

CRC Press is an imprint of the
Taylor & Francis Group, an **informa** business

First published 1980 by Marcel Dekker, Inc.

Published 2019 by CRC Press
Taylor & Francis Group
6000 Broken Sound Parkway NW, Suite 300
Boca Raton, FL 33487-2742

First issued in paperback 2019

ISBN-13: 978-0-367-45204-9 (pbk)
ISBN-13: 978-0-8247-6849-2 (hbk)

Visit the Taylor & Francis Web site at
http://www.taylorandfrancis.com

and the CRC Press Web site at
http://www.crcpress.com

Library of Congress Cataloging in Publication Data

Bond, Alan Maxwell, [Date]
    Modern polarographic methods in analytical chemistry.

    (Monographs in electroanalytical chemistry and
electrochemistry)
    Includes index.
    1. Polarograph and polarography. I. Title.
QD115.B6   543'.0872   80-302
ISBN 0-8247-6849-3

To Tunde, Stepher, and Andrew

CONTENTS

Foreword, Allen J. Bard   ix

Preface   xi

1 The Renaissance of Polarography   1
   1.1 History   1
   1.2 Aims and Scope   5
   Notes   7

2 Theoretical, Instrumental, and General Considerations
   Relevant to the Systematic Use of Polarographic
   Methodology   8
   2.1 The DC Polarographic Experiment: Simple Concepts   8
   2.2 Electrode Processes and Their Nature   12
        2.2.1 Electrochemical and Chemical Reversibility   15
        2.2.2 Classification of Reactions   35
        2.2.3 Adsorption, Film Formation, and Other
               Surface Phenomena   43
   2.3 Instrumentation   44
        2.3.1 The Two-Electrode Polarograph   45
        2.3.2 The Three-Electrode Polarograph   47
        2.3.3 General Recommendations in Instrumentation   67
   2.4 Faradaic and Nonfaradaic Processes   71
        2.4.1 Faradaic Current   71
        2.4.2 Charging and Capacitance of an Electrode:
               The Charging Current   72
        2.4.3 Migration Current   76
   2.5 The Modern Polarographic Experiment   78
   Notes   79

3 Conventional DC Polarography: Limitations and Uses   84
    3.1 Difficulties Encountered in Conventional
         DC Polarography   84
         3.1.1 Resolution, Wave Shape, and Readout   84
         3.1.2 Time Taken to Record a DC Polarogram   85
         3.1.3 Sensitivity and Charging Current   88
         3.1.4 Automation of Instrumentation and Readout   90
         3.1.5 Dependence of Drop Time on Potential   91
    3.2 Systematic Use of DC Polarographic Theory   91
         3.2.1 Diffusion-Controlled Limiting Currents   92
         3.2.2 Equation for DC Polarogram with
                Diffusion-Controlled Limiting Current   96
         3.2.3 Kinetically Controlled Limiting Currents   106
         3.2.4 Limiting Currents Controlled by Adsorption
                or Other Surface Phenomena   112
         3.2.5 Charging Current in DC Polarography   117
    3.3 DC Polarography as an Absolute or
         Comparative Method of Analysis   118
    Notes   119

4 Advances in DC Polarography   122
    4.1 Rapid DC Polarography with Short Controlled
         Drop Times   122
         4.1.1 Rapidly Dropping Mercury Electrodes
                and Their Theory   128
         4.1.2 Recording, Damping, and Maximum Scan
                Rates of Potential   129
         4.1.3 Reproducibility   130
         4.1.4 Dependence of i_d on Various Parameters   130
         4.1.5 Determination of Numbers of Electrons
                in Consecutive Electrode Processes   135
         4.1.6 Other Applications   137
    4.2 Spinning Dropping Mercury Electrodes   138
    4.3 Current-Sampled DC Polarography   139
    4.4 Current-Averaged DC Polarograms   142
    4.5 Derivative DC Polarography   145
    4.6 Subtractive DC Polarography   157
    4.7 Linear Charging Current Compensation   160
    4.8 Charging Current Compensation with
         Superimposed AC Signal   163
    4.9 Static Mercury Drop Electrodes
         and Minimization of Charging Current   165
    Notes   166

5. Linear Sweep Voltammetry and Related Techniques   169
    5.1 Nomenclature   169
    5.2 Synchronization of Sweep Time
         at a Dropping Mercury Electrode   172
    5.3 Solid Electrodes   176

5.4   Theory for Faradaic Processes   178
        5.4.1   Reversible Charge Transfer   182
        5.4.2   Nonreversible Charge Transfer   185
        5.4.3   Effect of Uncompensated Resistance   188
        5.4.4   Influence of Coupled Chemical Reactions   190
        5.4.5   Influence of Adsorption   192
        5.4.6   Diagnostic Criteria for Various
                 Electrode Mechanisms   194
5.5   Linear Sweep Voltammetry in Analytical Applications   197
5.6   The Charging Current   204
5.7   Sensitivity   206
5.8   Comparisons with DC Polarography   207
5.9   Advances in Voltammetry   209
        5.9.1   Staircase Voltammetry   209
        5.9.2   Derivative Voltammetry   212
        5.9.3   Convolution or Semi-Integral Techniques   216
        5.9.4   Deconvolution or Semidifferential Techniques   219
        5.9.5   Subtractive Voltammetry   223
        5.9.6   Interrupted Voltage Ramps   225
5.10   Voltammetry with Forced Convection   226
        5.10.1   Stationary Electrodes in Flowing Solutions   227
        5.10.2   Rotated Disk Electrodes   228
        5.10.3   Rotating Ring-Disk Electrodes   228
Notes   231

6 Pulse Polarography   236
6.1   Nomenclature   236
6.2   Theory and Analytical Implications   243
        6.2.1   Normal Pulse Polarography   243
        6.2.2   Differential Pulse Polarography   249
        6.2.3   Influence of Resistance   253
6.3   Characterization of Electrode Reversibility
        by Pulse Polarography   254
6.4   Charging Current and Direct Current Effects
        in Pulse Polarography   258
        6.4.1   Differential Pulse Polarography   259
        6.4.2   Normal Pulse Polarography   265
6.5   Pulse Voltammetry at Stationary Electrodes   267
        6.5.1   Normal Pulse Voltammetry   267
        6.5.2   Differential Pulse Voltammetry   269
        6.5.3   Pulse Voltammetry at a Dropping
                 Mercury Electrode   270
        6.5.4   Pulse Voltammetry at Rotated Electrodes   271
6.6   Some Other Variations in Pulse Polarography   273
        6.6.1   Pseudoderivative Pulse Polarography   273
        6.6.2   Differential Double-Pulse Voltammetry   273
        6.6.3   Alternate Drop Differential Pulse Polarography   277
        6.6.4   Constant Potential Pulse Polarography   280
        6.6.5   Miscellaneous   281
6.7   Assessment of Analytical Usefulness   282
Notes   284

7 Sinusoidal Alternating Current Polarography   288
   7.1   Nomenclature   288
   7.2   Basic Principles   292
   7.3   Fundamental Harmonic AC Polarography   298
         7.3.1   Systematic Use of Total Fundamental
                 Harmonic Alternating Current Polarography   298
         7.3.2   Reversible AC Electrode Processes   300
         7.3.3   Quasi-Reversible AC Electrode Processes   305
         7.3.4   Irreversible AC Electrode Processes   313
         7.3.5   Electrode Processes with Coupled
                 Chemical Reactions or Adsorption   314
   7.4   Phase-Sensitive Fundamental Harmonic AC Polarography   316
         7.4.1   Background or Charging Current   317
         7.4.2   Dependence of Charging Current
                 and Faradaic Current on Frequency   317
         7.4.3   Dependence of Charging and Faradaic
                 Currents on AC Amplitude   321
         7.4.4   Dependence of Charging and Faradaic
                 Currents on Phase Angle   324
   7.5   Other Developments in Fundamental Harmonic
         AC Polarography   332
         7.5.1   Short Controlled Drop Times   332
         7.5.2   Current-Sampled AC Polarograms   339
         7.5.3   AC Polarography in the Subtractive Mode   340
         7.5.4   Fast Sweep AC Voltammetry
                 at a Dropping Mercury Electrode   341
         7.5.5   AC Cyclic Voltammetry   349
         7.5.6   Pulsed DC Potentials in AC Polarography   349
         7.5.7   Analytical Applications of High-Frequency
                 AC Polarography   353
   7.6   Second Harmonic AC Polarography   356
   7.7   Intermodular and Faradaic Rectification
         AC Polarography Based on Sine Waves   361
   7.8   Tensammetry   363
   7.9   Examples and Trends in the Use of
         Sinusoidal AC Polarography   374
   Notes   386

8 Miscellaneous Polarographic Methods
   Used in Analytical Chemistry   390
   8.1   Square-Wave Polarography   391
   8.2   Radio-Frequency Polarography   399
   8.3   Chronopotentiometry   407
   8.4   DC Polarography with Controlled Current   416
   8.5   Chronopotentiometry with Controlled
         Alternating Current   419
   8.6   Charge-Step Polarography   424
   Notes   429

9 Stripping Voltammetry 435
   9.1   Introduction 435
   9.2   Electrodes Used in Stripping Voltammetry 439
   9.3   Theory and Techniques 448
   9.4   Comparison of Some of the Techniques 459
       9.4.1 Preformed and In Situ-Deposited Films 460
       9.4.2 Measurement Modes with Mercury
               Thin Film Electrodes 460
       9.4.3 Measurement Modes with Hanging
               Mercury Drop Electrodes 461
       9.4.4 Relative Sensitivities and Limits
               of Detection at MTFE and HMDE 464
       9.4.5 Resolution at MTFE and HMDE Electrodes 465
       9.4.6 Practical Examples of the HMDE
               and MTFE Techniques 468
   Notes 471

10 Computers and Digital Data Processing in Polarography 473
   10.1 Microprocessor- and Minicomputer-Controlled
        Polarographs 473
   10.2 The Computer-Controlled Polarograph 478
   Notes 491

Author Index 493

Subject Index 505

The rebirth of interest in electroanalytical methods in the 1970s
can be traced to a number of factors. Improvements and simplifi-
cations in the design of instrumentation with the growth of solid
state electronics and operational amplifiers led to versatile
commercial electrochemical instruments for such techniques as pulse
polarography and stripping methods. Advances in electroanalytical
theory, based on the initial work of Heyrovský and his coworkers,
but extended by the application of numerical methods and digital
simulations, provided a firm foundation and guide for the development of these methods. Interest in the determination of low levels
of metals and organic materials, and especially the desire to
ascertain the actual form of the substance of interest in the
sample, for example in environmental analysis, led to more widespread applications. Moreover, a growing appreciation of the power
of electrochemical methods as complements to spectroscopic techniques
has greatly increased the use of such methods as cyclic voltammetry
in organic and inorganic chemical investigations.

While the classic and widely read electroanalytical chemistry
books of the 1950s and 1960s aided in the development of this field,
they clearly are not useful as guides to the application of modern
electrochemical methods to chemical problems. Professor Bond gives
in this monograph an up-to-date discussion of modern polarographic

methods (broadly defined), complete with examples, experimental
details, and, most importantly, his own insight and opinions on
the relative merits and advantages of the different techniques.
Hopefully this book, designed for the practicing analyst, will be
a factor in bringing this reincarnated area of analytical chemistry
into a new and healthy maturity.

Allen J. Bard
Department of Chemistry
University of Texas
Austin, Texas

It is more than a decade since I accepted the publisher's suggestion to write a book on modern polarographic methods.* The inkling of a renaissance of an historically well-known, but not too widely used, method of instrumental analysis, was in the wind in some quarters ten years ago. However, at the same time, to many practicing analytical chemists, the technique was considered to be almost extinct. Thus, when I began the task of writing a book, I envisaged the main aim would be to promulgate the advances in polarographic analysis which, in the years ahead, could make the techniques competitive with existing instrumental methods of analysis commonly available in a well-equipped analytical laboratory. However, modern methods of polarography have now become well entrenched in all parts of the world as part of the arsenal of approaches available in trace analysis. The present need for an evangelical approach therefore has diminished considerably.

Writing this book over a period which has witnessed a technique moving from what might be described as a renaissance to substantial maturity (once again) has from time to time caused me considerable

_____

*The majority of this book was written when the author was at the Department of Inorganic Chemistry, University of Melbourne. The author is now in the Chemical and Physical Sciences Division at Deakin University, Waurn Ponds, Victoria, Australia.

difficulty in deciding on a correct emphasis and perspective with
respect to the audience that might be interested in the topic considered. Indeed, some sections have been rewritten many times
because of this problem. In reality, not being able to solve this
dilemma, I have ended up presenting a relatively personal account
of the status, attributes, and defects of modern polarographic
techniques as I have experienced them, employing as my guide the
belief that the areas and topics which interest me, might interest
others. The book and its content are therefore selective and highlight areas I regard as important or interesting. No attempt has
been made to exhaustively review all methods of polarographic analysis or to provide extensive literature reviews; rather, critical
accounts of the areas covered are provided in the belief that this
is the most useful aspect of the techniques I can write about. In
the analytical context, therefore, I have found it necessary to
dismiss some well-known modern polarographic techniques in a few
paragraphs as academically elegant, perhaps better than DC polarography, but not competitive with other modern approaches. To learned
colleagues who disagree with my assessment in some areas, I submit
an invitation to write and discuss any pertinent matters so that
due consideration may be given to revision of any of my ideas at a
later occasion, should the opportunity arise.

The writing of this book has required the assistance of numerous
people, who have shown remarkable patience in fulfilling my many
requests. To colleagues, librarians, secretaries, technical assis-
tants, and publisher I offer my sincere thanks. To journals, authors,
and instrument manufacturers who have willingly granted permission
to reproduce figures, and provided invaluable information used in
the book, I express my gratitude. To name every person who has
assisted with the writing of this book would have required a separate
chapter. These people, I hope, will therefore accept my thanks
anonymously. However, I would personally like to acknowledge the
considerable contribution of Professor Allen J. Bard, who has read
and commented on all chapters of the book and willingly held

discussions with me on many aspects, on many occasions, in different parts of the world. Finally, my family has also provided enormous support over the many years it has taken to complete the work. I have dedicated this book to them as an acknowledgment of my appreciation.

A. M. Bond

Modern Polarographic Methods in Analytical Chemistry

Chapter 1

THE RENAISSANCE OF POLAROGRAPHY

1.1 HISTORY

Direct current (dc) polarography, the original and still a very
commonly used form of polarography, was discovered by Jaroslav
Heyrovský more than 50 years ago. During the last three decades a
most impressive variety of new polarographic methodology has come
into existence. Most of these new techniques are associated with
certain theoretical advantages, which, if realizable in practical
applications, should have made the original dc form virtually obsolete by now. However, this has not eventuated, and it is only recently that many of the new methods are beginning to gain real practical acceptance as methods of trace analysis.

Ideally, of course, as soon as their advantages became apparent,
these new techniques should have been encouraged in preference to dc
polarography. However, during the 1950s and 1960s, most large analytical laboratories and teaching institutions in English-speaking
countries had only one simple dc polarograph, with few people familiar with its operation, and thus there was a high probability that
the instrument was simply gathering dust. During this period, a
most conservative and generally uninspiring approach to teaching
this method of analysis inhibited the advancement of polarography;
a wide gap was created between capabilities reported from research-
orientated electrochemical institutions using the newer methods and

those attributed to polarography in analytical laboratories which
still retained ideas formulated from the use of conventional dc polarography. The relative decline in the routine use of polarography,
compared with the upsurge in interest of other techniques, is therefore not difficult to understand. By contrast, in Eastern European
and other countries where electroanalytical chemistry has traditionally enjoyed much wider acceptance, the educational problem is not
evident and polarographic methods have more than held their own in
popularity since the inception of the technique.

Concomitant with the educational problem, a lack of attention
to instrument design in commercially available instruments was evident. By way of comparison, in other areas of trace instrumental
analysis, innovations and alterations to instrumentation were being
made almost routinely. This difference in attitude by instrument
manufacturers also presumably was symptomatic of a general lack of
enthusiasm for the polarographic technique.

Florence [1], in an article entitled "Is Polarography Dead?"
suggests that polarography reached its nadir around 1967 and was
at that time almost defunct. The environmental crisis had only
just awakened the public conscience, and atomic absorption spectrophotometry was at its peak of vigor. Importantly, modern solid state
operational amplifier circuitry had yet to make its full impact on
commercial polarographic instrumentation. During the past few years,
however, a sudden increase in interest in electroanalytical techniques,
including polarography, has become apparent. Papers discussing the
renaissance of polarography are appearing from many areas [1-3].
The resurgence of interest undoubtedly can be attributed partially
to the appearance of vastly improved, commercially available instruments, making the newer methodology available to all laboratories at
relatively low cost. Simultaneously, environmental scientists have
been making heavy demands for a large number of heavy metal and organic analyses. Ironically, even the polarographic determination of
cadmium has now assumed importance in practical applications. What
author of a new polarographic technique or writer of a review during
the history of polarography has not presented a polarogram for the
reduction of cadmium when making his claims or assessments? For

years polarographers were among the heaviest users of cadmium and spent an inordinate amount of time publishing polarograms of cadmium. Unfortunately, until recently the general analytical chemist was not impressed one iota because he had no interest whatsoever in cadmium. It actually has been suggested facetiously in some quarters that polarographers invented the environmental crisis, because unless they could discover a necessity for large-scale determinations of cadmium, they were doomed to early retirement.

The above discussion is of course a considerable oversimplification of the so-called renaissance, for in another sense polarography was never in danger of dying, if the contents of research journals of this period are examined closely. During the 1950s and 1960s, when the future use of polarography in trace analysis was being questioned, enormous advances in the theory and instrumentation of alternating current, pulse, linear sweep, and other polarographic techniques were taking place simultaneously. Admittedly, the major drive for this work was in areas of kinetic investigations and theoretical electrochemistry. However, it is the formulation of ideas and results arising out of this work that in fact has led to substantial improvement in the "state of the art" in the analytical area. The instrumentation evolved from the age of the manual polarograph to the stage where complete on-line computer control and automation of the experiment was feasible. Theoretical polarography far outstripped practical applications during this period, but this work has now given us an excellent foundation on which to exploit systematically the techniques to great advantage in the laboratory situation. Indeed, very few instrumental techniques are now founded on such an excellent theoretical understanding.

Writing in 1966, Smith [4], in a most authoritative review of some modern polarographic methods, made the following comment:

> The past decade has seen an inordinate amount of effort directed towards overcoming limitations of existing methods by applying electronic "trickery" to evolve more sophisticated techniques.... Many of these recent innovations have been investigated only briefly. It is likely that this emphasis on technique improvement will soon be replaced by emphasis on utilization of existing techniques within their useful realm.

This prediction has certainly proved correct and serves to summarize the research activities of this era. At the present time, those advances which Smith was describing now appear in the commercially available instrumentation being used by analytical chemists.

Flato [2] summarized his assessment of the overall situation: "The decade from 1955 to 1965 might be characterised as the one single period during which the greatest advancement in the technical aspects of polarography took place, while simultaneously, the greatest decline in the practical every day usage of these techniques occurred." Flato's appraisal, made with hindsight, complements the prognosis of Smith, suggesting that the dormancy experienced in the domain of trace analysis was indeed really a preparation or reorientation period necessary for the formulation of the revival currently being experienced.

Modern polarography is now a sensitive, rapid technique applicable to analysis in the inorganic, organic, geochemical, biochemical, medical, and pharmaceutical fields and indeed in most areas of analytical chemistry. It is probably the most versatile of all trace analysis methods. Pulse, phase-selective ac, and linear sweep polarography can compete successfully with atomic absorption spectrophotometry for the determination of several elements. Parts-per-billion and even lower levels are amenable to determination for many electroactive species. For the determination of trace levels of organic compounds, polarography has no real rival. A modern polarograph can give a linear response between concentration and current from 10^{-8} to 10^{-2} M, i.e., a range of six orders of magnitude. With most instruments and methods, spectrophotometry covers an absorbance range of 10^2 to 10^3. The great analytical range of polarography has obvious advantages. However, despite all these attributes there still has been considerable reluctance to use the techniques widely [5]. Any problem of its acceptance now appears to be largely one of education. Not only has the subject been taught inadequately in most tertiary chemistry courses until recently, but so few older analytical chemists have had experience in practical polarographic analysis other than the conventional dc form that they are somewhat inoculated

against the polarographic method, making its propagation more difficult.

1.2 AIMS AND SCOPE

In writing this book, I have endeavored to convey a general and
systematic approach to the use of modern polarographic methods. In
my laboratories, conventional dc polarography is rarely used or
recommended as an analytical technique on the grounds that any determination that can be done by conventional dc polarography can be done
faster, more accurately, or more conveniently by a modern polarographic
technique. With the ready availability of high-quality, inexpensive,
commercially available instrumentation, or alternatively, with the
relative simplicity of constructing one's own instrumentation, the
modern well-equipped analytical laboratory or tertiary teaching insti-
tution seriously contemplating polarographic analysis should have
access to the newer techniques. It is therefore clear that these
techniques, rather than conventional dc polarography, should be ad-
vocated under all circumstances for routine analytical work. Conventional dc polarography refers to the use of the well-known current-
voltage curves obtained by applying a dc potential between a dropping
mercury electrode (dme) with gravity-controlled drop times of approximately 2 to 10 sec and a reference electrode. This version of polarography will be presented solely as a convenient teaching medium
and reference point on which to base subsequent discussions. For
those people wishing to read the detailed history, theory, and practice of conventional dc polarography, the extensive literature on the
subject may be consulted [6-12]. The presentation of modern polarographic techniques is intended to have considerable didactic content
and is aimed at bringing the recent developments to the attention of
the well-trained analytical chemist. Extensive mathematical formula-
tions generally will be avoided in descriptions of techniques, and
results will be presented simply without derivation to facilitate
concise presentation of discussions of direct practical importance.

Almost by definition, it can be assumed that any modern polarographic technique should have certain specific advantages over dc polarography. These are usually fairly obvious. However, a more difficult fundamental question confronting an analytical chemist contemplating the use of polarography nowadays might be whether his problem would be best tackled by ac or pulse polarography, for example. Even then, having decided to use ac polarography, the chemist's problems remain: Should the fundamental or second harmonic variation be used, with or without phase-selective detection? If it is decided to use pulse polarography, should it be in the normal, derivative, or differential modes? New polarographic techniques and variations are published almost weekly, many appearing highly attractive in principle, others apparently having little to recommend them from either theoretical or practical considerations when compared with existing modern techniques, as distinct from comparing them solely with conventional dc polarography. The criterion adopted for publication of a method and discussion of techniques in reviews in the past, generally has been that they should offer an improvement on conventional dc polarography. In this book the modern polarographic techniques are compared critically with each other, and not just with dc polarography, in the belief that this is a requisite and responsibility of a reviewer of this period of development of the field of polarography.

The scope of what comes within the category of a polarographic method for the purposes of this book has presented somewhat of a dilemma. For example, rigorous definition of polarography based on the use and interpretation of current-constant potential-time curves at a dropping mercury electrode, while encompassing the most widely used current-potential techniques [13], would result in the omission of inverse or stripping methods as well as linear sweep voltammetry (LSV) at a dropping mercury electrode. The broader fields of voltammetry or electroanalytical chemistry of which polarography is a particular category or subcategory, respectively, would encompass too large a range to consider in a single book. As a compromise, the author has settled upon the expediency of a loose definition of polarography, which encompasses all the techniques correctly called *polarographic* and a selection of other voltammetric and electroanalytical techniques that can logically and conveniently be considered interwoven with polarographic research, development, and methodology.

NOTES

1. T. M. Florence, Proc. Roy. Australian Chem. Inst., *39*, 211 (1972).

2. J. B. Flato, Anal. Chem., *44*, (11) 75A (1972).

3. E. Jacobsen and T. Rojahn, Kjemi, *33*, 31 (1973).

4. D. E. Smith, in *Electroanalytical Chemistry* (A. J. Bard, ed.), Dekker, New York, 1966, vol. I, chap. 1.

5. E. J. Maienthal, Anal. Chem., *45*, 644 (1973).

6. I. M. Kolthoff and J. J. Lingane, *Polarography*, 2nd ed., Interscience, New York, 1952

7. G. W. C. Milner, *The Principles and Applications of Polarography and other Electroanalytical Processes*, Longmans, London, 1957.

8. L. Meites, *Polarographic Techniques*, 2nd ed., Interscience, New York, 1965.

9. O. H. Muller, *The Polarographic Method of Analysis*, 2nd ed., Chemical Education Publishing, Easton, Pa., 1951.

10. P. Zuman, *Organic Polarographic Analysis*, Macmillan, New York, 1964.

11. M. Brezina and P. Zuman, *Polarography in Medicine Biochemistry and Pharmacy*, Interscience, New York, 1958.

12. J. Heyrovsky and J. Kuta, *Principles of Polarography*, Academic, New York, 1966, and reference list therein of textbooks on the subject.

13. R. B. Fischer, Anal. Chem., *37*, 27A (1965).

Chapter 2

THEORETICAL, INSTRUMENTAL, AND
GENERAL CONSIDERATIONS RELEVANT TO THE
SYSTEMATIC USE OF POLAROGRAPHIC METHODOLOGY

Subsequent chapters contain individual and comprehensive accounts of advances in dc, ac, pulse, linear sweep, and other polarographic techniques, with much specialized emphasis and discussion relevant to the particular technique. However, many of the polarographic and related electroanalytical techniques actually have a basic formulation with respect to theory, instrumentation, and other characteristics. Recognition of the general principles and problems enables both an understanding of the evolution of polarographic methodology to be reached and predictions of future trends in research, development, and application to be made. Additionally, from generalized discussion, bases on which to later judge the usefulness and limitations of individual techniques arise, and a systematic approach to actually using the polarographic method can be developed.

2.1 THE DC POLAROGRAPHIC EXPERIMENT: SIMPLE CONCEPTS

A basic knowledge of conventional dc polarography is assumed as stated in Chap. 1, and the major emphasis given to this polarographic method is to illustrate general principles. However, in no other chapter will this role be performed so extensively; so to ensure that an initially familiar basis is indeed available on which to discuss theoretical, instrumental, and other phenomena relevant to the

general topic of electroanalytical chemistry, a very brief and
simplistic review of dc polarography is given immediately below.
More detailed discussion of the theory relevant to practical analytical applications is given in Chap. 3.

Figure 2.1 shows a schematic diagram of a dc polarogram. Applications of dc polarography are based on the measurement and interpretation of current-potential (i-E) curves as recorded at a dropping
mercury electrode (dme). The variation of current with a continuously changing potential can be measured instrumentally to give a
dc polarogram. The potential axis is defined with respect to a
reference electrode, traditionally a saturated calomel reference
electrode (sce), but it can be any electrode giving a stable and
reproducible potential under the conditions of measurement. In the
presence of substances which undergo reduction or oxidation at a
dropping mercury electrode, an increase in cathodic (positive) or
anodic (negative) current is observed over a particular potential
range of the current-vs.-potential plot. Subsequent to this potential range, a region is reached where the current is independent of
potential and has a limiting value. The S- or sigmoidal-shaped i-E
curve is called a *polarographic wave*.

The difference between the limting current and the current
before the wave rise, called the *wave height*, usually depends on the
concentration of the electroactive substance in solution. Most

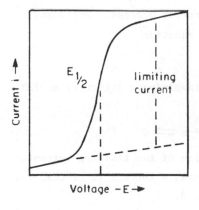

Voltage −E⟶

FIGURE 2.1 Schematic diagram of a
dc polarogram.

analytical applications of both organic and inorganic polarography
are based on the increase in wave height with concentration. The
limiting current can be diffusion-controlled (diffusion caused by a
concentration gradient between electrode surface and bulk solution
is rate-determining), kinetically controlled (chemical step is rate-
determining), or adsorption-controlled (process involving adsorption
is rate-determining).

Another important parameter is the half-wave potential $E_{1/2}$,
that is, the potential on a polarographic curve at which the current
reaches half of its limiting value. Although the wave height depends
on concentration, $E_{1/2}$ is frequently almost independent of the concentration of electroactive species. In some instances the value
of $E_{1/2}$ is closely related to the standard redox potential $E°$, and
in general, the $E_{1/2}$ value is a characteristic of the compound undergoing reduction or oxidation. Because $E_{1/2}$ depends on the nature
of the electrolyzed substance and therefore the composition of the
solution, etc., it is a quantity that can be used for qualitative
characterization of substances. The analogy with infrared (ir)
spectroscopy is obvious; for example, carbonyl and other functional
groups have ir bands at characteristic wavelengths that vary with
the structure of the compound. Likewise, compounds have characteristic half-wave potentials.

The shape of the wave is also an important ingredient in the
overall characterization or classification of a polarogram. This
is determined by the nature of the electrode process. For example,
the electrode process described by the equation

$$A + e \rightleftharpoons B$$

would be different in shape from that involving a different stoichiometry, such as

$$C + 3e \rightleftharpoons D \qquad or \qquad E + 2F + e \rightleftharpoons G$$

Complete theoretical rationalization of the limiting current,
$E_{1/2}$, and wave shape is available in notes 6 through 12 of Chap. 1.
Further details are also given in Chap. 3.

A convenient summary of the understanding of dc polarography required by the general analytical chemist at this stage may be gained by presenting the analogy existing between dc polarography and absorbance spectroscopy (Table 2.1).

In the more general context, all modern polarographic techniques are characterized by two parameters, one which is directly proportional to concentration and the other which is equivalent to $E_{1/2}$ and is a function of the structure and medium. The two kinds of parameter are used extensively in all analytical work, and the systematic use of each polarographic method requires an understanding of their significance. This knowledge is gained by consideration of theoretical descriptions of electrode processes, and much of the following work, therefore, has direct relevance in this area.

For the remainder of this chapter, formulation of theoretical concepts are couched in general terms applicable to most branches of electrochemistry, rather than dc polarography in particular. However, where an example is required to illustrate a particular principle, it is now assumed that dc polarography can be called upon. Illustration of the same principle with respect to other polarographic techniques is given in the chapters in which these methods are considered.

TABLE 2.1 Analogy Between Polarography and Absorption Spectroscopy

DC polarography	Absorbance spectroscopy
Readout consists of a plot of current vs. potential.	*Readout* consists of a plot of absorbance vs. wavelength.
Wave height (i_1) depends on *Concentration*.	Absorbance (A) depends on *Concentration*.
Half-wave potential $E_{1/2}$ = f (structure, medium).	Maximum wavelength λ_{max} = f (structure, medium).
Shape of i–E curve depends on nature of electrode process.	Shape of A–λ curve depends on nature of electronic transition.

Taken in part from Note 12.

2.2 ELECTRODE PROCESSES AND THEIR NATURE

The systematic use of modern electroanalytical methods in all areas, including trace analysis, invokes the need for an understanding of electrode processes. This presents some initial conceptual difficulties because the nature of any current-potential (i-E), current-time (i-t), or potential-time (E-t) curve encountered in a polarographic experiment is governed by interrelationships involving a wide spectrum of phenomena occurring in physical chemistry.

Even in the frequently reported so-called simple polarographic experiment in which cadmium ions are reduced to cadmium metal by applying an increasingly negative dc potential between a dme and reference electrode, many events occur to produce the dc polarogram seen in Fig. 2.2. The exact composition of the cadmium ions needs to be considered and depends on the solvent. In aqueous 1 M $NaClO_4$, cadmium may be considered to exist as $Cd(H_2O)_x^{2+}$, for example. The product of the electrode process, nominally cadmium metal, is soluble in mercury, so that an amalgam is actually produced. Writing the overall process as

$$Cd(II) + 2e \rightarrow Cd(0) \tag{2.1}$$

therefore masks a great deal of complexity. At some stage during the experiment, coordinated water molecules have to be removed. Diffusion of cadmium ions toward the electrode and then cadmium metal into a mercury drop, after electron transfer across the solution-mercury interface, then completes the process. Rewriting the electrode process hypothetically as

$$Cd(H_2O)_x^{2+} \underset{k_2}{\overset{k_1}{\rightleftharpoons}} Cd^{*2+} + xH_2O \tag{2.2}$$

$$Cd^{*2+} + 2e \underset{k_4}{\overset{k_3}{\rightleftharpoons}} Cd \text{ (amalgam)} \tag{2.3}$$

illustrates some of the difficulties in completely understanding the

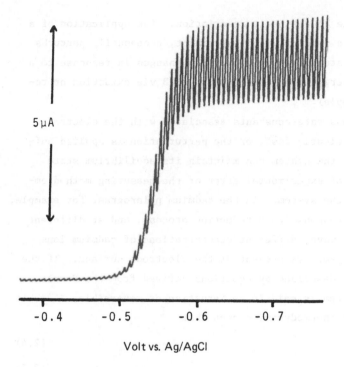

5 μA

-0.4   -0.5   -0.6   -0.7

Volt vs. Ag/AgCl

FIGURE 2.2   DC polarogram for reduction of cadmium(II) in 1 M NaClO$_4$.

electrode process; k_1 and k_2 are chemical rate constants (homogeneous kinetics or equilibria); Cd^{*2+} is a dehydrated form of cadmium postulated as an intermediate; k_3 and k_4 are rate constants describing the rate of the electron transfer steps (heterogeneous kinetics or equilibria).

A complete understanding of this electrode process requires a knowledge of homogeneous and heterogeneous kinetics, thermodynamics, diffusion, surface phenomena, solution chemistry, and basic electrochemical principles. Fortunately, the analytical chemist undertaking a polarographic experiment need not despair at the probable complexity of the rigorous theoretical treatment. All that needs to be recognized at this stage of the discussion is that a polarogram is the result of an electrode process occurring when a potential is applied to a cell and that the resulting i-E or other curve is generally amenable to a kinetic description. The application of a potential, as in a polarographic experiment, necessarily perturbs the system under study because the system changes in response to a change in free energy from state A to state B via oxidation or reduction (for example).

If the various rate constants associated with the electrode process are sufficiently fast, or the perturbation is applied sufficiently slowly, the system can maintain its equilibrium state within the limit of experimental error of the measuring method employed to follow the system. In the cadmium polarogram, for example, current flows in response to a reduction process, and at different potentials on the wave, different concentrations of cadmium ions and cadmium (amalgam) are present at the electrode surface. If the polarogram can be described by equations derived from the Nernst equation, the system is said to be Nernstian or reversible, i.e., it obeys the laws of thermodynamics such that

$$\Delta G = -nFE \tag{2.4}$$

$$\Delta G^{\circ} = nFE^{\circ} \tag{2.5}$$

$$E \approx E^{\circ} + \frac{RT}{nF} \ln \frac{[Cd^{2+}]a_{Hg}}{[Cd\ (amalgam)]} \tag{2.6}$$

Any process is of course only strictly *thermodynamically reversible* when an infinitesimal change in the direction of a driving force causes the direction of the process to reverse. All electrode processes and reactions occur at a finite rate and therefore do not proceed with thermodynamic reversibility. Thus, a *practical definition of reversibility* is really being used in defining a reversible electrode process as one obeying the Nernst equation, the implication being that departures from thermodynamic reversibility are too small to measure with the particular technique under consideration. In view of the approximation inherent in a thermodynamic description, the most general and useful descriptions of an electrode process are therefore formulated in terms of absolute rate theory or kinetic

equations and the conditions (e.g., rate constants) required for
practical reversibility noted. Electrode processes that do not meet
the specified conditions for reversibility can then be said to be
nonreversible or non-Nernstian.

One of the most important prerequisites to using modern polarographic methods systematically is the ability to classify electrode
processes as being reversible or otherwise and to define what is
meant by reversible in terms of rate constants with respect to a
particular technique. Different polarographic methods involve different time domains and measure different responses of the electrode
process, so that an electrode process may be reversible with respect
to, say, dc polarography but not ac polarography. Throughout this
book, statements similar to the following appear: "Second harmonic
ac polarography can be used to determine reversibly reduced species
down to 10^{-8} M. However, if the electrode process is irreversible,
a less favorable limit of detection will apply." In each polarographic technique, the question of reversibility needs to be defined
explicitly within the realms of its own time domain, and the ability
of the analyst to recognize and utilize the different time scales
is an invaluable asset. With conventional dc polarography, the
drop time (2 to 8 sec, approximately) controls the time domain, so
that variations over only an order of magnitude or less are involved.
The time domain aspect, therefore, could be neglected effectively
when dc polarography was the predominant technique. The range of
techniques now available, and discussed in this book, encompass
many orders of magnitude in time scale, and the response that can
be obtained from a given electrode process can be altered substantially in the required manner by judicious choice of technique
and/or experimental conditions.

2.2.1 Electrochemical and Chemical Reversibility

In its most general form, an electrode process consists of a charge
transfer step (heterogeneous kinetics), the chemical reactions
coupled to that charge transfer step (homogeneous kinetics), and

diffusion, as indicated on page 13. Additionally, adsorption and other surface reactions may frequently need to be considered. It is convenient, therefore, to consider the discussion of reversibility or otherwise of an electrode process not just in the overall context but in terms of both the electrochemical charge transfer step(s) and the chemical step(s). The statement that an electrode process is not reversible is most uninformative because in any kinetic process the information really required is the rate-determining step(s). An electrode process occurring across a solution-electrode interface can be irreversible because of either a slow electron transfer step (heterogeneous rate constant) or a slow chemical step (homogeneous rate constant) or adsorption, and a given polarographic technique can respond to an electrode process in a decidedly different fashion depending on the cause of the nonreversibility or kinetic complication. If the cause of the irreversibility is the electron transfer step, this will frequently dictate the use of a particular polarographic method in preference to another. Similarly, if the irreversibility is chemical in nature, there are considerable implications as will become clearer in later discussions.

Electrochemical Reversibility. Consider an electrode process such as $[Fe(III)(oxalate)_3]^{3-} \underset{-e}{\overset{+e}{\rightleftharpoons}} [Fe(II)(oxalate)_3]^{4-}$. The only process occurring in this example can be considered to be the electron transfer step [2]. Rearrangement of ligands in either oxidation state (or adsorption) is assumed to be absent at high oxalate concentrations, and the electron can be transferred in either direction. Figure 2.3 shows a dc polarogram of the reduction of Fe(III) in oxalate medium. By definition, the direction of current flow for reduction is considered positive. The oxidation of Fe(II) in the same medium gives a similar wave (see Fig. 2.4) except that the direction of current is reversed and the current is defined as negative. A mixture of Fe(III) and Fe(II) in the same oxalate medium gives a mixture of cathodic and anodic currents corresponding to the relative concentration of each oxidation state as shown in

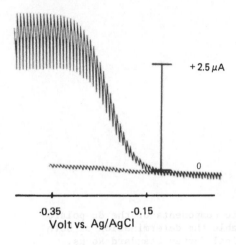

FIGURE 2.3 Reduction of
1×10^{-3} M Fe(III) in oxalate
media. [Reproduced from Anal.
Chem. *47*, 479 (1975). © American
Chemical Society.]

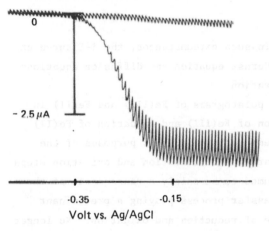

FIGURE 2.4 Oxidation of 1×10^{-3} M Fe(II) in oxalate media.
[Reproduced from Anal. Chem. *47*, 479 (1975). © American Chemical
Society.]

Fig. 2.5. By definition this electrode process is therefore electrochemically reversible because the same energetics are involved
whether reduction of Fe(III) or oxidation of Fe(II) is being considered; i.e., the path connecting the two states is thermodynamically reversible and the rate of electron transfer in both directions is sufficiently fast, so that at all points along the path the

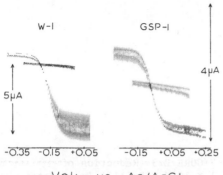

FIGURE 2.5 The cathodic and anodic components of the dc polaro-
gram for iron in oxalate media enable the determination of both
Fe(III) and Fe(II) in U.S. Geological Survey Standard Rocks.
[Reproduced from Anal. Chem. *47*, 479 (1975). © American Chemical
Society.]

system is in equilibrium. In such circumstances, the i-E curve can
be defined in terms of the Nernst equation and diffusion equations
relating current to concentration.

Figure 2.6 shows pulse polarograms of Fe(III) and Fe(II) in
another media [3]. Reduction of Fe(III) and oxidation of Fe(II)
now occur at totally different potentials. For purposes of the
present discussion, the separation of reduction and oxidation steps
in this instance can be assumed to be solely a consequence of the
kinetics of the electron transfer process playing a predominant
role in determining the rate of reduction and oxidation. No longer
are the reduction and oxidation pathways between the two states
equivalent, nor is the pathway between them an equilibrium one.
The pathways are now actually defined by kinetic considerations.

Consider the electrode reaction in terms of rate constants:

$$A + ne \underset{k_b}{\overset{k_f}{\rightleftharpoons}} B \qquad\qquad (2.7)$$

where k_f and k_b are heterogeneous rate constants (cm sec^{-1}) describing the rates of the forward and reverse reactions. The rate

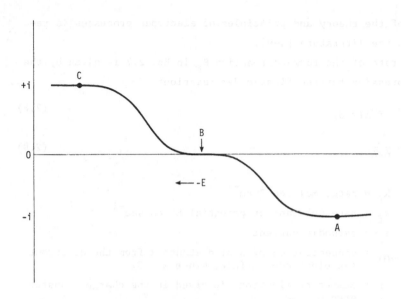

FIGURE 2.6 Schematic diagram of pulse polarographic behavior of
the irreversible Fe(III)/Fe(II) redox couple in pyrophosphate medium
(pH 8). Stepping from rest potential B (-0.6V vs. sce) to potential
C (-1.4 V vs. sce) gives a signal for Fe(III) reduction, while
stepping to potential region A (-0.2 V vs. sce) gives a signal for
Fe(II) oxidation. Note how in contrast to the oxalate medium in
Figure 2.5, oxidation and reduction waves are observed at different
potentials. [Reproduced from Anal. Chem. 45, 458 (1973). © American Chemical Society.]

of an electrode process may be described by the kinetic equation
derived by Butler [4] and in terms of models based on potential
energy curves, transition state theory, or absolute rate theory.
It is not the purpose of this book to discuss the detailed theory
of electrode kinetics. All that is required for the systematic use
of polarographic methodology, as indicated previously, is an understanding of the type of rate constants best used to define reversibility. That is, we need to know what magnitude of the rate constants for the electron transfer step enables us to call an electrode process reversible with respect to the technique under consideration. Consequently, only a nonrigorous illustrative discussion
of the electron transfer step is presented here, without the derivation of the equations. The reader who is interested in the detailed

aspects of the theory and principles of electrode processes is referred to the literature [4-9].

The rate of the forward reaction R_f in Eq. 2.7 is given by the
usual expression for the first-order reaction:

$$R_f = k_f C_{A(x=0)} \tag{2.8}$$

$$= \frac{i_c}{nFA} \tag{2.9}$$

where

$$R_f = \text{rate, mol sec}^{-1} \text{ cm}^{-2}$$

$$k_f = \text{rate constant at potential E, cm sec}^{-1}$$

$$i_c = \text{cathodic current}$$

$$C_{A(x=0)} = \text{concentration of A at distance x from the electrode (at electrode surface, since x = 0)}$$

$$n = \text{number of electrons involved in the charge transfer step}$$

$$A = \text{area of the electrode}$$

$$F = \text{Faraday}$$

The rate of the backward reaction R_b is similarly given by
Eqs. (2.10) and (2.11), with i_a as the anodic current.

$$R_b = k_b C_{B(x=0)} \tag{2.10}$$

$$= \frac{i_a}{nFA} \tag{2.11}$$

The net rate R_n, which determines the magnitude of the measured current i, is given by the difference between forward and reverse
rates, so that

$$R_n = R_f - R_b = k_f C_{A(x=0)} - k_b C_{B(x=0)} = \frac{i}{nFA} \tag{2.12}$$

Thus,

$$i = nFA (k_f C_{A(x=0)} - k_b C_{B(x=0)}) = i_c - i_a \tag{2.13}$$

Note the difference between heterogeneous and homogeneous reactions:

1. Rate R is given in mol sec^{-1} cm^{-2} and rate constant k in cm^{-1} because reactions are normalized to unit area.
2. Concentration is specified at the reaction site (i.e., C is defined at the electrode surface, where x = 0).
3. Rate constant is a function of potential: k = f(E).

To establish criteria for defining heterogeneous rate constants vs. reversible electrode processes approximations, the functions relating k and E are needed. The appropriate functions can be derived from transition state theory. These concepts require the introduction of a dimensionless parameter called the *transfer coefficient* α, which denotes the fraction of the potential influencing the rate of electroreduction. In describing an electrode process in terms of free energy-vs.-reaction coordinate diagrams, as is usually done in transition state theory, it can be proposed that the height of the activation energy of the forward reaction is altered by some fraction α of the total free energy change resulting from the potential difference at the electrode-solution interface when the reduction occurs. The same assumption requires that the activation energy for the reverse reaction is altered by 1 - α of the total free energy change. From this simple concept,

$$0 < \alpha < 1 \tag{2.14}$$

The transfer coefficient α appears in many polarographic equations and has great significance in the theoretical discussion of electrode kinetics. However, this parameter is not a rate constant and therefore does not concern us with respect to defining the reversibility or otherwise of an electrode process. We may simply regard it as an adjustable parameter that, when used in conjunction with the heterogeneous charge transfer constant, facilitates a complete mathematical description of the electrode process. Apart from noting that α performs this function and writing it into many equations, no further discussion of this parameter is needed in describing analytical applications of polarography.

The functions relating potential and rate constant, after establishing the appropriate theoretical model, turn out to be

$$k_f = k_f^{\circ} e^{-\alpha n F E / RT} \tag{2.15}$$

and

$$k_b = k_b^{\circ} e^{(1-\alpha) n F E / RT} \tag{2.16}$$

where k_f° and k_b° are the values of k_f and k_b when $E = 0$, on an arbitrary potential scale. A far more useful potential at which to define the rate constants is the standard redox couple E°, where E° has the usual thermodynamic significance.

When $E = E^{\circ}$, $C_{A(x=0)} = C_{B(x=0)}$ and $R_f = R_b$. Thus, $k_f = k_b$ and

$$k_f^{\circ} e^{-\alpha n F E^{\circ} / RT} = k_b^{\circ} e^{(1-\alpha) n F E^{\circ} / RT} = k_s \tag{2.17}$$

k_s denotes the heterogeneous rate constant in cm sec^{-1} of the electrode process at a standard potential E°. Hence,

$$k_f = k_s e^{-\alpha n F (E - E^{\circ}) / RT} \tag{2.18}$$

and

$$k_b = k_s e^{(1-\alpha) n F (E - E^{\circ}) / RT} \tag{2.19}$$

The complete current-potential equation is then

$$i = n F A k_s (C_{A(x=0)} e^{-\alpha n F (E - E^{\circ}) / RT} - C_{B(x=0)} e^{(1-\alpha) n F (E - E^{\circ}) / RT}) \tag{2.20}$$

This equation and variations derived from it are used in all treatments of electrode kinetics. Thus, despite the fact that we have yet to consider specific polarographic equations, the above treatment enables an understanding of how a practical definition of reversibility can be derived.

Under genuine equilibrium conditions,

$$R_f = R_b = R^{\circ} \qquad\qquad E = E_{eq} \qquad\qquad i = 0$$

$$i_c = i_a = i^{\circ} \qquad C_{A(x=0)} = C_A \qquad C_{B(x=0)} = C_B$$

where i° is called the *exchange current*. Substitution into previous equations therefore gives

$$nFAk_s C_A e^{-\alpha nF(E_{eq}-E^{\circ})/RT} = nFAk_s C_B e^{(1-\alpha)nF(E_{eq}-E^{\circ})/RT} = i^{\circ}$$

$$(2.21)$$

or

$$E_{eq} = E^{\circ} + \frac{RT}{nF} \ln \frac{C_A}{C_B}$$

$$(2.22)$$

as required by a correct kinetic treatment. Equation (2.22) is of course the well-known Nernst equation.

"Reversible" polarographic waves obviously involve a net flow of current, but if k_s is sufficiently large, the corresponding departure from equilibrium potential is negligible and the Nernst equation holds quite accurately. It can be seen from Eq. (2.20) that if $k_s = \infty$,

$$\frac{i}{nFAk_s} = 0$$

and

$$C_{A(x=0)} e^{-\alpha nF(E-E^{\circ})/RT} = C_{B(x=0)} e^{(1-\alpha)nF(E-E^{\circ})/RT}$$

$$(2.23)$$

or

$$E = E^{\circ} + \frac{RT}{nF} \ln \frac{C_{A(x=0)}}{C_{B(x=0)}}$$

$$(2.24)$$

Hence, the Nernst equation is rigorously valid at the electrode surface under the conditions $k_s = \infty$. All electrode processes occur at a finite rate, so that $k_s < \infty$, and a decision has therefore to be made as to what upper limit of k_s will be acceptable when defining a reversible electrode process. Previous discussion as to the need for defining polarographic reversibility in terms of the technique being used now becomes clear, and a definition of what is meant by *reversible* is now possible. Similarly, an understanding as to why a practical rather than thermodynamic definition of reversibility is necessary becomes apparent.

The time domain of a dc polarographic experiment is governed by the drop time, so that the value of k_s required to define a reversible electron transfer step needs to be stated with respect to a specific drop time. However, the time domain in ac polarography is, for example, usually governed by the frequency of the alternating potential rather than drop time, so that reversibility would need to be redefined in terms of a stated frequency rather than drop time when employing this technique. On the other hand, in a linear sweep voltammetric experiment (see Chap. 5), the scan rate of potential would need to be stated when defining explicitly what is meant by a reversible electrode process. Delahay [10] was the first to show that if $k_s \geq 2 \times 10^{-2}$ cm sec^{-1}, a dc electrode process can be considered reversible for a drop time of approximately 3 sec. However, a value of $k_s \geq 1$ might be needed in an ac polarographic experiment of moderate frequency to define the same electrode process as reversible in the ac sense.

If the magnitudes of k_f and k_b are comparable and k_s lies within the limits $2 \times 10^{-2} \geq k_s \geq 5 \times 10^{-5}$ cm sec^{-1}, dc polarographic waves [10,11] are sometimes called *quasi-reversible* (drop time $\approx$ 3 sec). For a totally irreversible electrode process, the backward reaction can be neglected. Such a condition corresponds to a value of $k_s \leq 5 \times 10^{-5}$ cm sec^{-1} in dc polarography. In general, therefore, a reversible electrode process can also be defined as one in which diffusion control rather than the electron transfer step is the determining factor. This class of electrode process is characterized by large values of k_s, and the current magnitude at all potentials is considered to be independent of k_s, k_f, or k_b within the limit of experimental error of the method of measurement. Quasi-reversible electrode processes are then defined as those where complete mathematical description requires inclusion of terms involving both k_f and k_b. Totally irreversible electrode processes, on the other hand, are governed solely by k_f and have small values of k_s.

The position of the irreversible polarographic wave on the potential axis relative to the reversible case can now be understood.

Consider a dc polarographic reduction process. It will be shown later that if the electrode process in Eq. (2.7) is reversible, $E_{1/2}$ occurs at a potential very near to $E°$. For the totally irreversible process, k_s is very small and the rate is governed by k_f. However, k_f is a function of potential. At $E = E°$, $k_f = k_s$, but k_s is so small that the rate of reduction is negligible and no current flows. At more negative potentials, k_f increases until a finite and measurable reduction is observed, as monitored by a flow of current. At this potential, $k_f \gg k_s$ and the rising portion of the S-shaped polarographic wave commences. At sufficiently negative potentials, k_f becomes so large that the electron transfer step no longer remains rate determining and the limiting current region corresponds to diffusion control, as is also the case for the limiting current region of a reversible electrode process. The $E_{1/2}$ value of an irreversible reduction wave is therefore considerably more negative than the half-wave potential $E_{1/2}^r$ of a reversible reduction wave. The differences between reversible and irreversible electrode processes with respect to dc polarography are diagrammatically presented in Fig. 2.7. Exactly the reverse arguments hold for oxidation, and $E_{1/2}$ for irreversible oxidation waves are more positive than $E_{1/2}^r$, as also shown in Fig. 2.7. Thus, the electrode

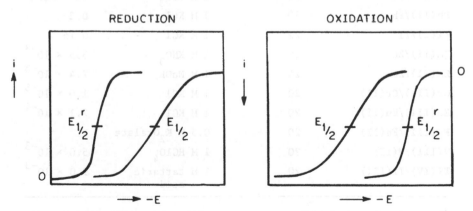

FIGURE 2.7 Differences between reversible ($E_{1/2}^r$) and irreversible ($E_{1/2}$) dc polarographic waves.

process is defined as reversible on the observation that $E_{1/2}^{r}$ for oxidation of Fe(II) and reduction of Fe(III) in oxalate media are equivalent (see Figs. 2.3 to 2.5). On the other hand, the large separation in $E_{1/2}$ values found for the reduction and oxidation processes of iron in pyrophosphate media (Fig. 2.6) is also explained in terms of the occurrence of an irreversible electrode process.

The position and shape of a polarographic wave are therefore determined by the kinetics of the electron transfer step. The important parameter we need to know in discussing the electrochemical reversibility of an electrode process is the heterogeneous charge transfer rate constant k_s. Table 2.2 gives values of k_s for some typical electrode processes.

TABLE 2.2 Electrochemical Rate Constants (k_s) for Some Electrode Processes Measured at Mercury Electrodes[a]

System	Temperature, °C	Supporting electrolyte	k_s (cm sec^{-1})
Bi(III)/Bi	25	1 M HClO$_4$	3.0×10^{-4}
Bi(III)/Bi	26	1 M HCl	2
Cd(II)/Cd	22	1 M NaClO$_4$	2.3
Cd(II)/Cd	22	0.5 M HCl	0.17
Cu(II)/Cu	25	1 M KNO$_3$	4.5×10^{-2}
Pb(II)/Pb	22	1 M KCl	0.2
Tl(I)/Tl	22	1 M KCl	0.15
Zn(II)/Zn	20	1 M KNO$_3$	3.5×10^{-3}
Zn(II)/Zn	25	1 M NaOH	7.4×10^{-5}
Cr(III)/Cr(II)	20	1 M KCl	1.0×10^{-5}
Eu(III)/Eu(II)	20	1 M KCl	2.1×10^{-4}
Fe(III)/Fe(II)	20	0.5 M K$_2$Oxalate	>1
V(III)/V(II)	20	1 M HClO$_4$	5.0×10^{-3}
Ti(IV)/Ti(III)	20	1 M tartaric acid	9.0×10^{-3}

[a]The majority of the values are taken from those tabulated in Note 14.

Chemical Reversibility. The question of chemical reversibility arises when a homogeneous chemical reaction precedes or follows the charge transfer step. This may occur, for example, when the product of the electron transfer step reacts with the solvent or when the electroactive species is formed from a nonelectroactive one. Frequently, substances that react with the product of the electron transfer reaction or coordinate with the electroactive species or its product are deliberately added to the solution, since one of the consequences of a coupled chemical reaction is a shift in the position of the polarographic wave. An obvious example of this would occur in dc polarography when two species have similar $E_{1/2}$ values in a particular medium. To achieve resolution, a reagent that preferentially complexes one of the species might be added.

Homogeneous chemical kinetics (or equilibria in the limiting case) coupled to an electron charge transfer process are usually treated theoretically via the normal concepts of solution kinetics, despite the fact that the reactions occur at an electrode surface.

By way of example, consider the electrode reaction in which A is reduced to B reversibly, but B is unstable and rearranges to give C, that is,

$$A + e \rightleftharpoons B \underset{k_2}{\overset{k_1}{\rightleftharpoons}} C \tag{2.25}$$

The treatment of the heterogeneous (electron transfer) step has been considered previously. For the chemical step

$$B \underset{k_2}{\overset{k_1}{\rightleftharpoons}} C \tag{2.26}$$

the usual expressions apply. Thus, the rate of disappearance of B, $-dB/dt$, or the formation of C, dC/dt, is given by

$$-\frac{dB}{dt} = \frac{dC}{dt} = k_1[B] - k_2[C] \tag{2.27}$$

At equilibrium,

$$K = \frac{[C]}{[B]} = \frac{k_1}{k_2} \qquad \frac{dB}{dt} = \frac{dC}{dt} = 0 \qquad\qquad (2.28)$$

and

$$\Delta G^\circ = -RT \ln K$$

The same considerations apply as for the electron transfer step, and reversibility is defined as a limiting condition with respect to the technique of measurement. That is, k_1 and k_2 may be sufficiently fast to define the system as reversible under dc polarographic conditions, but this does not mean that a fast scan rate linear sweep or high-frequency ac polarographic method would "see" the system as being an equilibrium one.

If the chemical step is reversible and coupled to a reversible charge transfer step, then the overall electrode process can be considered as reversible and amenable to a thermodynamic treatment as follows:

For the electron transfer step,

$$E = E^\circ + \frac{RT}{nF} \ln \frac{[A]}{[B]}$$

and for dc polarography, $E^\circ \approx E_{1/2}$, as will be shown in Chap. 3. Thus,

$$E = E_{1/2} + \frac{RT}{nF} \ln \frac{[A]}{[B]} \qquad\qquad (2.29)$$

In the presence of a reversible follow-up chemical reaction,

$$K = \frac{[C]}{[B]}$$

therefore

$$E = E^\circ + \frac{RT}{nF} \ln K + \frac{RT}{nF} \ln \frac{[A]}{[C]} \qquad\qquad (2.30)$$

Because the overall electrode process is reversible, the formal description as a two-stage electron transfer, coupled chemical reaction, and the overall reaction

$$A + e \rightleftharpoons C \tag{2.31}$$

are exactly equivalent.

For Eq. (2.31),

$$E = E^{\circ\prime} + \frac{RT}{nF} \ln \frac{[A]}{[C]} \tag{2.32}$$

From Eqs. (2.30) and (2.32) it follows that

$$E^{\circ\prime} = E^{\circ} + \frac{RT}{nF} \ln K \tag{2.33}$$

or

$$E'_{1/2} = E_{1/2} + \frac{RT}{nF} \ln K \tag{2.34}$$

in dc polarographic terms, and the ability of the chemical reaction to shift the position of the polarographic wave (characterized by $E_{1/2}$ or $E'_{1/2}$) is readily seen.

If the follow-up chemical reaction had been totally irreversible, the electrode process would be described by the reaction scheme

$$A + e \rightleftharpoons B \xrightarrow{k_1} C \tag{2.35}$$

In general, from analagous considerations, the presence of an irreversible chemical step implies

$$E'_{1/2} = E_{1/2} \pm f \text{ (rate constant)} \tag{2.36}$$

and the presence of a reversible chemical reaction implies

$$E'_{1/2} = E_{1/2} \pm f \text{ (equilibrium constant)} \tag{2.37}$$

The ± notation takes into account whether a following or preceding chemical reaction is occurring.

In organic polarography [12,13], many charge transfer steps are preceded or followed by a chemical step involving an acid-base reaction. Thus, for the reversible case, and writing similar equations as previously, would give a result of the kind

$$E'_{1/2} = E_{1/2} \pm \frac{RT}{nF} \ln K_a \pm \frac{RT}{nF} pH \tag{2.38}$$

where K_a is the acidity constant. Consequently $E'_{1/2}$ varies in a
predictable manner as a function of pH [14].

Dialkyl phenacyl sulfonium salts, for example, show acid-base
properties; the $E_{1/2}$ value for the polarographic reduction of the
C-S bond in dialkyl phenacyl sulfonium ions is strongly dependent on
pH [15]. Figure 2.8 shows the shift in $E_{1/2}$ as a function of pH
along with the well-known potentiometric titration curve. The
electrode process being considered is rather complex [15]; however,
it may be written for simplicity as

$$C_6H_5 - CO - \overset{(-)(+)}{CHSR_2} + H^+ \overset{K_a}{\rightleftharpoons} C_6H_5 - CO - \overset{(+)}{CH_2SR_2}$$

$$\downarrow 2e$$

$$C_6H_5 - \underset{\overset{||}{O}}{C} \cdots CH_2 + SR_2 \tag{2.39}$$

The reason for the frequent need to buffer in organic polarographic
analysis, i.e., to ensure that calibration and unknown solutions are
at the same pH, is now obvious.

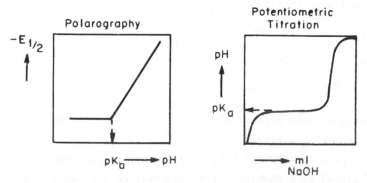

FIGURE 2.8 $E_{1/2}$ for reduction of dialkyl phenacyl sulfonium ions
shows pH dependence which indicates an acid-base equilibrium is involved in the overall electrode process [15]. pK_a value is between
7 and 8 depending upon the nature of the alkyl group. Potentiom-
etry proved the presence of this equilibrium.

In inorganic polarography, coordination numbers and stability constants of metal ion complexes play a similarly important role in determining the position and nature of the polarographic wave [14]. Figure 2.9 shows a dc polarogram of Sn(II) in 1 M NaF. A two-electron reduction wave and a two-electron oxidation wave can be seen. Note that the difference in sign of the current distinguishes the reduction and oxidation processes.

The reduction and oxidation waves can be described by Eqs. (2.40) and (2.41), respectively.

$$Sn(II) + 2e \rightleftharpoons Sn \ (amalgam) \tag{2.40}$$
$$Sn(II) \rightleftharpoons Sn(IV) + 2e \tag{2.41}$$

Table 2.3 shows the $E_{1/2}$ values as a function of fluoride concentration for both electrode processes. Data are taken from note 16 and this work can be consulted for experimental details.

In fluoride media, Sn(II) exists predominantly as SnF_3^-. Thus the reduction can be written as

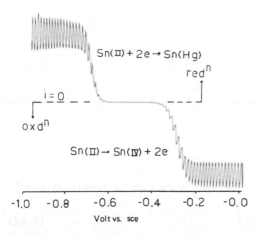

FIGURE 2.9   In NaF, tin(II) gives both reduction and oxidation waves.

TABLE 2.3 $E_{1/2}$ Values for Reduction and Oxidation of Sn(II) in Fluoride Media

Sn(II) + 2e ⇌ Sn (amalgam)		Sn(II) ⇌ Sn(IV) + 2e	
Fluoride concentration (M)	$-E_{1/2}$ (V vs. sce)	Fluoride concentration (M)	$-E_{1/2}$ (V vs. sce)
0.005	0.506	0.05	0.217
0.010	0.528	0.07	0.224
0.025	0.558	0.09	0.217
0.10	0.611	0.10	0.242
0.15	0.627	0.15	0.263
0.20	0.140	0.20	0.285
0.25	0.650	0.25	0.309
0.30	0.656		
0.35	0.663		
0.40	0.669		
0.45	0.673		
0.50	0.683		
0.55	0.683		
0.60	0.688		
0.65	0.686		
0.70	0.691		
0.75	0.693		

Data from Note 16.

$$\left.\begin{array}{c} SnF_3^- \overset{\beta_3^{-1}}{\rightleftharpoons} Sn^{2+} + 3F^- \\[2em] Sn^{2+} + 2e \rightleftharpoons Sn \text{ (amalgam)} \end{array}\right\} \equiv SnF_3^- + 2e \rightleftharpoons Sn \text{ (amalgam)} + 3F^-$$

$$(2.42)$$

where

β_3 = the stability constant for the formation of SnF_3^- and

$$\beta^{-1} = \frac{[Sn^{2+}][F^-]}{[SnF_3^-]} = \frac{1}{\beta}$$

Thus, neglecting differences in diffusion coefficients,

$$(E_{1/2})_{SnF_3^-/Sn(Hg)} = (E_{1/2})_{Sn^{2+}/Sn(Hg)} - f(\beta_3) - f([F^-])$$

and the higher the concentration the more negative the $E_{1/2}$. In terms of free energies, the shift in $E_{1/2}$ can be explained in terms of the lowering of free energy resulting from complexation. That is, SnF_3^- has a lower free energy than Sn^{2+} and its reduction is more difficult and occurs at more negative potentials.

The predominant fluoro complex of tin in oxidation state IV is SnF_6^{3-}. Thus the oxidation process can be written as

$$Sn(II) \begin{Bmatrix} Sn(II)F_3^- \overset{\beta_3^{-1}}{\rightleftharpoons} Sn^{2+} + 3F^- \\ 2e \, \big\Updownarrow \, 3F^- \\ Sn(IV)F_6^{2-} \overset{\beta_6^{-1}}{\rightleftharpoons} Sn^{4+} + 6F^- \end{Bmatrix} \equiv Sn(II)F_3^-$$

$$+ 3F^- \rightleftharpoons Sn(IV)F_6^{2-} + 2e$$

$$(2.43)$$

The direction in which $E_{1/2}$ shifts with increasing concentration of fluoride in this instance depends on the relative magnitudes of β_3 and β_6 and the coordination numbers. The oxidation-state-IV fluoro complex is in fact more stable than oxidation state II and also has a higher coordination number. Hence, a negative shift in $E_{1/2}$ occurs on addition of fluoride, in accordance with the usual free energy relationships.

In summary, for totally reversible systems, in which all the equilibria (electron transfer and chemical) are established instantly, the polarographic $E_{1/2}$ values are practically equal to overall standard oxidation-reduction potentials measured potentiometrically. Consequently, the half-wave potentials are a function of the equilibrium constants for the chemical reactions and the $E°$ for the charge transfer step, and little other than a shift in $E_{1/2}$ occurs on altering the solution conditions (e.g., pH, concentration of ligand).

Similarly, with ac, pulse, and other modern polarographic techniques,
shifts in wave position occur in a manner expected from the usual
thermodynamic relationships, and wave heights and shapes, etc., are
almost unaltered. No kinetic phenomena are operative, and the differences in time scale between different polarographic methods of
measurement do not enter into the discussion. The direction and
magnitude of the shift in wave position can be calculated from the
usual free energy relationships $\Delta G° = -RT \ln K$ and $\Delta G° = -nFE°$.
Figure 2.10 summarizes the situation with respect to dc polarography.
Generally, the shift toward the more positive potential of a cathodic
wave (reduction process) indicates that the reduction proceeds more
readily (increase in free energy). A shift toward more negative
potentials indicates that the reduction proceeds with greater difficulty (lowering of free energy). Conversely, for anodic waves corresponding to an oxidation process, the shift toward more negative
values indicates that the oxidation proceeds more readily, etc.

However, as soon as either the charge transfer step or chemical
reaction or both are considered to be nonreversible, the distinction
in time scale between the various polarographic methods becomes all

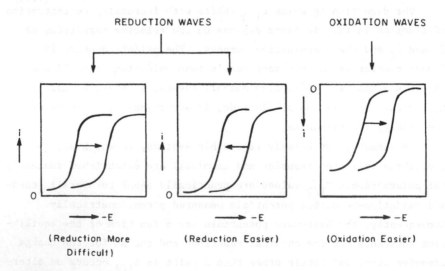

FIGURE 2.10 Shifts in position of a polarographic wave indicate how
reduction or oxidation occur.

important, and this must be clearly recognized. The following discussion should assist further in understanding the relevance of the
electrode process classification used in this book.

2.2.2 Classification of Reactions

It is convenient to classify the different possible reaction schemes
by using letters to signify the nature of the step: E signifies an
electron transfer step and C signifies a chemical step [17]. A
sequence in which a chemical reaction follows the electron transfer
would be designated as an "EC" reaction. In the equations that follow, substances designated A and B are assumed to be electroactive
in the potential range of interest, whereas substances Y and Z are
assumed not to be electroactive. Reductions will generally be considered, but naturally the same principles apply to oxidation waves.

Preceding Reaction, CE.

$$Y \underset{k_2}{\overset{k_1}{\rightleftharpoons}} A \qquad\qquad\qquad (C)$$

$$A + ne \rightleftharpoons B \qquad\qquad\qquad (E)$$

In this case the electroactive species A is generated by a preceding chemical reaction. An example of this case is said to be the reduction of formaldehyde in aqueous solutions [18–20], where formaldehyde
can exist as the nonreducible hydrated form methyleneglycol, $H_2C(OH)_2$,
and the reducible form formaldehyde, H_2CO. The chemical reaction

$$\underset{H}{\overset{H}{>}}C\underset{OH}{\overset{OH}{<}} \underset{k_2}{\overset{k_1}{\rightleftharpoons}} \underset{H}{\overset{H}{>}}C = 0 + H_2O \qquad\qquad (2.44)$$

precedes the electrode reaction, which is the reduction of H_2CO.
Other examples of this class are the reduction of weak acids and
their anions and the reduction of some metal complexes [18].

At the electrode surface, only A is reduced. This disturbs the
equilibrium at the electrode surface and additional A is formed.
Thus, the amount of A reduced and, therefore, the current flowing
through the cell are functions of the chemical rate constant k_1.

In dc polarography, for example, the magnitude of the limiting current of a polarogram in such circumstances is said to be kinetically controlled rather than the more common diffusion control found in the absence of a preceding chemical reaction. A formal description of this dc case is given in Chap. 3.

In view of the small magnitude of kinetic currents relative to diffusion-controlled limiting currents, the sensitivity (current per unit concentration) of such electrode processes is rather poor, and their use in analytical chemistry is restricted on these grounds at least. Furthermore, being governed by a rate-determining chemical step, kinetic currents are potentially extremely sensitive to even subtle changes in conditions, and unless it can be rigorously ensured that the media for recording the calibration curves and the solution to be determined are exactly identical, there is considerable danger involved in using this class of electrode process. That is, any foreign entity in an unknown solution can potentially influence the rate constant, giving rise to erroneous results when referred to a calibration curve prepared on pure solutions of the compound being determined. Similar considerations apply to the use of CE electrode processes with all modern polarographic methods, and the use of a purely kinetically controlled electrode process is not recommended if it can be avoided. This recommendation again emphasizes that the ability to classify electrode processes is an integral part of the systematic use of polarographic methodology; means for doing this will be presented with respect to the commonly used methods.

Following Reaction, EC

$$A + ne \rightleftharpoons B \qquad\qquad\qquad\qquad (E)$$

$$B \underset{k_2}{\overset{k_1}{\rightleftharpoons}} Y \qquad\qquad\qquad\qquad (C)$$

This is one of the most common mechanisms in electrochemistry in which the product of the electrode process reacts to produce a nonelectroactive species. For example, the reduction of many organic species are followed by protonation reactions, reaction with the solvent or dimerization, etc. [21].

$$HO-\langle\!\!\!\bigcirc\!\!\!\rangle-NH_2 \rightleftharpoons O=\langle\!\!\!\bigcirc\!\!\!\rangle=NH + 2H^+ + 2e$$

$$(E) \quad (2.45a)$$

$$O=\langle\!\!\!\bigcirc\!\!\!\rangle=NH + H_2O \longrightarrow O=\langle\!\!\!\bigcirc\!\!\!\rangle=O + NH_3$$

$$(C) \quad (2.45b)$$

$$RR'C = O + e + H^+ \rightleftharpoons RR'\overset{\cdot}{C}OH \qquad (E) \quad (2.46a)$$

$$2RR'\overset{\cdot}{C}OH \longrightarrow \underset{\underset{OH\ OH}{|\ \ |}}{RR'C-CRR'} \qquad (C) \quad (2.46b)$$

The product of the electrode process may also isomerize or rearrange to its most stable form. For example, complexes of the form $[M(CO)_2(DPE)_2]^{x+}$ obviously can exist as geometric cis or trans isomers. (DPE = 1,2 bisdiphenylphosphinoethene), and the oxidation of *cis*-Mo(CO)$_2$(DPE)$_2$ gives rise to the following electrode process [22]:

$$cis\text{-Mo(CO)}_2(DPE)_2 \rightleftharpoons cis\text{-}[Mo(CO)_2(DPE)_2]^+ + e \qquad (E) \quad (2.47a)$$

$$cis\text{-}[Mo(CO)_2(DPE)_2]^+ \longrightarrow trans\text{-}[Mo(CO)_2(DPE)_2]^+ \qquad (C) \quad (2.47b)$$

In dc polarography, the presence of a follow-up chemical reaction has no effect on the limiting current, and the limiting current remains diffusion controlled with the value expected for the A + ne $\rightleftharpoons$ B step. Analytical use of this class of electrode process provides no new difficulties. However, the position of the wave is a function of the rate constant k_1. As already discussed, the greater the influence of the follow-up chemical reaction, the more positive becomes the $E_{1/2}$ value. The influence of the time scale of any polarographic experiment, however, now also becomes extremely important. The shorter the time scale, obviously, the smaller the influence of the follow-up reaction. In dc polarography, for example, the use of very short drop times, as shown in Chap. 4, may enable the follow-up reaction to be quenched or eliminated completely. If this can be achieved, the description of the electrode process reverts to a simple charge transfer step. On the other hand, the deliberate use of extremely long drop times provides more time for equilibrium to be obtained, and complete chemical reversibility may sometimes be achieved in this manner.

The influence of this class of electrode process in the analytical use of linear sweep voltammetry, ac polarography, and other techniques is generally much more important than in dc polarography. In some polarographic techniques, the peak height used in constructing an analytical calibration curve is a function of the rate constant, unlike the limiting current in dc polarography, and some caution has therefore to be exercised. Indeed in such instances, to provide the best analytical procedure, one would ideally endeavor to either decrease the time scale of the experiment to outrun the chemical reaction or increase the time scale to allow chemical equilibrium to be reached. In either of these limiting cases, the waves revert to the reversible case (assuming reversible charge transfer), and the analytical procedure is simplified. Discussion in this area will be amplified considerably when examining specifically the various polarographic methods.

Disproportionation, Catalytic, and Other Regenerative Mechanisms. In this class of electrode process the starting material is regenerated by a chemical reaction occurring subsequent to the initial charge transfer step (C'). Thus, the electroactive material is effectively reduced more than once, and starting material is produced at the electrode surface by both diffusion and the chemical step.

In dc polarography, the limiting current flowing through the cell is therefore larger than the diffusion-controlled value, the enhancement being a function of the rate constant pertaining to step C'.

The reduction of U(VI) provides an example of a disproportionation mechanism in mineral acid and other media [23]. The electrode process can be summarized as

$$U(VI) + e \rightleftharpoons U(V) \qquad\qquad\qquad (E) \quad (2.48a)$$

$$\underset{\Big\lfloor}{} 2U(V) \xrightarrow{k} U(VI) + U(IV) \qquad\qquad (C') \quad (2.48b)$$

The U(VI) generated by step C' is immediately available for further
reduction.

In some media, step C' is so slow that effectively only step E
occurs and a reversible one-electron step is found. In other media,
step C' is so rapid that the disproportionation virtually goes to
completion within the lifetime of the polarographic experiment. In
the limiting current region of a dc polarogram, the electrode process
would then be equivalent to

$$U(VI) + 2e \rightleftharpoons U(IV) \qquad\qquad\qquad (2.49)$$

and the wave height would correspond to a two-electron reduction.
In other media, the wave height lies intermediately between an apparently one- and two-electron step, depending on the value of k in
Eq. (2.48b). In the determination of uranium by dc polarography,
the recommended procedure [23], therefore, is to choose a medium
where either the disproportionation step is negligible or else it is
so fast that effectively a simple two-electron reduction occurs.
The same recommendation applies to ac, pulse, linear sweep, and all
other polarographic methods. In general, it can be seen that wher--
ever possible, avoidance of a chemical rate constant is recommended,
on the grounds that the potential for interference is greatly enhanced.

The catalytic mechanism can be summarized as

$$A + ne \rightleftharpoons B \qquad\qquad\qquad\qquad (E)$$

$$\underset{\Big\lfloor}{} B + Z \xrightarrow{k} A \qquad\qquad\qquad\qquad (C')$$

Here the product B of the electrode reaction reacts with a nonelectroactive substance Z to regenerate A. Since Z is capable of oxidizing B, it should also be reducible at the electrode according to
the usual thermodynamic considerations. However, if the electrode
process for Z is kinetically very slow, electrochemical reduction

can be far more negative than the value predicted from thermodynamic
considerations. The chemical redox process, however, need not exhibit
a kinetically slow step, and step C' can be extremely fast, giving
rise to a catalytic process.

The limiting current in a dc polarogram, like the disproportionation case, obviously will be enhanced compared with the diffusion-
controlled value, but to a much greater extent. If k is fast, remarkable increases in the limiting current can occur and very high
currents for very low concentrations of A can sometimes be obtained.
The catalytic wave represents one of the cases where the current
increase being a function of a chemical rate constant may be an
asset. Despite the fact that catalytic waves do not generally exhibit a high degree of specificity in the sense that the extent of
catalysis may frequently be altered by the presence of almost any
entity, the remarkably high sensitivity may be extremely valuable
in certain analytical applications where solution conditions and
experimental work can be carefully controlled.

By way of example the kinetic wave observed for tungsten in the
presence of hydrogen peroxide and oxalate can be considered [24].
Peroxide induces a catalytic current which is proportional to the
tungsten concentration. Table 2.4 shows the effect of diverse ions,
and the nonspecificity of this class of electrode process is clearly
demonstrated.

Blažek and Koryta [25,26] have investigated the dc polarographic
catalytic wave for reduction of the titanium(IV) oxalate complex in
the presence of hydroxylamine. Hydroxylamine oxidizes the
titanium(III) formed by reduction at the electrode back to
titanium(IV). Equations (2.50a) and (2.50b) summarize the electrode
reaction.

$$\text{Ti(IV)} + e \rightleftharpoons \text{Ti(III)} \qquad\qquad\qquad\qquad (E) \quad (2.50a)$$

$$\text{Ti(III)} + \text{NH}_2\text{OH} \xrightarrow{k} \text{Ti(IV)} + \text{OH}^- + \overset{\cdot}{\text{N}}\text{H}_2 \qquad (C') \quad (2.50b)$$

Further different, but more complicated, examples are contained in
Notes 27 and 28, and the observation of high sensitivity coupled
with nonspecificity can be seen to be characteristic of catalytic
mechanisms.

TABLE 2.4 Effect of Diverse Ions on the Polarographic Determination of Tungsten Using a Catalytic Wave[a]

Ion	Concentration, M	Relative error, %	Interference
Vanadium	3.0×10^{-6}	−6.8	Yes
	3.0×10^{-5}	−16.7	Yes
	3.0×10^{-4}	−19.5	Yes
Chromium(VI)	3.0×10^{-6}	−1.8	No
	3.0×10^{-5}	−4.1	Yes
	3.0×10^{-4}	−16.7	Yes
Iron(III)	3.0×10^{-6}	−4.1	Yes
	3.0×10^{-5}	−16.7	Yes
	3.0×10^{-4}	−46.2	Yes
Molybdenum(VI)	3.0×10^{-6}	+0.9	No
	3.0×10^{-5}	+11.7	Yes
	3.0×10^{-4}	+79.6	Yes
Titanium(IV)	3.0×10^{-6}	−24.4	Yes
	3.0×10^{-5}	−24.4	Yes
	3.0×10^{-4}	No peak	Yes

[a]3.0×10^{-5} M WO_4^{2-}, 8.0×10^{-2} M H_2O_2, 1.0×10^{-2} M $H_2C_2O_4$
Data taken from Note 24.

ECE Reaction

$$A_1 + n_1 e \rightleftharpoons B_1 \qquad (E_1)$$

$$B_1 \rightarrow A_2 \qquad (C)$$

$$A_2 + n_2 e \rightleftharpoons B_2 \qquad (E_2)$$

Two types of ECE reactions are possible. In one, the substance A_2 is reducible at the same or more positive potentials than A_1, and thus it is immediately reduced at the electrode. The oxidation of aromatic hydrocarbons in nonaqueous solvents [29] usually proceeds via this mechanism.

$$A \rightleftharpoons A\overset{+}{\circ} + e \qquad\qquad (E_1) \quad (2.51a)$$

$$A\overset{+}{\circ} + Z \overset{k}{\rightarrow} B \qquad\qquad (C) \quad (2.51b)$$

$$B \rightarrow C + e \qquad\qquad (E_2) \quad (2.51c)$$

Z can be the solvent, small amounts of water, or other impurities.

Many other examples of this mechanism occur in organic polarography [30]. For example, the reduction of p-nitrophenol occurs as follows:

In the ECE reaction of this kind, the limiting current in dc polarography again depends on the rate of step C. In the case where $n_1 = n_2 = 1$, the overall electrode process can appear as a one-electron or two-electron step or as an intermediate between these two extremes. Analytically, the ideal experiment should use conditions where $(n)_{apparent} = 1$ or 2, so that the limiting current is independent of the chemical rate constant, for reasons given previously.

The second kind of ECE electrode process occurs when the second reduction E_2 occurs at more negative potentials than the first, E_1. Two polarographic waves result in such a circumstance. The first

wave is the EC mechanism given previously. The second or more negative wave is then obviously controlled by the kinetics of step C.

Many other reactions schemes are possible. Variations of the
above also depend upon the reversibility or irreversibility of the
electron transfer and chemical steps. For example, the EC case
might involve four possible subcases involving reversible (R) or
irreversible (I) reactions: RR, RI, IR, II. While the examples
above usually were presented for the case of electrochemical reductions, obviously they apply to oxidations as well.

In developing or using a modern polarographic method, the most
systematic approach results when the nature of the electrode process
is understood. If this is the case, the choice of the best possible
polarographic techniques can be readily ascertained. The above examples of possible electrode processes are by no means exhaustive.
They are given to show that in polarography we are dealing with a
kinetic problem in which the electron transfer and diffusion-control
or chemical steps, etc., can be rate-determining.

2.2.3 Adsorption, Film Formation, and Other Surface Phenomena

An electrode process occurs at a solution-solid interface. Thus,
possible influences in polarography arising from surface phenomena
such as adsorption or film formation, from say an insoluble product,
also need to be considered in giving a complete description of the
electrode process. Such phenomena can be treated as additional
nuances to those describing the kinetics of the electron transfer
step and coupled chemical reactions. Adsorption and/or film formation can alter the shape and position of a polarographic wave, control the limiting current in dc polarography, and, in general, exert
most of the influences attributable to kinetic phenomena. Products
or reactants may be adsorbed and reactions with adsorbed ligands,
etc., may be postulated in any given electrode process.

Thus, an electrode process may be written as

$$A \rightleftharpoons A(ads) + e \rightleftharpoons B \qquad (2.53)$$

or

$$A + e \rightleftharpoons B \rightleftharpoons B(ads)$$

or

$$A + Z(ads) \rightleftharpoons AZ(ads) + e \rightleftharpoons B$$

There are probably no more despised words in polarography than *adsorption* or *film formation*, as they are frequently associated with "bad news." In dc polarography, maxima, erratic drop behavior, non-linear calibration curves, inhibition of the electrode process, split or drawn-out waves, and other analytically undesirable phenomena are frequently attributed to or associated with adsorption/film formation. Notes 31 to 38 provide some examples of the influence of surface phenomena. Specific examples are presented and discussed in subsequent chapters.

However, many modern polarographic techniques minimize, eliminate, or even exploit the adsorption phenomena that give analytically undesirable data in dc polarography [33]. In presenting a discussion of individual polarographic techniques, the need to consider not only their response to the homogeneous and heterogeneous kinetic aspects of the electrode process, but their response to surface phenomena, can now be appreciated.

2.3 INSTRUMENTATION

At the outset it needs to be emphasized that in discussing instrumentation the author assumes that the reader is using, or is considering the use of, polarography as a modern instrumental method of analysis. The underlying assumption, therefore, is that if the polarographic method is chosen, it will necessarily have been done so on the basis that it is competitive with other commonly used analytical methods. On this assumption, the criterion that the technique should be amenable to a substantial degree of automation, and therefore suitable for routine use, is considered to be of

paramount importance. Similarly, in line with the use of other
modern instrumental methods, it is assumed that a considerable degree of sophistication will be required and utilized in performing
and interpreting the experiment. Manual recording polarographs, for
example, are therefore excluded from discussions in the section on
instrumentation, despite the fact that they can serve several useful
functions in other contexts [1]. Based on this criterion, other
areas traditionally covered in discussions of polarographic instrumentation are also omitted.

2.3.1 The Two-Electrode Polarograph

Classically, dc polarographs have been represented by a circuit of
the kind shown in Fig. 2.11. Similarly, the description of a polarographic measurement accompanying this diagram used to be something
like the following:

> In polarography, a dropping mercury electrode is dipped into
> an electrolysis cell containing about 5 to 50 ml of solution.
> The inner diameter of the capillary is about 0.05 to 0.08 mm
> and the capillary is connected by tubing to a mercury reser-
> voir, which is placed 30 to 80 cm, above the capillary orifice.

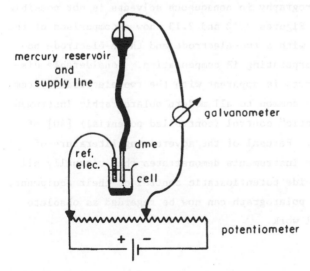

FIGURE 2.11 Circuit typically used to represent a two-electrode
polarograph.

By varying the height of the mercury head, the pressure of the
mercury column can be adjusted to give a drop time of about 2
to 8 sec. Atmospheric oxygen is removed from the solution by
bubbling an inert gas such as nitrogen, hydrogen, or argon
through the solution. A reference electrode is present in the
solution along with the dme, and the electrochemical circuit is
formed by connecting the cell to a potentiometer, which permits
the application of any voltage across the dme and reference
electrode. The resulting current flowing at potential E, is
measured directly by means of a galvanometer or indirectly by
measuring the potential drop across a standard resistor in-
serted in the circuit. A graphical plot of mean current versus
potential then gives the dc polarogram.

The modern polarograph does not even remotely resemble the
preceding description, and virtually all aspects of the polarographic
experiment have been altered substantially in the last two decades.
For those chemists historically inclined, the evolution of the
polarograph has recently been presented by Ewing [39]. One of the
greatest drawbacks in the classical two-electrode polarograph is that
the potential is applied across the entire cell, rather than across
the working electrode (dme in polarography)-solution interface.
Data recorded with such a system are considerably in error if solution resistances and the resultant ohmic, iR, drop across them are significant. Thus, polarography in nonaqueous solvents is not possible
with such a circuit. Figures 2.12 and 2.13 show a comparison of the
cells and polarograms with a two-electrode and three-electrode potentiostat system incorporating iR compensation. Considerable distortion of the i-E curves is apparent with the two-electrode system.
One key characteristic common to all modern polarographic instrumentation is "potentiostatic" control (controlled potential) [40] of
the working potential. Perusal of the advertizing literature of
commercially available instruments demonstrates that virtually all
manufacturers now provide potentiostatic control of their equipment,
and the two-electrode polarograph can now be regarded as obsolete
for routine analytical work.

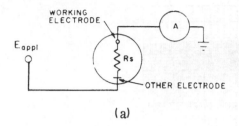

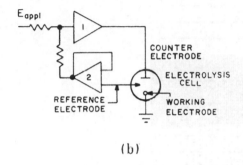

FIGURE 2.12 (a) Simple two-electrode cell including resistance. (b) Potentiostatic three-electrode system. [Reproduced from Anal. Chem. *44*, 75A (1972). © American Chemical Society.]

2.3.2 The Three-Electrode Polarograph

Potentiostatic Control. In the two-electrode experiment, if the reference electrode is constructed properly, its potential relative to the solution will not change with current flow. The total potential applied to a cell will consist of the constant potential of the reference electrode, a potential between the solution and the working electrode (dme in polarography), and a potential through the solution owing to its resistance. Under these conditions the total current flowing through the cell will depend on the oxidation or reduction processes taking place at the working electrode at a particular potential, as well as in the capacitive charging current required to charge the double layer (see pp. 72–76). Since the current flowing through the cell depends on the potential between the solution and the working electrode, it is important that this potential be well known and controlled. However, in the two-electrode arrange-

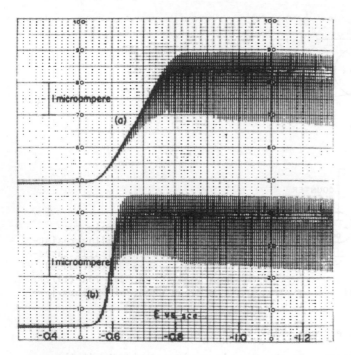

FIGURE 2.13a Polarograms of 5×10^{-4} M Cd (II) in 0.1 M HCl recorded
with a 60 kΩ fiber-type calomel reference electrode. The same reference electrode was used to obtain both curves, but in the lower one
a platinum auxiliary electrode was added and the polarogram recorded
with a three-electrode cell. [Reproduced from J. Chem. Educ. *41*,
202 (1964).]

ment, the potential which is actually controlled is that across both
the working electrode and the reference electrode. The potential of
the reference electrode is constant. Hence, the potential of the
working electrode varies in the same way as the total cell voltage,
provided that the ohmic resistance of the solution does not cause an
appreciable potential drop. This requires that the solution have
either a low specific resistance or that the two electrodes be placed

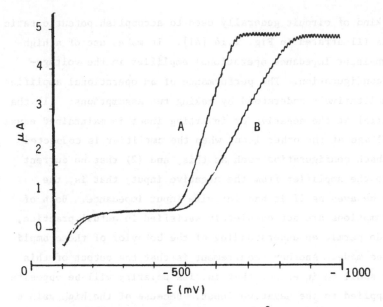

FIGURE 2.13b Comparison of waves for 10^{-3} M Sn(II) with two- and three-electrode systems in nonaqueous media: solvent, isopropyl alcohol; supporting electrolyte, 5×10^{-2} M LiCl. (A) Three-electrode systems give the true $E_{1/2}$ value of -570 mV. (B) Two-electrode systems give an apparent $E_{1/2}$ value of -675 mV. [Reproduced from Bull. Soc. Chem. Fr. *1966*, 811.]

very close together. It was to avoid the problems of solution resistance that three-electrode potentiostatic systems were developed. The three-electrode system provides greater flexibility in the location of the reference and working electrodes and minimizes the effect of solution iR drop. In addition, it has the advantage that virtually no current passes through the reference electrode.

The general requirements for a three-electrode control system in polarographs are that it consist of the dme, a reference electrode, and an auxiliary electrode, with the external circuit arranged so that potential control is maintained between the dme and the reference electrode, but so that cell current passes between the dme and the auxiliary electrode [Fig. 2.12(b)].

There are almost as many circuits as potentiostats. However, the circuits and discussion given here should enable a general understanding of a potentiostat to be gained.

The kind of circuit generally used to accomplish potentiostatic control is illustrated in Fig. 2.14 [41]. It makes use of a high-gain, high-input impedance operational amplifier in the voltage-follower configuration. The performance of an operational amplifier can be qualitatively understood by making two assumptions: (1) that the potential at the negative or inverting input is maintained equal to the voltage at the other input when the amplifier is connected in a feedback configuration such as this, and (2) that no current flows into the amplifier from the negative input; that is, the amplifier behaves as if it had infinite input impedance. Both of these assumptions are not completely satisfied in actual practice, but they do permit an understanding of the behavior of these ampli-fiers to be made. Another requirement is that the output of this amplifier shall be inverted; that is, its polarity will be opposite to that applied to the negative input. Because of the high gain of the amplifier, a very small difference in potential between the two inputs is necessary to obtain an output.

Now, consider the effect of this circuit, as shown, when the output of the amplifier goes to the auxiliary electrode; the ref-erence electrode is connected to the negative input; the dme is con-nected directly to ground; and the positive input is connected to the source voltage, which is to be applied to the cell. If a

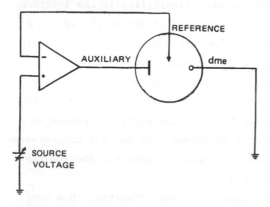

FIGURE 2.14 Three-electrode potentiostatic control system for polarography. (By courtesy of Melabs [41]).

positive potential is applied to the noninverting or positive input
of the operational amplifier by the source voltage, the output of the
amplifier will become more positive; hence, the auxiliary electrode
will also become more positive. When current passes through the
solution, the potential of the solution itself will become more positive relative to ground. The potential of the reference electrode
will in turn become more positive, since the potential between it
and the solution remains constant. But the potential at the reference electrode is applied directly to the inverting or negative input
of the operational amplifier, and if this becomes more positive, then
the output of the amplifier will become negative. Therefore, we have
a feedback system which is self-stabilizing and will act to automatically make the potential of the reference electrode equivalent
to that applied by the source voltage to the positive input of the
amplifier. Under these conditions it can be said that the potential
between the reference electrode and ground is equivalent at all times
to the potential applied by the source voltage to the positive input
of the follower amplifier. But since the dme is connected directly
to ground in the configuration shown, the voltage across the cell
from the reference electrode to the dme is just equal to the source
voltage.

The current, on the other hand, does not pass through the reference electrode, but instead passes only through the auxiliary
electrode and the dme. Hence, the purpose of maintaining potential
control between two electrodes without permitting any current to
pass through the reference electrode has been accomplished. This
is one of the major reasons for the three-electrode system. In
addition, the follower amplifier supplies current to the dme adequate
to maintain potential control, regardless of the solution resistance.
Therefore, in solutions of very high resistance, the voltage drop
between the auxiliary electrode and the dme can become very high
(up to the limit of the amplifier's output capability). When this
happens, a voltage gradient exists through the solution, and it is
important that the reference electrode be located in the field in a

position which minimizes the potential between it and the dme. In
general, this is on the side of the mercury drop opposite the auxiliary electrode and as close to the drop as possible. Thus, the
flexibility of position of the reference electrode can be used to
minimize the problem of iR drop through the solution.

Other circuit configurations used to achieve three-electrode
potential control in an electrochemical cell may be analyzed in the
same way as this one (that is, to indicate that two loops exist, one
the potential control loop and the other the current loop, and that
these both pass through the dme but are independent of each other
at the other electrode). For additional and detailed discussion on
the three-electrode potentiostat and related aspects of this kind
of instrumentation, Notes 42 to 70 are recommended for further
reading.

Voltage Input, Current Readout. To be operational, the three-
electrode polarograph basically requires these components: an input
voltage, a potentiostat, a control system, and a current readout.
Figures 2.15 and 2.16 show two representations incorporating each
of these features to give the complete polarograph.

Obviously, three electrodes are employed as required for potentiostat control. In Fig. 2.16 [43], the reference electrode is in
the summing loop and the auxiliary electrode is in the feedback loop,
as required. For versatility, the input voltage E_{in} is usually composed of three voltage sources. These are E_i, which can be set to
a fixed value; E_s, which can be varied linearly with time; and E_r,
the reference electrode voltage. An external voltage E_{ex} can also be
applied if required, as in ac and differential pulse polarography, for
example, as will be seen in Chaps. 6 and 7, respectively. These voltages (E_i, E_s, E_r, and E_{ex}) are additive in the summing loop. E_{in} is
sensed at the summing point of the amplifier. The amplifier then generates current i_f in the feedback loop through the cell by means of the
working and auxiliary electrodes, so that the voltage at the working
electrode surface is equal and opposite to E_{in}. Thus, the potential at
the surface of the working electrode is selected by adjusting E_{in} and

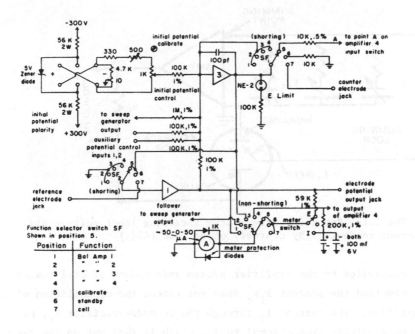

Function selector switch SF
Shown in position 5.

Position	Function
1	Bal Amp 1
2	" " 2
3	" " 3
4	" " 4
5	calibrate
6	standby
7	cell

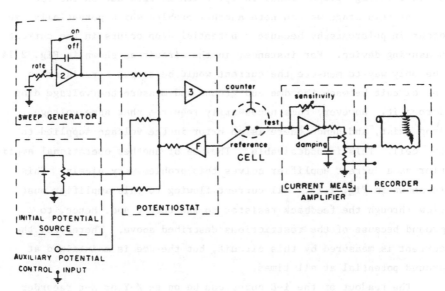

FIGURE 2.15 Block diagram of controlled-potential polarograph:
Upper diagram gives circuit of potential control section, lower
gives complete polarograph. [Reproduced from J. Chem. Educ. *41*,
202 (1964).]

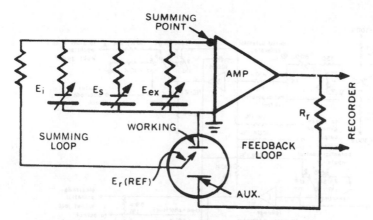

FIGURE 2.16 Potentiostatic system including input voltage and current readout. (By courtesy of Beckman [43].)

is controlled by the amplifier system independent of cell resistance R_c provided the product $i_f R_c$ does not exceed the capabilities of the amplifier. The current i_f through the recorder resistor R_r, produces a voltage proportional to i_f, which is read out on the recorder.

At this stage we can note another problem which potentially can occur in polarography because a potential drop occurs in the current measuring device. For instance, in the circuit as shown in Fig. 2.14 the only way to measure the current would be to put a resistor in the circuit between the dme and ground and measure the voltage drop across it. However, this necessarily requires that some voltage drop exist, and hence there is an error in the voltage supplied to the cell. This is undesirable. The use of another operational amplifier as a current amplifier solves this problem very nicely. This is shown in Fig. 2.17. All current flowing to this amplifier must flow through the feedback resistor to the output and, hence, to ground because of the restrictions described above. Therefore, the current is measured by this circuit, but the dme is maintained at ground potential at all times.

The readout of the i-E curve can be on an X-Y or X-t recorder or an oscilloscope. Alternatively, digital readout and/or computer printout of data are also possible after analog-to-digital

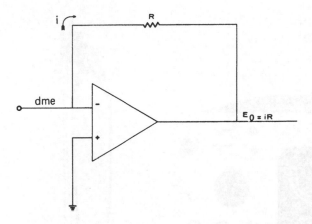

FIGURE 2.17 Measurement of current in a three-electrode system by a current amplifier. (By courtesy of Melabs [41].)

conversion. Figure 2.18 shows a commercially available instrument. Oscilloscope and digital readout are obviously available. Figure 2.19 shows a range of other commercially available instruments with X-Y recorder and other kinds of display and features. A high degree of automation is now available. With the arrival of the minicomputer and microprocessor in the laboratory, the experiment is likely to come more and more under computer control in the future (see Chap. 10) and the current sophistication of polarographic instrumentation is in line with developments in other areas of instrumental analysis.

Uncompensated Resistance. Positive Feedback Circuitry. The importance of resistance effects in polarography has been well recognized for a long time [1]. Indeed, along with the unwanted charging current contribution discussed elsewhere, uncompensated resistance ranks as the chief villain in polarography, and a considerable amount of research has been undertaken over the last 20 years or so, specifically directed toward eliminating or minimizing resistance effects. Resistance effects are mainly manifested in any polarographic technique as ohmic potential losses (iR drops) and even with the advent of the three-electrode potentiostat problems associated with iR drop remain a source of much concern in electrochemistry.

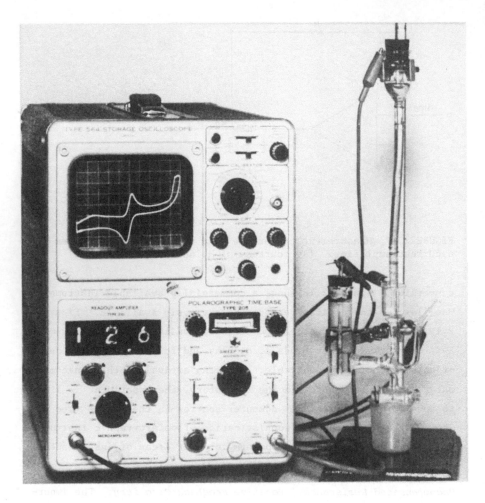

FIGURE 2.18   CHEMTRIX Polarographic Analyzer System including digital
readout.

Previous discussion was somewhat idealized, and although the three-
electrode potentiostat removes many of the sources of iR drop operative in the common polarographic methods, it does not eliminate
all the sources of iR drop [44,45,50-54,66,70-74]. To a good approximation, the only uncompensated resistance is the ohmic drop between
the tip of the reference electrode and the summing point of the
current-measuring amplifier. That is, the uncompensated resistance
basically includes the resistance of the dme itself and the solution
resistance between the dme and the reference electrode.

FIGURE 2.19a   Princeton Applied Research Corporation Model 170
Electrochemistry System enables a wide range of polarographic techniques to be obtained from the one unit.

FIGURE 2.19b   Princeton Applied Research Corporation Model 174A
Polarographic Analyzer. To perform polarographic techniques other
than dc or pulse, a modular approach is available with this instrumentation. Compare this with Fig. 2.19a and c, where many techniques
are available in the one unit.

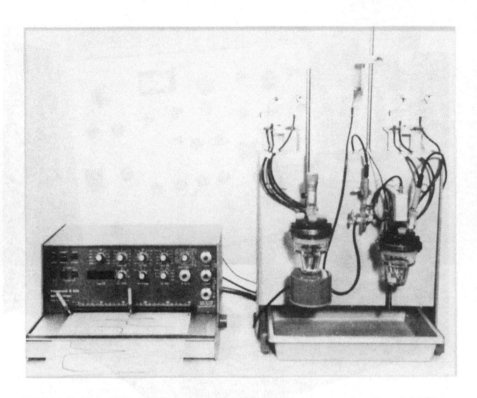

FIGURE 2.19c   Metrohm Polarecord E506 and Polarography Stand E505.
This instrumentation can perform dc polarography, fundamental and
second harmonic ac polarography, a range of pulse techniques, and
many other methods described in this book.

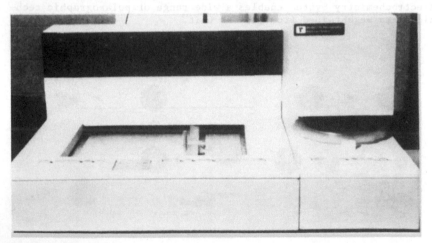

FIGURE 2.19d   Computerized (microprocessor controlled) polarograph
with automatic cell changer.   Princeton Applied Research Corporation
Model 374 Polarographic Analyzer System (see also Chap. 10).

Additional electronic circuitry to eliminate the remaining uncompensated resistance in phase-selective ac polarography has been presented by Hayes and Reilley [53]. A method of general applicability has been suggested by Booman and Holbrook in which an additional positive feedback loop is incorporated into the three-electrode potentiostat [51]. Several successful applications of this technique have been reported, and the modification is now available in commercially available instrumentation. The positive feedback circuitry derives a signal from the measured current flow and feeds it back to the point of the current-measuring amplifier. This provides a means of overcompensating the three-electrode potentiostat.

Too much positive feedback will result in instability (observed on an oscilloscope as oscillation of the summing amplifier), but the application of an amount of feedback, just under that required for instability, will compensate for the iR drop across the uncompensated resistance. In most work employing positive feedback, instability of the potentiostat is encountered unless something less than the ideal of 100% iR compensation is used. However Brown et al. [60] showed that 100% iR compensation can be realized over a widespread range of operating conditions via the use of high-performance operational amplifiers and appropriate stabilization techniques, and ideal results with very high resistance solutions could be obtained. In principle, therefore, modern polarographic instrumentation can provide accurate readout of the faradaic component, unencumbered by effects of iR drop over a wide range of operating conditions. At present, however, despite this possibility in analytical work, positive feedback circuitry would normally only be used in high-resistance nonaqueous solutions, as under most operations conditions, uncompensated resistance terms present with a three-electrode system exert negligible influence on the determination. This conclusion is reached on the grounds that with most positive feedback systems, the adjustment of the circuitry to give close to 100% compensation, while avoiding oscillation, still remains a nontrivial task, and therefore, such circuitry would be employed by an analytical chemist only if essential. Indeed, at one stage such circuits have been the

subject of outspoken criticism [75]. Recent suggestions that electrolyte resistance can be completely removed without producing oscillation [72] and that dynamic compensation of the cell resistance is
possible via a self-adjusting positive feedback circuit [73,74,76]
could mean, however, that in the future, practical objections will
no longer be valid. Under such circumstances the analytical chemist
would then be in the ideal situation of conveniently being able to
work in the absence of iR drop.

An alternative approach to compensate for iR drop is the basis
of a digital rather than analog pontentiostat [77] and a current
interruption technique [78] which is closely related. In contrast
to the preceding methods, the potential of the working electrode
with respect to the reference electrode is measured with no current
flowing through the cell (i.e., with zero iR drop). If the measured
value is different from that required, a charge of correct size is
injected to make them equal. The total charge per unit time gives
the average current through the cell when using this approach.

Mechanical Control of the Drop Time in Polarography. As will be seen
in subsequent chapters, most modern polarographic techniques incorporate the use of mechanically controlled drop times with the dme
rather than the gravity-control or normal drop time methods conventionally depicted with polarographic instrumentation. Study of
manufacturers' literature shows the widespread nature of mechanical
drop knockers or hammers which tap, vibrate, or displace the dme
periodically and enable the mercury drop to be dislodged at prese-
lected and controlled drop times. Figure 2.20 shows a commercially
available drop knocker. With many modern polarographic techniques,
a controlled drop time is an integral part of the experiment; in
others, while not essential, it invariably provides considerable
advantage, and more and more analytical polarographic work is incorporating this approach to obtain a controlled and regulated drop
time. The use of controlled drop times with the dme is now becoming
the standard mode of operation, and theoretical discussions in terms
of controlled drop time is given with all techniques in this book.

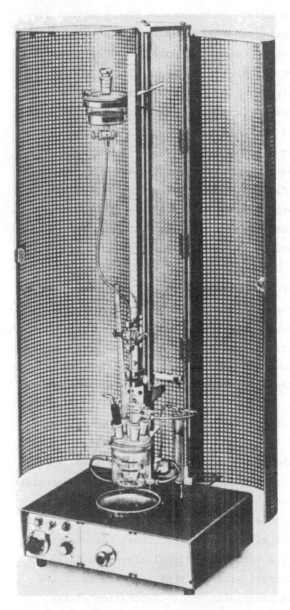

FIGURE 2.20 The Tacussel Electromagnetic Hammer enables the drop
time of the mercury electrode to be controlled between 0.01 and
100 sec. Note Faraday cage surrounding cell to achieve electrical
shielding. (By courtesy of Tacussel Electronique.)

A General Description of a Modern Polarograph. From previous discussion, a general description of a modern polarograph is virtually
impossible to provide because of the wide spectrum of options available at every stage. However, Figs. 2.21 and 2.22 [79,80] show
block diagrams of two instruments to give a representation of what
components the complete polarographic instrumentation might contain,
including computer control in Fig. 2.22. Trends toward multifunctional instrumentation are now apparent [81,82]. The schematic
representation of the analog instrumentation in Fig. 2.21 resembles
closely that provided in the most recent commercially available
multifunctional instrumentation [e.g., Fig. 19(a)-(c)] now being
used in many analytical laboratories. However, computerized data
acquisition or digital readout is becoming more frequent in polarographic instrumentation. Chapter 10 is devoted to the use of micro-
processes and minicomputers in polarography.

Reference and Auxiliary Electrodes. The use of a potentiostat has increased the scope of reference electrodes available for use in polarography. With the two-electrode system, the choice has been restricted
to those having both a low resistance and a reference potential which
is unaffected by the passage of current. Now any electrode with a
reasonable impedance that gives a reproducible potential under potentiometric (zero current) conditions is suitable. In nonaqueous
solvents, a wide range of reference electrodes are available also,
and the construction or choice of a reference electrode now presents
very few problems [83]. Furthermore, many are available commercially
[84]. The usual precaution that the chemistry of the reference
electrode should be compatible with the chemistry of the test solution being studied is still given of course. Thus, because of
the insolubility of $KClO_4$, a saturated calomel electrode (sce)
is not placed in direct contact with the test solution to record
a polarogram of cadmium in perchloric acid. Separation of the
test solution via an "inert" salt bridge or use of, say, a
Ag/AgCl(NaCl) reference electrode would be required. Similarly,
the still too frequent use of the aqueous sce to record polarograms

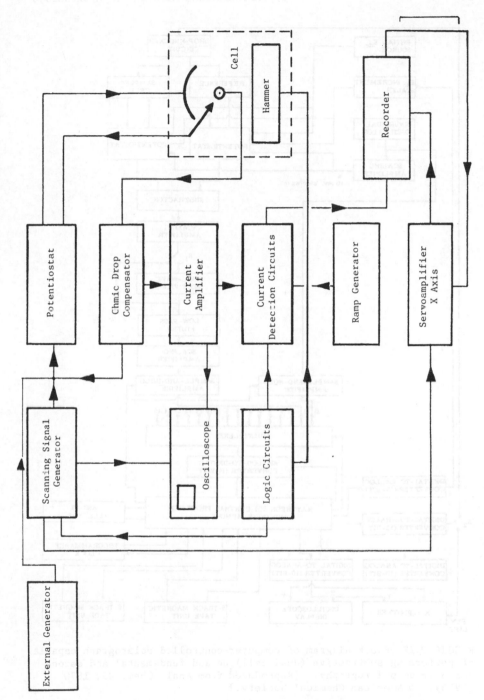

FIGURE 2.21 Block diagram of basic apparatus for polarograph. (By courtesy of Tacussel Electronique.)

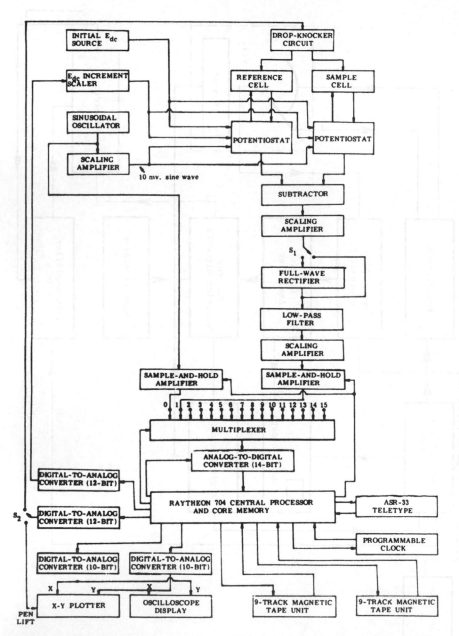

FIGURE 2.22 Block diagram of computer-controlled polarograph capable
of performing subtractive (dual cell) dc and fundamental and second
harmonic ac polarography. [Reproduced from Anal. Chem. 45, 1870
(1973). © American Chemical Society.]

in nonaqueous solvents is a potential hazard. KCl is often insoluble in nonaqueous solvents, and undesirable phenomena at the aqueous-nonaqueous interface (e.g., time-dependent junction potentials) can arise. Furthermore, contamination of the nonaqueous solvent with water may also be undesirable. The three-electrode potentiostat has made the use of most solvents possible in polarographic analysis, but this does not mean the traditionally used aqueous sce has to be maintained as an integral part of the polarographic experiment as it all too frequently seems to have been in many applications.

The third or auxiliary electrode, through which the current passes and where the electrode process counter to that being used in the polarographic analysis occurs, does not need to have a constant or reproducible potential and usually can be any metal or other inert electrode. Mercury pool, platinum wire, graphite, and a host of other auxiliary electrodes have been used. In polarographic work, the quantity of material produced at the auxiliary electrode (e.g., products of oxidation or reduction of the solvent) is too small to cause any difficulties, and the auxiliary electrode is usually placed directly in contact with the test solution, rather than being separated via a salt bridge or equivalent arrangement. Furthermore, the time scale of the experiment is usually too small for the products produced at the auxiliary electrode to diffuse to the working electrode and cause interference via this reaction. However, in a large-scale controlled-potential electrolysis experiment, possible interaction or contamination from product(s) produced at the auxiliary electrode definitely needs to be considered. In an experiment of this kind, where interference can occur, the auxiliary electrode would usually be separated from the test solution [42].

As a final caution regarding the care required in making the choice and application of reference electrodes (both working or auxiliary), the following example from the literature is given. Some years back [85] metallic molybdenum was suggested as a useful reference electrode for use in polarography. Later a method for the

determination of silver using the molybdenum reference electrode was
reported [86]. Subsequently, however, it was shown [87] that, in fact,
the silver ions displaced molybdenum from the reference electrode
and the polarographic waves reported were due to molybdenum now
present in solution. Thus, actually a determination of molybdenum
rather than silver had been performed unknowingly. Obviously
molybdenum would make an ideal auxiliary or reference electrode
under many conditions, but there were hazards not appreciated by
the authors when presenting their method for the determination of
silver.

The use of the three-electrode potentiostat also obviously
eases restrictions placed on the positioning of the working reference and auxiliary electrodes compared with two-electrode work,
and generally this is no longer a critical problem in polarographic
analytical work. However, it is still not advisable to have the
electrodes separated by too great a distance, particularly in nonaqueous solvents. A commonly used arrangement is to have the auxiliary and reference electrodes approximately equidistant from the
dme, the three electrodes being in the same plane.

Galvanostatic Control. In most polarographic techniques, controlled-
potential conditions are required. However, in certain related electrochemical experiments, such as chronopotentiometry, controlled
current or galvanostatic conditions are needed. The function and
operation of the three-electrode galvanostat, with respect to controlling the current, is entirely analogous to that of the three-
electrode potentiostat, with respect to controlling the potential.
Indeed, some manufacturers of multipurpose electrochemical apparatus
include alternative potentiostatic or galvanostatic control with
their instruments, since the basic electronic components of both
circuits are the same.

Figure 2.23 shows a three-electrode galvanostatic circuit and
experimental arrangement, which should be compared with the potentiostatic one given in Fig. 2.16. The analogies and similarities are
obvious. In this case E_{in} is generated by a variable voltage source,

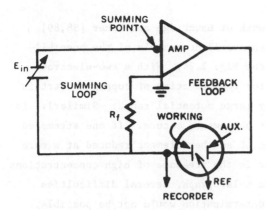

FIGURE 2.23 Galvanostatic system. (By courtesy of Beckman [43].)

and all electrodes are in the feedback loop. E_{in}, in the summing loop, causes the amplifier system (AMP) to generate the current i_f in the feedback loop; i_f also flows through it. Furthermore, as long as E_{in} and R_f are constant, i_f is constant and a constant current passes through the cell. The polarity and magnitude of the controlled current can be altered by changing E_{in} or R_f. Finally, the current will not be affected by the cell resistance R_c, provided that the product $i_f(R_c + R_f)$ does not exceed the power capabilities of the amplifier. The voltage at the surface of the working electrode is monitored (as a function of time in chronopotentiometry) by placing a reference electrode close to its surface.

2.3.3 General Recommendations in Instrumentation

The author strongly recommends the use of the three-electrode system when using a modern polarographic technique for routine analysis. "Resistance effects" in these techniques are frequently far more severe than in dc polarography, and unless potentiostatic-galvanostatic control is used, experiment-vs.-theory correlations will not be obtained readily, and the systematic use of the polarographic method will be severely hindered. Furthermore, many insidious instrumental and other artifacts, apart from the well-known one of iR drop, are even now being found to be minimized or eliminated with the three-electrode system. The work of Hawkridge and Bauer [88,89] with respect to dc maxima is an excellent example of the possibilities in this area, as seen from Fig. 2.24. With a two-electrode system, the maximum observed for the reduction of copper in certain media occurs over an extremely large potential range. Similarly, in ac polarography, an equivalent phenomenon occurs. If one attempted to undertake the determination of another element reduced at a more negative potential than copper in the presence of high concentrations of copper with a two-electrode polarograph, several difficulties would be encountered and the determination would not be possible. However, with a three-electrode potentiostat, the copper maximum is confined to a small potential range, and the determination of a more negatively reduced species is now achieved without difficulty. It has been the author's continued experience that a considerable number of "interferences" reported in polarography, particularly with respect to modern polarographic methods, are no more than instrumental artifacts arising from iR drop effects encountered with the two-electrode polarograph. The use of a three-electrode system is generally assumed in subsequent chapters.

For recording the i-E, i-t, or E-t curve, an X-Y recorder is still the most frequently used device in polarography. For the application of many modern polarographic techniques, a fast-response recorder is often required. Furthermore, the recorder is used with little or no damping, unlike the approach most commonly used in dc polarography in bygone days. Recorders with a response time of 0.5 to 1 sec or better are now commonly in use, and these have the capability to measure directly maximum currents in dc polarography, in contrast to average values of the current obtained from slow response or heavily damped recorders usually provided with earlier dc polarographic instrumentation. The tendency in modern techniques of polarography, for theoretical and practical reasons, is now more and more toward measuring currents at the end of the drop life, as will be seen in subsequent chapters. High-speed data readout or acquisition

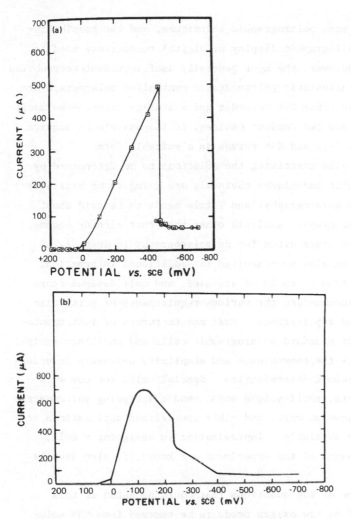

FIGURE 2.24 Differences between two-electrode and three-electrode systems are often considerable: (a) acute maximum obtained with two-electrode apparatus; (b) maximum obtained with a three-electrode circuit. Wave is reduction of 8×10^{-3} M copper in 1 M Na_2SO_4. [Reproduced from Anal. Chim. Acta *58*, 203 (1972).]

is required with some polarographic techniques, and the possibility
of utilizing oscillographic display or digital readout may need to
be considered. However, the most generally useful readout accessories
to go with a potentiostatic-galvanostatic controlled polarograph are
probably a fast-response X-Y recorder and a storage cathode-ray oscilloscope. With these two readout devices, it is possible to acquire
a variety of i-E, i-t, and E-t curves in a suitable form.

The actual cells containing the solutions to be determined by
modern polarographic techniques obviously are going to be basically
the same as in dc polarography, and little needs to be said about
this area of polarographic analysis other than that already encompassed in the literature cited for dc polarography in Chap. 1.
Notes 90 and 91 can also be consulted for additional information.
Solution volumes from <1 to 50 ml are used, and cell designs range
far and wide to accommodate the various requirements of particular
determinations and applications. Most manufacturers of instrumentation now also make standard polarographic cells and ancillary equipment which provide the convenience and simplicity necessary in undertaking a polarographic determination. Special cells for use with
automatic analyzers, small-volume work, anodic stripping voltammetric
experiments, nonaqueous work, and other specialized applications are
also commercially available. Improvization in designing a cell,
once the requirements of the experiment are known, is also usually
not difficult.

Finally, some of the questions that could be asked at this
stage, such as if or how oxygen needs to be removed from the solution prior to polarographic analysis, will not be dealt with here.
In books on dc polarography, the removal of oxygen is generally
recommended with good reason. However, with more modern polarographic techniques this problem needs to be reviewed after consideration of the theory. Other pertinent areas of general polarographic methodology are also better left until the theory behind
the various techniques has been established.

2.4 FARADAIC AND NONFARADAIC PROCESSES

2.4.1 Faradaic Current

Two distinctly different types of processes occur at electrodes.
One type includes those in which electrons are transferred across
the electrode-solution interface. In these processes, oxidation or
reduction occurs, and because they obey Faraday's law, they are
called *faradaic processes*. The magnitude of the faradaic current is
(as indicated previously) governed by the electrode mechanism or
mass transfer process, the technique being used, and whether the rate
of electrolysis is limited by diffusion, electron transfer, chemical
kinetics, or adsorption, etc. With a few exceptions, in analytical
applications of polarography we will be concerned with the use of
faradaic processes. Previous discussion has been concerned only
with this class of process, and it has usually been assumed that all
of the current flowing through the electrochemical cell is derived
from the redox couple. However, this is not so, and one of the major
problems in polarography, that of the nonfaradaic current component,
has yet to be formally introduced.

At all potential regions, including those where charge transfer
reactions do not occur because they are thermodynamically or kinetically unfavorable, the structure of the electrode-solution interface can change with changing potential or solution concentration,
and processes not involving electrolysis, such as adsorption or desorption, can occur. Because no electron transfer is involved,
these processes are called *nonfaradaic*; however, they can contribute
current flow to the electrochemical cell, as can the faradaic ones.
In the presence of a faradaic process, the nonfaradaic process
still occurs, albeit sometimes modified, but usually the assumption
that they are independent is invoked. Thus, the total current i_T
flowing through the cell can be considered as the sum of faradaic
i_F and nonfaradaic i_{NF} contributions, or

$$i_T = i_F + i_{NF} \tag{2.54}$$

The analytical use of most polarographic techniques, more specifically than stated previously, actually requires a plot of $f(i_F)$
vs. concentration to be made rather than $f(i_T)$ vs. concentration.
Hence, the nonfaradaic current contribution must be subtracted from
the total current or, even better, eliminated instrumentally from
the readout.

Just as a large body of polarographic research in the last 20
years has been directed toward correcting the potential for iR drop,
an equivalent amount of attention has been paid to minimizing or
eliminating the charging current contribution from the readout,
this generally being the most important of the nonfaradaic components.
Indeed, most of the new polarographic techniques are designed to
achieve this goal, and therefore, it is obviously most important that
the theory for the charging current be understood. The ability of a
particular technique to discriminate against the charging current in
fact usually governs the limit of detection or sensitivity of the
method. This arises because during the electrochemical reduction or
oxidation of small concentrations of electroactive species, the
charging current may become larger than the faradaic current and
completely mask or hide the required parameter i_F. Furthermore, in
applying theoretical relationships, the ability to correctly subtract
or eliminate i_{NF} is assumed, and the systematic use of modern polarography requires that this can be done. The importance of the nonfaradaic aspect of the overall electrode process, therefore, cannot
be overemphasized.

2.4.2 Charging and Capacitance of an Electrode: The Charging Current

In the absence of an electroactive species, charge cannot cross the
interface via electron transfer, and the behavior of the electrode-
solution interface must remain analogous to that of a capacitor.

A capacitor is an electrical circuit element whose behavior is
governed by the equation

$$\frac{q}{V} = C \qquad\qquad\qquad (2.55)$$

where

  q = charge on the capacitor, coulombs

  V = potential across the capacitor, volts

  C = capacitance, farad

When a potential is applied across a capacitor, it will charge until q attains a value governed by Eq. (2.55). During this charging process, a current will flow (the charging current).

The charge of a capacitor consists of an excess of electrons on one plate and a deficiency of electrons on the other. The electrode-solution interface can be described in a similar manner. At a given potential there will exist a charge on the electrode, q_E, and a charge in the solution, q_s. The charge in the solution is characterized by a specific orientation of the cations and anions [1,14]. Whether the electrode is charged negatively or positively depends upon the potential; however, $q_E = -q_s$.

The capacitance of the electrode-solution interface is a function of potential. The charge on the electrode represents an excess or deficiency of electrons, and resides on a thin layer of the electrode surface. The charge in the solution is made up of an excess of either cations or anions in the vicinity of the electrode. The whole array of charged species and orientated ions existing in the metal-solution interface is called the *electrical double layer* and represents an important branch of electrochemistry [92]. However, for the purposes of this book, only results from considering the interface as a capacitor are required directly.

Figure 2.25 shows dc and ac polarograms of the charging current. These two polarographic methods are used to show the procedure for calculating the approximate value of the charging current.

For an electrode of area A, growing with time t, the charge q required to bring the double layer up to any potential E is given by Eq. (2.56).

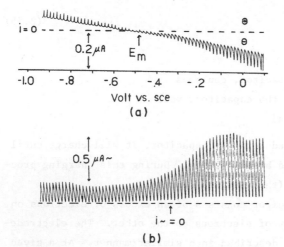

FIGURE 2.25   (a) Polarogram of the charging current in dc polarography.   (b) Polarogram of the charging current in ac polarography.
Medium is 1 M NaF.

$$q = C'_{F(E)} A(E_m - E) \tag{2.56}$$

$C'_{F(E)}$ is the capacity of the double layer per unit area and E_m is
the potential where $q = 0$.

During the life of any single mercury drop, the variation of
potential in dc polarography, even as applied with modern instrumentation, is so small that the potential may be considered to be a
constant.   Since $i = dq/dt$, the value of the charging current at
time t sec is therefore given by Eq. (2.57).

$$i_c = C'_{F(E)} \frac{dA}{dt} (E_m - E) \tag{2.57}$$

From this equation it can be seen that i_c is positive when E is more
negative than E_m, zero when $E = E_m$, and negative when E is more positive than E_m as required by Fig. 2.25(a).   However, the value of
$C'_{F(E)}$ is not independent of potential, and so a linear plot of i_c vs. E
is not observed.   Detailed discussion of these equations is available
in Notes 93 and 94.

In an ac experiment, E varies as a function of time (see Chap.
7), and $E = E_{dc} - \Delta E \sin \omega t$, where E_{dc} is the dc potential, and ΔE

and ω are the amplitude and frequency of the alternating potential, respectively. The total charging current can be written as the sum of dc and ac components, and after taking into account the relationship $dE/dt = -\Delta E \, \omega \cos \omega t$, it can be shown that

$$i_c^{\sim} = AC_{dl} \, \Delta E \, \omega \cos \omega t \qquad (2.58)$$

where C_{dl} is the differential capacity of the double layer per unit area $= dq'/d(E_m - E)$, q' is the charge density or charge per unit area and $i_c^{\sim}$ is the ac charging current.

Equations equivalent to (2.57) and (2.58) can be written for all polarographic techniques, and while only approximate, they enable us to understand how a particular variation in polarographic methodology enables the required discrimination against charging current to be achieved in a large number of circumstances.

It is apparent from Fig. 2.25(a) that the dependence of the dc charging current on potential has two different slopes, depending on the potential relative to E_m. At potentials more positive than E_m, part of the electrical double layer in the solution is formed by anions, while cations form part of the double layer at potentials more negative than E_m. The value of $C'_{F(E)}$ for anions exceeds that for cations. Thus, the change in slope around E_m is qualitatively explained. Figure 2.25 and the preceding discussion, however, assume the absence of adsorption and that values of $C'_{F(E)}$ change far more markedly and nonuniformly when surface-active substances are adsorbed or desorbed.

Just as adsorption of an electroactive species can modify the faradaic electrode process, as stated earlier, adsorption can also alter the nonfaradaic processes in a most complex and often not well-understood manner. At the adsorption and desorption potentials, sudden changes in differential capacity occur and give rise to sharp peaks [94]. In the potential region in which the substance is adsorbed on the electrode surface, the capacity of the double layer is lowered. Adsorption in polarography can also alter the drop time of the dme. In ac polarography, tensammetric waves (see Chap. 7) can be produced at potentials where adsorption or desorption occurs.

This clearly provides an example of an analytical application of a nonfaradaic process, as will be shown in Chap. 7 on ac polarography.

In discussing the manner in which various techniques discriminate against nonfaradaic charging current, it must be realized that in using equations such as (2.57) and (2.58), the absence of adsorption (and iR drop) is assumed. Frequently this is unjustified; so some nonideality must be expected, and the inability to correctly compensate for the charging current, using methods recommended with modern polarographic techniques, often can be attributed to the phenomena of adsorption.

2.4.3 Migration Current

The magnitude of the faradaic current is governed by the mass transfer mechanism as implied previously when discussing electrode processes. During electrolysis, the decrease of reactant concentration at the electrode surface can be governed by diffusion (coupled with electrode kinetics), convection (stirring of solutions or rotating of electrode), and migration.

These are the three basic mechanisms of mass transfer affecting both the electrolysis potential and current. The diffusion and convection components are normally included in the mathematical description of the faradaic electrode process and are the phenomena from which the basic concepts of polarographic analysis arise. Thus, the migration current is assumed to be zero or negligible in most descriptions of electrode processes, and it is important to recognize that this assumption exists when performing a polarographic experiment. To achieve this condition the deliberate addition of an inert supporting electrolyte is usually undertaken in polarography.

Prior to the advent of the three-electrode potentiostat, the supporting electrolyte was also added to increase the conductance of the solution and minimize iR drop effects. This consideration is no longer as important in certain instances, but elimination of migration current is still required and the presence of a supporting electrolyte still remains an integral part of most polarographic experiments.

In any electrolysis experiment of the kind performed in polarography, where an electroactive species is reduced or oxidized at the working electrode, an appropriate redox reaction also occurs simultaneously at the reference (or auxiliary) electrode. The net current flow observed occurs as a result of the current being conducted through the solution via the movement or migration of ions. Cations move toward the cathode and anions toward the anode to achieve this current flow, and if the species being reduced or oxidized is also charged, transfer or movement of these ions other than via diffusion or convection occurs. That is, the mass transfer process of the electroactive species will be modified via migration, and the migration current can be positive, zero, or negative, depending on the charge of the electroactive species.

The value of the migration current depends on the transference number T of the species being reduced or oxidized; the larger the transference number, the greater the fraction of the current carried by the ion. In particular,

$$T_j = \frac{C_j \lambda_j}{\sum_i C_i \lambda_i} \tag{2.59}$$

where C_j is the concentration of the electroactive entity and λ_j is its equivalent conductance. Subscript i refers to any charged species present in solution. The addition of a supporting electrolyte, whose ions do not contribute to the current because they cannot be oxidized or reduced, causes the transference number of the electroactive species to decrease. If the concentration of the supporting electrolyte is high, e.g., 100 times that of the electroactive species, the transference number of the species being oxidized or reduced becomes practically zero. Consequently, most polarographic determinations covering the 10^{-3} to 10^{-5} M range are carried out in the presence of 0.1 M or greater concentrations of supporting electrolyte. However, with many modern techniques having detection limits below 10^{-7} M, concentrations of supporting electrolyte around 10^{-3} M can be used in trace analysis, provided the iR drop is not a problem.

In dc polarography, the concentration of supporting electrolyte necessary to meet the above requirement is experimentally observed as the concentration where the measured limiting current is essentially independent of supporting electrolyte concentration. Table 2.5 shows some data for the reduction of lead in KNO_3.

Analogous results are obtained in ac, pulse, and other polarographic techniques. Note 1 can be consulted for further details of the general theory of the migration current in dc polarography.

2.5 THE MODERN POLAROGRAPHIC EXPERIMENT

The preceding discussion, coupled with the usual considerations applicable to all branches of analytical chemistry, enables guidelines to be established for the systematic use of modern polarographic methods. Obviously, the decision to use one polarographic method in preference to another depends on knowing how the various techniques respond to the electrode process under consideration. The ability of each technique to discriminate against charging current and its response to different kinds of faradaic processes will therefore be of fundamental importance. The speed, resolution, simplicity of

TABLE 2.5 Effect of Adding KNO_3 on dc Polarographic Limiting Current for Reduction of Pb(II)[a]

[KNO_3] M	Limiting current (μA)
0	17.6
0.0001	16.2
0.0002	15.0
0.0005	13.4
0.001	12.0
0.005	9.8
0.1	8.45
1.0	8.45

[a]Pb(II) = 9.5×10^{-4} M.
Data taken from Note 95.

apparatus, and operation of each method will also be important, since gains in one area and losses in another will be encountered frequently in changing techniques. Chapters subsequent to the initial resumé of dc theory in Chap. 3 are intended to provide pertinent information in each of these areas.

NOTES

1. L. Meites, *Polarographic Techniques*, 2nd ed., Interscience, New York, 1965.

2. J. J. Lingane, J. Amer. Chem. Soc. *68*, 2448 (1946); L. Meites, Anal. Chem. *20*, 895 (1948).

3. E. P. Parry and D. P. Anderson, Anal. Chem. *45*, 458 (1973).

4. J. A. V. Butler, Trans. Faraday Soc. *19*, 729 (1924); *28*, 379 (1932).

5. B. E. Conway, *Theory and Principles of Electrode Processes*, Ronald Press, New York, 1965.

6. K. J. Vetter, *Electrochemical Kinetics*, Academic Press, New York, 1967.

7. J. Koryta, J. Dvořak, and V. Boháčková, *Electrochemistry*, Methuen, London, 1970.

8. J. O'M Bockris and A. K. N. Reddy, *Modern Electrochemistry*, Plenum Press, New York, 1970.

9. H. Bauer, *Electrodics*, Georg Thieme Verlag, Stuttgart, 1972.

10. P. Delahay, J. Amer. Chem. Soc. *75*, 1430 (1953); Advan. Polarogr. *1*, 26 (1960), and references cited therein.

11. H. Matsuda and Y. Ayabe, Z. Elektrochem. *63*, 1164 (1959).

12. P. Zuman, Chem. Eng. News *46*, 94 (1968).

13. P. Zuman, *Topics in Organic Polarography*, Plenum Press, New York, 1970.

14. J. Heyrovský and J. Kůta, *Principles of Polarography*, Academic Press, New York, 1966.

15. P. Zuman and S. Tang, Collect. Czech. Chem. Commun. *28*, 1524 (1963).

16. W. B. Schaap, J. A. Davis and W. H. Nebergall, J. Amer. Chem. Soc. *76*, 5226 (1954).

17. A. C. Testa and W. H. Reinmuth, Anal. Chem. *33*, 132 (1961).

18. Note 14, pp. 341-380.

19. K. Vesely and R. Bridicka, Collect. Czech. Chem. Commun. *12*, 313 (1947).

20. R. Bieler and G. Trumpler, Helv. Chem. Acta *30*, 706, 791, 1109, 1286, 1534, 2000 (1947).

21. Note 14, pp. 394-400.

22. F. L. Wimmer, M. R. Snow and A. M. Bond, Inorg. Chem. *13*, 1617 (1974).

23. G. L. Booman and J. E. Rein, *Treatise on Analytical Chemistry*, Interscience, New York, 1962, vol. 9, part II, pp. 115-128.

24. T. A. O'Shea and G. A. Parker, Anal. Chem. *44*, 184 (1972).

25. A. Blazek and J. Koryta, Collect. Czech. Chem. Commun. *18*, 326 (1953).

26. J. Koryta, Chem. Zvesti *8*, 723 (1954).

27. H. Sour and K. Wienhold, J. Electroanal. Chem. *35*, 219 (1972).

28. Note 14, pp. 386-388.

29. R. N. Adams, Current Accounts Chem. Res. *2*, 175 (1969).

30. R. N. Adams, *Electrochemistry at Solid Electrodes*, Dekker, New York, 1969, and references cited therein.

31. D. R. Canterford, A. S. Buchanan, and A. M. Bond, Anal. Chem. *45*, 1327 (1973).

32. I. M. Kolthoff and C. S. Miller, J. Amer. Chem. Soc. *63*, 1405 (1941).

33. R. Kalvoda, W. Anstine, and M. Heyrovský, Anal. Chim. Acta *50*, 93 (1970), and references cited therein.

34. C. N. Riley and W. Stumm, *Progress in Polarography*, Interscience, New York, 1962, vol. 1, chap. 5, and references cited therein.

35. I. M. Kolthoff and Y. A. Okinaka, J. Amer. Chem. Soc. *82*, 3528 (1960).

36. J. Kuta, *Modern Aspects of Polarography*, Plenum Press, New York, 1966, pp. 62-70.

37. K. Tsuji, *Modern Aspects of Polarography*, Plenum Press, New York, 1966, pp. 233-242.

38. B. B. Damaskin, O. A. Petrii, and V. V. Batrakov, *Adsorption of Organic Compounds on Electrodes*, Plenum Press, New York, 1971.

39. G. W. Ewing, Chemtech, June 1973, pp. 326-330.

40. A. Hickling, Trans. Faraday Soc. *38*, 27 (1942).

41. *Some Aspects of Modern Polarography*, supplied by courtesy of Melabs, Palo Alto, Calif.

42. J. E. Harrar, in *Electroanalytical Chemistry* (A. J. Bard, ed.), Dekker, New York, 1975, vol. 8, chap. 1.

43. R. J. Joyce, *An Introduction to Electroanalytical Chemistry*, Beckman Instruments, Inc., Fullerton, Calif., 1966.

44. D. E. Smith, C. R. C. Crit. Rev. Anal. Chem. *2*, 247 (1971), and references cited therein.

45. D. E. Smith, in *Electroanalytical Chemistry* (A. J. Bard, ed.), Dekker, New York, 1966, vol. 1, chap. 1, and references cited therein.

46. D. E. Smith, in *Computers in Chemistry and Instrumentation* (H. B. Mark, Jr., J. S. Mattson, and H. C. MacDonald, eds.), Dekker, New York, 1972, vol. 2, chap. 12, and references cited therein.

47. W. H. Reinmuth, Anal. Chem. *36*, 211R (1964).

48. W. L. Underkofler and I. Shain, Anal. Chem. *37*, 218 (1965).

49. D. E. Walker, R. N. Adams, and J. R. Alden, Anal. Chem. *33*, 308 (1961).

50. E. R. Brown, T. G. McCord, D. E. Smith, and D. D. Deford, Anal. Chem. *38*, 1119 (1966).

51. G. L. Booman and W. B. Holbrook, Anal. Chem. *35*, 1793 (1963).

52. G. L. Booman and W. B. Holbrook, Anal. Chem. *37*, 795 (1965).

53. J. W. Hayes and C. N. Reilley, Anal. Chem. *37*, 1322 (1965).

54. D. Pouli, J. R. Huff, and J. C. Pearson, Anal. Chem. *38*, 382 (1966).

55. G. Lauer and R. A. Osteryoung, Anal. Chem. *38*, 1106 (1966).

56. J. W. Hayes and H. H. Bauer, J. Electroanal. Chem. *3*, 336 (1962).

57. M. E. Peover and J. S. Powell, J. Polarogr. Soc. *12*, 64 (1966).

58. H. Gerischer and K. E. Staubach, Z. Elektrochem. *61*, 789 (1957).

59. P. Valenta and J. Vogel, Chem. Listy *54*, 1279 (1960).

60. E. R. Brown, D. E. Smith, and G. L. Booman, Anal. Chem. *40*, 1411 (1968).

61. E. R. Brown, H. L. Hung, T. G. McCord, D. E. Smith, and G. L. Booman, Anal. Chem. *40*, 1424 (1968).

62. R. Bezman and P. S. McKinney, Anal. Chem. *41*, 1560 (1969).

63. J. R. Tacussel, Electrochim. Acta *11*, 449 (1966).

64. W. E. Thomas, Jr., and W. B. Schaap, Anal. Chem. *41*, 136 (1969).

65. H. C. Jones, W. L. Belew, R. M. Stelzner, T. R. Mueller, and D. J. Fisher, Anal. Chem. *41*, 772 (1969).

66. W. B. Schaap and P. S. McKinney, Anal. Chem. *36*, 29 (1964).

67. D. E. Smith, Anal. Chem. *35*, 1811 (1963).

68. G. L. Booman, Anal. Chem. *29*, 213 (1957).

69. M. T. Kelley, D. J. Fisher, and H. C. Jones, Anal. Chem. *31*, 1475 (1959); 1262 (1960).

70. L. Namee, J. Electroanal. Chem. *9*, 166 (1964).

71. A. M. Bond, Anal. Chem. *44*, 315 (1972).

72. C. Lamy and C. C. Herrmann, J. Electroanal. Chem. *59*, 113 (1975), and references cited therein.

73. C. Yarnitzky and Y. Friedman, Anal. Chem. *47*, 876 (1975), and references cited therein.

74. C. Yarnitzky and N. Klein, Anal. Chem. *47*, 880 (1975).

75. A. Bewick, Electrochim. Acta *13*, 825 (1968).

76. J. Dévay, B. Lehghel, and L. Mezarus, Acta Chim. Acad. Sci. Hung. *66*, 269 (1970).

77. W. W. Goldsworthy and R. G. Clem, Anal. Chem. *44*, 1360 (1972).

78. R. Bezman, Anal. Chem. *44*, 1781 (1972).

79. J. R. Tacussel, *The Voctan: An Apparatus for Electrochemical Techniques Involving Voltage/Current/Time Relationships*, S.O.L.E.A., Villeurbanne, France.

80. D. E. Glover and D. E. Smith, Anal. Chem. *45*, 1869 (1973).

81. J. B. Flato, Amer. Lab., February 1969, p. 10.

82. G. G. Willems and R. Neeb, Z. Anal. Chem. *269*, 1 (1974).

83. D. J. G. Ives and G. J. Janz (eds.), *Reference Electrodes*, Academic Press, New York, 1961.

84. R. D. Caton, Jr., J. Chem. Educ. *50*, A571 (1973); *51*, A7 (1974).

85. V. T. Athavale, S. V. Burangey, and R. G. Dhaneshwar, *Proc. SAC Conf.*, Nottingham, 1965, pp. 446–454.

86. V. T. Athavale, M. R. Dhaneshwar, and R. G. Dhaneshwar, Analyst *94*, 855 (1969).

87. D. S. Allan, B. Lamb, and D. Teasdale, Analyst *97*, 409 (1972).

88. F. M. Hawkridge and H. H. Bauer, Anal. Chim. Acta *58*, 203 (1972).

89. F. M. Hawkridge and H. H. Bauer, Anal. Chem. *43*, 768 (1971).

90. Z. P. Zagorski, *Progress in Polarography*, Interscience, New York, 1962, vol. II, chap. 27, and references cited therein.

91. D. R. Crow, *Polarography of Metal Complexes*, Academic Press, New York, 1969, chap. 6.

92. P. Delahay, *Double Layer and Electrode Kinetics*, Interscience, New York, 1965, and references cited therein.

93. Note 1, chap. 3.

94. Note 14, pp. 53–60, and references cited therein.

95. J. S. Lingane and I. M. Kolthoff, J. Amer. Chem. Soc. *61*, 1045 (1939).

Chapter 3

CONVENTIONAL DC POLAROGRAPHY:
LIMITATIONS AND USES

Many of the major advances and innovations in the analytical use of
polarography have resulted from the consideration of means for over-
coming or minimizing drawbacks, restrictions, and limitations found
in conventional dc polarography. Figures 3.1 to 3.5 reveal some of
the disadvantages inherent in classical dc polarography. An understanding of the inadequacies of conventional dc polarography provides
an excellent basis for appreciation of the advances in polarography
as an analytical method.

3.1 DIFFICULTIES ENCOUNTERED IN
     CONVENTIONAL DC POLAROGRAPHY

3.1.1 Resolution, Wave Shape, and Readout

Even the idealized sigmoidal-shaped polarogram (e.g., Fig. 2.2),
classically presented in many publications, provides difficulties
in analytical work. Calculation of the mean limiting current and
the $E_{1/2}$ value requires evaluation of the average current at a large
number of potentials. After including the difficulty of subtracting
the charging current at each potential, as is necessary where low
concentrations are to be determined, the complete evaluation
procedure can be seen to be a nontrivial process. Obtaining the
required parameters from a dc polarogram is therefore a time-
consuming and tedious procedure compared with the usually simple

readout of absorbance required in ultraviolet (uv)-visible spectrophotometry, for example. Furthermore, the ideal-shaped curve is all
too infrequently obtained. Maxima and other distortions (Figs. 3.1
and 3.2) are often observed in conventional dc polarography, introducing the necessity of using semi-empirical or even arbitrary procedures for evaluation of the wave. Sometimes these distortions can
be eliminated by judicious choice of solution conditions, but this
is by no means generally true. Nonidealities add considerably to
the difficulties of obtaining the required parameters.

Any approach improving the ease and convenience of readout should
be useful. Obviously the production of a peak-shaped curve with
direct readout of peak current and peak potential as alternatives to
measurement of i_1 and $E_{1/2}$ would be advantageous in many situations.
Derivative readout provides this kind of curve. Similarly, production of a smooth trace devoid of the serrations resulting from the
periodic nature of the growth and fall of the mercury drop would
facilitate measurement of the i-E curve. This is readily achieved
with the aid of modern electronics, as will be seen in Chap. 4.

Concomitant with the drawbacks encountered with respect to wave
shape and readout previously listed is the restriction the wave shape
places on resolution. Figure 3.3 shows, for example, that calculation of the limiting current of a second electrode process occurring
in the limiting current region of another electrode process must
present severe difficulties in all but the most ideal situation.
Obviously, a readout of two peaks referred to a common base line
(e.g., derivative dc plot) would provide for superior resolution in
most instances. Most modern techniques of polarography have the
advantage of considerably improved resolution over dc polarography.

3.1.2 Time Taken to Record a DC Polarogram

The time taken for the determination of a given entity is often of
prime consideration in modern analytical chemistry. Conventional
dc polarography is necessarily relatively slow because the time
taken to record a polarogram is restricted by the slow rate at which

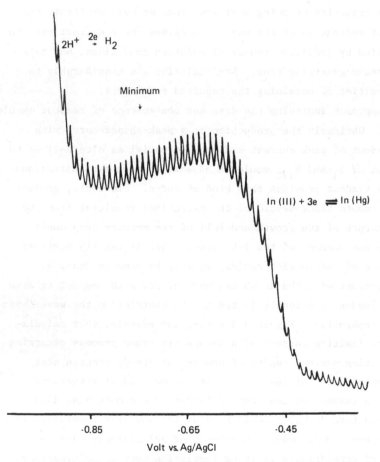

FIGURE 3.1 Maxima, minima, and other irregularities encountered in dc polarography caused difficulties in the evaluation of polarographic parameters because the limiting current region is nonlinear: 4.6×10^{-4} M In(III) in 0.8 M $HClO_4$/0.5 M NaCl.

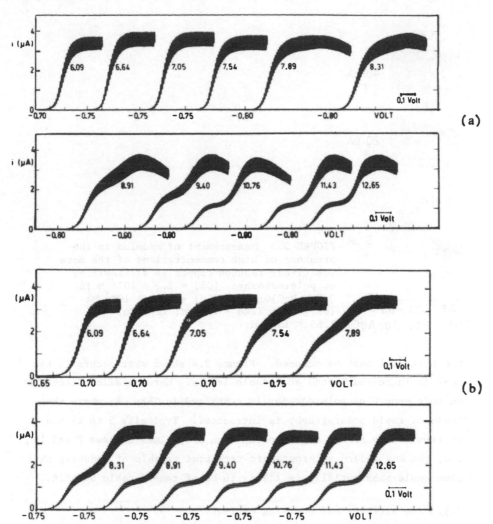

FIGURE 3.2 Polarograms frequently depart from the ideal sigmoidal shape. Note how wave shape and definition of limiting current can change drastically with change in solution conditions in this example. (a) 3-Diazocamphor reduction in aqueous buffers of pH shown: concentration = 2.8×10^{-4} M. (b) 3-Diazocamphor reduction in aqueous buffers of pH shown plus 2×10^{-3} M phenyltrimethylammonium bromide. [Reproduced from J. Electroanal. Chem. *23*, 399 (1969).]

(a)

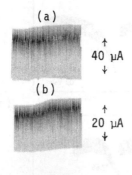

↑
40 μA
↓

(b)

↑
20 μA
↓

(c)

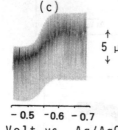

↑
5 μA
↓

FIGURE 3.3 Measurement of cadmium in the
presence of high concentrations of the more
positively reduced copper is difficult by
dc polarography: [Cd] = 1.3×10^{-2} M in
0.5 M $NaClO_4$; (a) [Cu] = 4.00×10^{-3} M;
(b) [Cu] = 7.46×10^{-3} M; (c) [Cu] =
1.65×10^{-3} M.

-0.5 -0.6 -0.7
Volt vs. Ag/AgCl

the potential must be scanned. Figure 3.4 shows what occurs if the
scan is increased beyond acceptable levels. More detailed discussion
on this aspect of polarography is contained in Chap. 4, where the
topic of rapid polarography is introduced. Typically 5 to 15 min
is required to record a polarogram with drop times between 2 and 10
sec, and any modern polarographic technique capable of reducing this
time scale substantially is likely to be of considerable benefit.

3.1.3 Sensitivity and Charging Current

In conventional dc polarography with diffusion-controlled limiting
currents and in aqueous media, the charging current generally masks
the faradaic current when the concentration of electroactive species
is in the 10^{-6} to 10^{-5} M region. Thus, the limit of detection is
also in this range. However, even in the 10^{-5} to 10^{-4} M concentration range, the magnitude of the charging current is significant,
and curves with sloping base lines (see Fig. 3.5) rather than the
idealized sigmoidal shaped curves are usually found in dc polarography. Any technique discriminating against the charging current

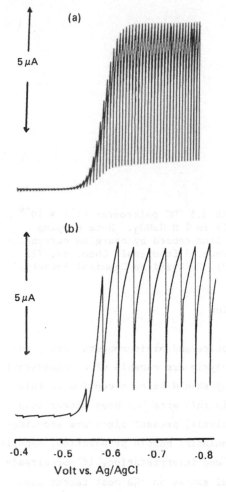

FIGURE 3.4 If the scan rate is too fast in dc polarography, insufficient data points are available to define accurately the polarogram, and the constant potential condition is violated. Solution is 1×10^{-3} M Cd(II) in 1 M HCl. (a) scan rate selected correctly; (b) scan rate too fast.

should increase the sensitivity of the determination and lower the limit of detection, all other things being equal. Alternatively, any method automatically compensating for or subtracting the charging current may be similarly advantageous. In nonaqueous media the charging current may be somewhat higher, increasing the limit of detection in certain solvents. For kinetically controlled limiting currents, limits of detection may be higher or lower than for the diffusion-controlled case, depending on the mechanism (see subsequent discussion in Sec. 3.2).

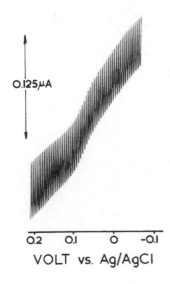

0.125μA

0.2   0.1   0   -0.1

VOLT vs. Ag/AgCl

FIGURE 3.5 DC polarogram of 5×10^{-6} M
Cu(II) in 1 M $NaNO_3$. Note sloping
base line caused by charging current.
[Reproduced from Anal. Chem. *44*, 721
(1972). © American Chemical Society.]

3.1.4 Automation of Instrumentation and Readout

With the advent of digital electronics and minicomputers, etc., most
modern instrumental methods of analysis are capable of a considerable
degree of automation. Polarography should be no exception to this
trend, and even though evolution in this area has been slower than
in other areas of instrumental analysis, present signs are encouraging. Automation at all stages, including sample preparation, experimental control, data acquisition, and interpretation, if not already
standard, are available as optional extras on the most recent commercially available instrumentation. Certain polarographic methods
lend themselves more readily to automation then the dc technique,
and this needs to be recognized as an advantage in modern analytical
chemistry. Chapter 10 is devoted to a discussion of the computer,
and some important concepts with respect to automation are to be
found in this context.

3.1.5 Dependence of Drop Time on Potential

In conventional dc polarography, the dependence of drop time on potential presents difficulties in using the method, particularly at
very negative potentials, where large changes in drop time occur.
The limiting current is a function of drop time as shown later in
this chapter and the potential dependence of this term may need to
be taken into account in certain applications of conventional dc
polarography. The potential dependence of drop time is frequently
overlooked, but it is not always negligible. The use of mechanically
controlled drop times, a feature of virtually all modern polarographic instrumentation, as noted in Chap. 2, obviates this difficulty. Further discussion on this aspect is presented in Chap. 4.

3.2 SYSTEMATIC USE OF DC POLAROGRAPHIC THEORY

The greatest advantage of dc polarography is probably the relative
simplicity of the theory and its use. As stated in Chap. 2, an
understanding of the nature of the electrode process is essential
in the systematic use of any polarographic analytical method. Hence,
even if one were to develop a square-wave polarographic method for
an analytical determination of a particular entity, preliminary work
using dc polarography would probably be undertaken to assign the
electrode process as reversible or irreversible in the dc sense.
Indeed, the future major use of conventional dc polarography may be
in this domain, rather than in actual practical analytical applications. The subtleties of the electrode process and the extension
to the time domain of the technique being actually employed are
facilitated using guidelines emanating from the dc study.

A summary of the well-known results of dc polarographic theory
commonly useful in analytical work is given below. The assumption
introduced in Chap. 2, that a supporting electrolyte is present to
eliminate migration current, needs to be emphasized, as the theoretical results are only valid under this condition.

3.2.1 Diffusion-Controlled
Limiting Currents

If the potential of the dme is at the limiting current region of
the dc polarogram and all electroactive species reaching the electrode can be reduced or oxidized immediately as they reach the electrode surface, the magnitude of the limiting current will be controlled solely by diffusion and not by the heterogeneous kinetics of
the electron transfer step or homogeneous solution kinetics. Since
it is assumed that all the ions or molecules initially at the electrode surface will be electrolyzed immediately as the potential is
applied in the limiting current region, the concentration of electroactive species will approach zero in a very thin layer of solution
near the electrode surface. The concentration gradient established
between the solution in the region of the electrode surface and the
bulk of the solution causes ions or molecules to diffuse toward the
electrode surface. Since the current flowing through the cell will
be proportional to the quantity of material that can be electrolyzed,
the limiting current is proportional to the rate at which the electroactive substance can diffuse toward the electrode surface. Under
this condition, the limiting current i_1 is therefore referred to as a
diffusion-controlled limiting current i_d.

The exact nature of the mass transfer process at a dme, in
terms of a mathematical formulation, represents an extremely complex
problem to solve rigorously. Solutions of partial differential
equations (boundary-value problems), taking into account the geometry
of the diffusion process (linear, cylindrical, spherical, etc.),
are required.

The diffusion current is determined by the concentration gra-
dient at the surface of the electrode, that is, $(\partial C/\partial x)_{(x=0)}$, which
is time dependent (as are the concentration gradients at all dis-
tances). Fick's laws of diffusion [1] are applicable to a diffusion-
controlled process. Thus,

$$i = nFAD \left(\frac{\partial C}{\partial x} \right)_{(x=0)} \tag{3.1}$$

where D = diffusion coefficient (other symbols are as used in Chap. 2). For linear diffusion,

$$\frac{\partial C}{\partial t} = D \frac{\partial^2 C}{\partial x^2} \tag{3.2}$$

Initial and boundary conditions can be invoked

$$C_{x,t} \begin{cases} C & \text{for } t = 0 \\ C_{(x=0)} & \text{for } t > 0 \end{cases}$$

where C is the concentration in the bulk solution and $C_{(x=0)}$ is the concentration at the electrode surface. For spherical diffusion

$$\frac{\partial C}{\partial t} = D \left[\frac{\partial^2 C}{\partial r^2} + \frac{2}{r} \frac{\partial C}{\partial r} \right] \tag{3.3}$$

where r is the radius of spherical electrode.

For a growing dropping electrode, corrections to the above models are required. For linear diffusion at a growing electrode, the Ilkovic model is

$$\frac{\partial C}{\partial t} = D \frac{\partial^2 C}{\partial x^2} + \frac{2x}{3t} \frac{\partial C}{\partial x} \tag{3.4}$$

For spherical diffusion at a growing electrode,

$$\frac{\partial C}{\partial t} = D \frac{\partial^2 C}{\partial r^2} + \frac{2}{r} \frac{\partial C}{\partial r} - \frac{dr}{dt} \frac{\partial C}{\partial r} \tag{3.5}$$

In principle, correction for stirring produced by the preceding drop falling off, the exact geometry of the electrode, including shielding effects, etc., should be taken into account [2-4]. However, for a model based on linear diffusion at a growing dropping electrode [2-6], the Ilkovic equation for the maximum current at the end of the drop life is

$$i = 0.732nF(C - C_{(x=0)})D^{1/2}m^{2/3}t^{1/6} \qquad\qquad (3.6)$$

where

$\qquad$ i = diffusion-controlled current at the end of the drop life, A

$\qquad$ n = number of electrons involved in charge transfer process

$\qquad$ F = Faraday

$\qquad$ C = concentration of electroactive species in the bulk
$\qquad\quad$ solution, mol cm^{-3}

$\qquad C_{(x=0)}$ = concentration of electroactive species at electrode
$\qquad\qquad\quad$ surface, mol cm^{-3}

$\qquad$ D = diffusion coefficient, cm^2 sec^{-1}

$\qquad$ m = flow rate of mercury, g sec^{-1}

$\qquad$ t = drop time, sec

$\quad$ $C - C_{(x=0)}$ represents the concentration gradient and at the
limiting current region $C_{(x=0)}$ = 0 and i = i_d; so

$$i_d = 0.732nFCD^{1/2}m^{2/3}t^{1/6} \qquad\qquad (3.7)$$

which is commonly referred to as the *Ilkovic equation*. Note the
units of concentration are not the more usual moles per liter but
moles per cubic centimeter as the unit of length used in deriving
the equation is the centimeter.

$\quad$ The Ilkovic equation in the form of Eq. (3.6) gives the magnitude of the current at all potentials of a reversible polarographic
wave because the concentration term $C_{(x=0)}$ is governed by the Nernst
equation.

$\quad$ Modern electronic methods are well suited to provide direct
measurement of maximum currents at the end of the drop life, and
these are increasingly being used in modern polarography rather than
the mean currents formerly used when long-period galvanometers and
heavily damped (slow response) X-Y recorders were in common use [3,
4]. If mean currents are used, the numerical coefficient 0.732 in
Eqs. (3.6) and (3.7) is replaced by 0.627 [2-4].

$\quad$ The validity or otherwise of the Ilkovic equation has been
argued at length by Meites [3], and the conclusion emerges that the
relatively good correlation with experiment is fortuitous (factors

nulling each other) rather than the more attractive alternative that
the excellent agreement results from the use of a rigorously valid
model. Many extensions have been made to the Ilkovic equation, such
as incorporation of spherical diffusion correction terms, to give
solutions of the form (mean current)

$$i = 0.627nF\{(C - C_{(x=0)})D^{1/2}m^{2/3}t^{1/6}[1 + f(D,m,t)]\} \qquad (3.8)$$

While these more rigorous equations may be academically more satis-
fying, the Ilkovic equation appears to be as useful as any other in
analytical work, in the sense that it provides an adequate explanation of the experimental results and no significant advance appears
to be accomplished by using more complex expressions. All equations
predict a linear dependence of i_d on C, and analytically, this is
one of the most important results. A feature of polarographic analysis generally is the widespread occurrence of linear current
parameters-vs.-C plots over extremely wide ranges of concentration.
This cannot be said of many other instrumental methods of analysis.

Equations (3.6) to (3.8) also predict that i_d is dependent on
n, D, and the capillary characteristics (m and t). Since D is a
function of temperature, thermostating is usually undertaken in
polarography. The experimental dependence of i_d on each of the
parameters is thoroughly discussed in Notes 2 to 4 and need not be
examined further.

It is obviously important in dc polarography to establish
whether the limiting current is diffusion-controlled, rather than
kinetically controlled, for example. Apart from the linear dependence on concentration, the other diagnostic criterion commonly
employed in analytical work to ascertain that the limiting current
is diffusion-controlled, is the linear dependence on the square root
of mercury column height. The mercury column height h is the distance between the top of the mercury reservoir and the tip of the
dme. The value should be corrected for back pressure if necessary
[4].

Since $t \propto 1/h$ and $m \propto h$,

$$i_d \propto m^{2/3} t^{1/6} \propto h^{1/2} \qquad (3.9)$$

Thus, a plot of i_d vs. $h^{1/2}$ should be linear, passing through the origin, if the limiting current is diffusion-controlled.

3.2.2 Equation for DC Polarogram with Diffusion-Controlled Limiting Current

For a reversible electrode process, the shape of the i-E curve at all potentials can be derived readily by combining the Nernst and Ilkovic equations.

If A is reduced reversibly to B, for example, say,

$$Cd(II) + 2e \rightleftharpoons Cd \text{ (amalgam)} \qquad \text{or} \qquad Ti(IV) + e \rightleftharpoons Ti(III)$$

or in general,

$$A + ne \rightleftharpoons B \qquad \text{(cathodic process, positive currents)}$$

then

$$E = E^\circ + \frac{RT}{nF} \ln \frac{[A]_{(x=0)}}{[B]_{(x=0)}} \qquad (3.10)$$

A diffuses toward the electrode and from the Ilkovic equation

$$i = 0.732nF([A] - [A]_{(x=0)})D_A^{1/2}m^{2/3}t^{1/6}$$

$$= i_d - 0.732nF([A]_{(x=0)})D_A^{1/2}m^{2/3}t^{1/6}$$

Thus

$$[A]_{(x=0)} = \frac{i_d - i}{0.732nFD_A^{1/2}m^{2/3}t^{1/6}} \qquad (3.11)$$

After being produced by electrolysis, B can diffuse either into the bulk solution or into the mercury to form an amalgam. In either case,

$$i = 0.732nF([B]_{(x=0)} - [B])D_B^{1/2}m^{2/3}t^{1/6}$$

But [B] = 0, so that

$$[B]_{(x=0)} = \frac{i}{0.732nFD_B^{1/2}m^{2/3}t^{1/6}} \tag{3.12}$$

Substituting Eqs. (3.11) and (3.12) into (3.10) gives

$$E = E° + \frac{RT}{nF} \ln \frac{i_d - i}{i} \left(\frac{D_B}{D_A}\right)^{1/2} \tag{3.13}$$

Since the diffusion coefficients of oxidized and reduced forms are often almost equal and moreover appear as the ratio of their square roots, $(D_B/D_A)^{1/2}$ may be set equal to unity, and the equation can be written as

$$E = E° + \frac{RT}{nF} \ln \frac{i_d - i}{i} \tag{3.14}$$

Now, when $i = i_d/2$, $E = E_{1/2}$ so that $E_{1/2} = E°$ and

$$E = E_{1/2} + \frac{RT}{nF} \ln \frac{i_d - i}{i} \tag{3.15}$$

The equation for a reduction wave was first derived by Heyrovský and Ilkovic [7], and Eq. (3.15) is often referred to as the Heyrovský-Ilkovic equation.

From Eq. (3.15) it follows that a graphical plot of E vs. log $[(i_d - i)/i]$ should be linear with slope $2.303RT/nF$. When $i = i_d/2$, $\log[(i_d - i)/i] = 0$ and $E = E_{1/2}$. Thus, a plot of this kind is frequently used to assess the reversibility or otherwise of a dc electrode process and to calculate $E_{1/2}$. Note that $E_{1/2}$ may be considered to be a constant, independent of concentration of the electroactive species for this class of electrode process.

In determining an unknown solution for say cadmium, ideally, in addition to simply measuring the limiting current and determining the concentration, and as part of the overall evaluation procedure, one would check that both $E_{1/2}$ and the slope of the "log" plot of E vs. $\log[(i_d - i)/i]$ are the same as for a known cadmium solution. However, construction of a large number of log plots is a tedious and time-consuming procedure, and an expedient alternative approach is to simply measure the difference $E_{1/4} - E_{3/4}$ from the polarogram, where $E_{1/4}$ and $E_{3/4}$ correspond to values of E at $(1/4)i_d$ and $(3/4)i_d$, respectively.

Since

$$E = E_{1/2} + 2.303 \frac{RT}{nF} \log \frac{i_d - i}{i}$$

$$E_{1/4} = E_{1/2} + 2.303 \frac{RT}{nF} \log 3$$

$$E_{3/4} = E_{1/2} + 2.303 \frac{RT}{nF} \log \frac{1}{3}$$

then

$$E_{1/4} - E_{3/4} = 2.303 \frac{RT}{nF} \log 9 = 0.95(2.303 \frac{RT}{nF}) \qquad (3.16)$$

and is similar to the slope of the log plot. Many authors tabulate values of $E_{1/4} - E_{3/4}$ instead of the slope of a log plot.

For the reversible oxidation process

$$B \rightleftharpoons A + ne$$

it is readily shown that

$$E = E_{1/2} - \frac{RT}{nF} \ln \frac{i_d' - i}{i} \qquad (3.17)$$

where i and i_d' are now negative, and similar criteria for reversibility based on log plots and values of $E_{1/4} - E_{3/4}$ are available.

If both A and B are present in the bulk solution, it logically follows that

$$E = E_{1/2} + \frac{RT}{nF} \ln \frac{i_d - i}{i - i_d'} \qquad (3.18)$$

Equation (3.18) incorporates the two special cases represented by (3.15) and (3.17). If B is absent, $i_d' = 0$ and Eq. (3.18) is equal to (3.15). Alternatively, if A is absent, $i_d = 0$ and Eq. (3.18) reduces to (3.17). The fundamental definition of reversibility is incorporated into this discussion. That is, the same $E_{1/2} \approx E°$

value is obtained whether the oxidation or reduction process is being
studied, and if both are simultaneously present in solution, only
one wave is observed.

For the irreversible process the Nernst equation is no longer
valid and rate theory must be used. A number of calculation procedures are available [8]. For the irreversible reduction

$$A + ne \rightarrow B$$

and using mean currents a solution is

$$E = E° + \frac{RT}{\alpha nF} \ln \frac{i_d - i}{i} + \frac{RT}{\alpha nF} \ln 0.886k_s \left(\frac{t}{D}\right)^{1/2} \tag{3.19a}$$

Thus

$$E_{1/2} = E° + \frac{RT}{\alpha nF} \ln 0.886k_s \left(\frac{t}{D}\right)^{1/2} \tag{3.20a}$$

and

$$E = E_{1/2} + \frac{RT}{\alpha nF} \ln \frac{i_d - i}{i} \tag{3.21a}$$

In Eqs. (3.19) to (3.21), i is no longer the diffusion-controlled
value but is governed by the electron transfer rate. However, the
limiting current is still the diffusion-controlled value.

Provided α is independent of potential, the log plot in this
case is a straight line, the slope of which is larger than that in
the reversible case because α lies in the range 0 to 1. Similarly,
values of $E_{1/4} - E_{3/4}$ are larger than those for the reversible case.
Note $E_{1/2}$ is now also a function of drop time and a rate constant,
unlike the reversible case. Distinction between reversible and
irreversible processes can be made, therefore, on the basis of wave
shape, wave position, and drop time dependence.

Unlike the situation for the reversible electrode process where
both mean and maximum current versions of the i-E curves are analogous,
alterations to the basic expressions are necessary for conversion of
the above mean current equations to the generally more useful equations derived from consideration of currents at the end of the
drop life. In terms of maximum currents

$$E = E° + 0.916 \frac{RT}{\alpha nF} \ln \frac{i_d - i}{i} + \frac{RT}{\alpha nF} \ln 1.359 k_s \left(\frac{t}{D}\right)^{1/2} \qquad (3.19b)$$

$$E_{1/2} = E° + \frac{RT}{\alpha nF} \ln 1.359 k_s \left(\frac{t}{D}\right)^{1/2} \qquad (3.20b)$$

and

$$E = E_{1/2} + 0.916 \frac{RT}{\alpha nF} \ln \frac{i_d - i}{i} \qquad (3.21b)$$

For the quasi-reversible case, corresponding to intermediate values of charge transfer rate constants, no simple solution is available, but as could be guessed, a plot of E vs. log $[(i_d - i)/i]$ in appropriate circumstances can be curved with limiting slopes of 2.303RT/nF and 2.303RT/αnF. Figure 3.6 shows an example for the reduction of zinc in fluoride media [9]. Consequently, log plots are widely used to assess the reversibility, quasi-reversibility, or irreversibility of electrode processes.

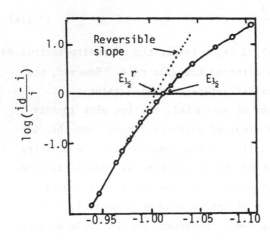

FIGURE 3.6 Logarithmic analyses are frequently used to assess the reversibility, or otherwise, of electrode processes. Example given is the quasi-reversible reduction of 4×10^{-4} M Zn(II) in 0.3 M NaF [9]. Note slope at foot (most-positive potentials) has Nernstian value and the reversible half-wave potential $E_{1/2}^r$ can be calculated.

At this point it should be noted that despite the fact that most reversible systems are of the A $\underset{-ne}{\overset{+ne}{\rightleftharpoons}}$ B type, where A and B are both soluble in solution or in mercury, stoichiometries other than 1:1 are found and log plots other than E vs. $\log[(i_d - i)/i]$ are possible. Whenever alternative stoichiometry occurs in the charge transfer step, the Nernst equation is modified, and so is the dc polarographic wave shape, compared with preceding equations. The all too routine use of $\log [(i_d - i)/i]$-type plots has caused several authors to provide misleading information on the reversibility of the electrode process. Particularly when the mercury electrode itself is involved in the charge transfer step, different kinds of stoichiometry are prevalent. For example, the oxidation of mercury in the presence of chloride, bromide, iodide, xanthates, and other species in nonaqueous solvents can occur via two steps [10-14] as shown in Fig. 3.7.

$$Hg + 3X^- \rightleftharpoons HgX_3^- + 2e \qquad \text{(wave 1)}$$

$$2HgX_3^- + Hg \rightleftharpoons 3HgX_2 + 2e \qquad \text{(wave 2)}$$

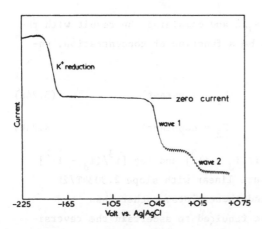

FIGURE 3.7 DC polarogram of 10^{-2} M potassium *o*-butylxanthate in acetone. [Reproduced from J. Electroanal. Chem. *48*, 71 (1973).]

The species X diffuses toward the electrode, and in writing the Nernst equation for wave 1, the fact that the activity of mercury is unity and 3 mol of X are involved needs to be considered. Similar considerations apply to wave 2.

The potential at the dme for wave 1 (E_1) is given by

$$E_1 = E_1^\circ + \frac{RT}{2F} \ln \frac{[HgX_3^-]_{(x=0)}}{[X]_{(x=0)}^3} \tag{3.22a}$$

and for wave 2 (E_2) by

$$E_2 = E_2^\circ + \frac{RT}{2F} \ln \frac{[HgX_2]_{(x=0)}^3}{[HgX_3]_{(x=0)}^2} \tag{3.22b}$$

The equations for the two waves may therefore be written

$$E_1 = C_1 + \frac{RT}{2F} \ln \frac{i}{(i_d - i)^3} \qquad (C_1 = \text{constant}) \tag{3.23a}$$

$$E_2 = C_2 + \frac{RT}{2F} \ln \frac{i^3}{(i_d - i)^2} \qquad (C_2 = \text{constant}) \tag{3.23b}$$

Solving for $E = E_{1/2}$ when $i = i_d/2$ and combining the result with the Ilkovic equation shows $E_{1/2}$ to be a function of concentration, unlike previous examples:

$$E_{1,1/2} = K_1 - \frac{RT}{F} \ln [X^-] \qquad (K_1 = \text{constant}) \tag{3.24a}$$

$$E_{2,1/2} = K_2 + \frac{RT}{2F} \ln [X^-] \qquad (K_2 = \text{constant}) \tag{3.24b}$$

Thus plots of E vs. $\log [i/(i_d - i)^3]$ and $\log [i^3/(i_d - i)^2]$ as shown in Figs. 3.8 and 3.9 are linear with slope $2.303RT/2F$ and $2.303RT/F$, respectively, and these plots, rather than E vs. $\log[(i_d - i)/i]$, would be required to ascertain the reversibility or otherwise of the electrode processes. Further, $|E_{1/4} - E_{3/4}|$ values would now be $2.303(RT/2F) \log 81$ for wave 1 and $2.303(RT/2F) \log 243$ for wave 2. This illustrates the point

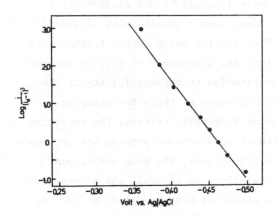

FIGURE 3.8 Logarithmic analysis for wave 1 for a 10^{-2} M solution
of potassium *o*-ethylxanthate in acetone. [Reproduced from J.
Electroanal. Chem. *48*, 71 (1973).]

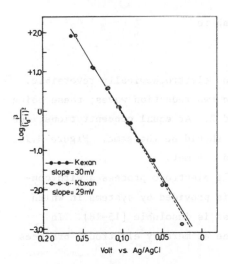

FIGURE 3.9 Logarithmic analysis
for wave 2 for 10^{-3} M solutions
of potassium *o*-ethyl(butyl)-
xanthate in acetone. [Repro-
duced from J. Electroanal. Chem.
48, 71 (1973).]

that values of $E_{1/4} - E_{3/4}$ greater than 2.303(RT/nF) log 9 need
not necessarily be an indication of irreversibility, as has sometimes been assumed. Finally, with a 10-fold increase in concentration of X, the $E_{1/2}$ value of wave 1 should become (2.303RT/F) V
more negative. On the other hand, wave 2 should shift (2.303RT/2F) V
in the opposite direction. That is, the waves become further separated with higher concentration, and dependence of $E_{1/2}$ on concentration is not necessarily a criterion of irreversibility if the
electrode process has the stoichiometry of these two examples.

While log plots and related diagnostic criteria for assessing
the reversibility or otherwise of an electrode process are extremely
useful and widely used in analytical work, the most unambiguous
method for assessing the reversibility is to study the polarography
of the assumed product of the electrode process noted, as in Chap.
20. In the above example, the complex HgX_3^- should give one oxidation wave corresponding to

$$2HgX_3^- + Hg \rightleftharpoons 3HgX_2 + 2e$$

and one reduction wave corresponding to

$$HgX_3^- + 2e \rightleftharpoons Hg + 3X^-$$

if the system is both chemically and electrochemically reversible.
On the other hand, HgX_2 should give two reduction waves, these being
the reverse of previous waves 1 and 2. At equal concentrations, the
$E_{1/2}$ values for X^-, HgX_3^-, and HgX_2 should be the same. Figure 3.10
confirms that criteria of this kind are met.

A further example of reversible electrode processes with concentration dependent $E_{1/2}$ values, is provided by systems in which
the product of the electrode process is insoluble [15-18]. In
aqueous media, oxidation of chloride at mercury electrodes produces
insoluble calomel:

$$2Cl^- + 2Hg \rightleftharpoons Hg_2Cl_2 \downarrow + 2e$$

Since the activity of Hg_2Cl_2 can be taken as unity,

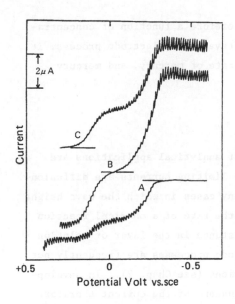

FIGURE 3.10 Polarograms of 10^{-3} M solutions of Et_4NCl (A), Et_4NHgCl_3 (B), and $HgCl_2$ (C) in DMF containing 0.1 M Et_4NClO_4 at 25°C. [Reproduced from Bull. Chem. Soc. Jap. *43*, 2046 (1970).

$$E = E° - \frac{RT}{F} \ln [Cl^-]_{(x=0)} \qquad (3.26)$$

Alternatively, writing the electrode process in a stepwise fashion as

$$2Hg \overset{2e}{\rightleftharpoons} Hg_2^{2+} + 2Cl^- \overset{K_s}{\rightleftharpoons} Hg_2Cl_2$$

where K_s is the solubility product, shows that Eq. (3.26) is equivalent to

$$E = E°_{Hg_2^{2+}/Hg} + \frac{RT}{F} \ln K_s - \frac{RT}{F} \ln [Cl^-]_{(x=0)} \qquad (3.27)$$

which is the equation for the potential of the calomel electrode.

Substitution of the Ilkovic equation gives

$$E = \text{constant} - \frac{RT}{F} \ln (i = i_d) \qquad (3.28)$$

and

$$E_{1/2} = E° - \frac{RT}{F} \ln \frac{[Cl^-]}{2} \qquad (3.29)$$

The half-wave potential is therefore a function of concentration of chloride, but unlike an irreversible electrode process, it
is independent of drop time, flow rate of mercury, and mercury
column height.

3.2.3 Kinetically Controlled
        Limiting Currents

In dc polarography, the majority of analytical applications are
based on electrode processes whose limiting currents are diffusion-
controlled. However, there are many cases in which the wave height
is partly or wholly determined by the rate of a chemical reaction
that produces an electroactive substance in the layer of solution
around the electrode. Such electrode processes are frequently not
as suitable in analytical applications (see Chap. 2). In developing any polarographic method, the nature of the current therefore
needs to be recognized, and this is easily accomplished *for the
limiting current region* in dc polarography, because kinetically
controlled currents exhibit considerably different behavior to that
described above for the diffusion-controlled case.

Preceding Reaction: CE Mechanism. In its simplest form, the CE
mechanism, using nomenclature from Chap. 2, can be written as

$$Y \underset{k_2}{\overset{k_1}{\rightleftharpoons}} A \tag{C}$$

$$A + ne \rightleftharpoons B \tag{E}$$

where k_1 and k_2 are first-order or pseudo-first-order rate constants
(sec^{-1}). In other cases, k_1 and k_2 could be second or higher order.
The limiting current of a kinetically controlled electrode
process will obviously depend on the ratio of the equilibrium concentrations of Y and A in the bulk solution and the values of k_1
and k_2. Under conditions where the equilibrium concentration of A
is low and k_1 is slow, then the limiting current will be purely
kinetically controlled and

$$i_1 = i_k \tag{3.30}$$

If the equilibrium concentration of A is significant, the observed
wave height can be approximated to the sum of the diffusion current,
which is proportional to the concentration of A and the kinetic
current arising from the transformation of Y into A at the electrode
surface, or

$$i_1 = i_d + i_k \tag{3.31}$$

A final possibility is that k_1 is so fast that the limiting current
is an appreciable fraction of the hypothetical diffusion-controlled
limiting current of A. In the limiting case where the transformation of Y into A occurs at a diffusion-controlled rate,

$$i_1 = i_d \tag{3.32}$$

The solution of the CE mechanism is based on a modified boundary-
value problem compared to that for the diffusion-controlled case, and
the literature can be consulted for details [2,3,19-29].

An expression derived by Koutecký for the mean current is

$$\frac{i_k}{i_d - i_k} = 0.886 \frac{k_1}{k_2^{1/2}} t^{1/2} \tag{3.33}$$

where i_d is the diffusion current that would be obtained from the
complete reduction of Y and i_k is the observed kinetic current.

In the case where the limiting current is purely kinetically
controlled [28], that is, $i_k \ll i_d$ (mean currents),

$$i_1 = i_k = 0.493nD^{1/2} [Y]m^{2/3}t^{2/3} \frac{k_1}{k_2^{1/2}} \tag{3.34}$$

From Eq. (3.34) a kinetic current is easily recognized because
of the independence of i_k on mercury column height, that is,
$m \propto h$, $t \propto 1/h$, so that $m^{2/3}t^{2/3}$ is independent of h. It was demonstrated previously that diffusion-controlled limiting currents show
a $h^{1/2}$ dependence on mercury column height; so the behavior of i_1
on variation of the mercury column height readily enables the two
types of limiting currents to be distinguished.

If Eq. (3.31) is valid, then

$$\frac{i_1}{m^{2/3}t^{1/6}} = 0.607nD^{1/2}[A] + 0.493nD^{1/2}[Y]t^{1/2}\frac{k_1}{k_2^{1/2}} \qquad (3.35)$$

From Eqs. (3.34) and (3.35) it can be seen that for kinetically
controlled currents the limiting current may be linearly dependent
on concentration of Y. The height of a kinetic wave also can be
influenced easily by the experimental conditions that affect k_1 and
k_2, e.g., temperature, pH, or ionic strength. These mathematical
results amplify conclusions reached from the qualitative discussion
in Chap. 2.

Following Reaction: EC Mechanism. The dc limiting current for the
EC mechanism is not influenced by the kinetics of a follow-up reaction; so equations applicable to diffusion control remain valid.
Chapter 2 adequately describes the dc polarography of the EC-type
mechanism, and electrode processes of this type are most suitable
for a polarographic analysis. However, this conclusion is not
generally valid for all methods, and the EC mechanism will be examined in detail when discussing other polarographic techniques.

Catalytic Mechanisms. Approximate theoretical descriptions of the
dc polarography of this class of electrode process, as for other
mechanisms, quantitatively support the discussion in Chap. 2.

Unlike the CE mechanism, where $i_1 = i_k \ll i_d$, or the EC mechanism, where $i_1 = i_d$, the catalytic current i_c is frequently much
greater than i_d, and the extremely large current per unit concentration may justify the use of this class of electrode process.

If the catalyzing reagent Z is in a considerable concentration
excess, the rate of the catalytic step can be written in terms of a
pseudo-first-order rate constant, $k_1[Z]_{(x=0)}$. This simplifies the
mathematical description of catalytic waves of the kind

$$A + ne \rightleftharpoons B \qquad B + Z \underset{k_2}{\overset{k_1}{\rightleftharpoons}} A$$

If $k_1[Z]_{(x=0)}$ is small, very little B will be reoxidized and the
wave height will be equal almost to the diffusion current of A as
measured in the absence of Z. However, as $k_1[Z]_{(x=0)}$ increases, the
chemical reaction may produce much more reducible material at the
electrode surface than that arising by diffusion from the bulk solution, so that $i_c \gg i_d$.

The mean catalytic current [19,30,31] is given by

$$i_c = 0.493nD^{1/2}[A]m^{2/3}t^{2/3}\{(k_1 + k_2)[Z]_{(x=0)}\}^{1/2} \qquad (3.36)$$

Like kinetic currents, this kind of catalytic current can be
identified by the fact that it does not vary with mercury column
height and apart from the magnitude of the limiting current, both
kinetic and catalytic currents have many similar characteristics.
The limiting current for the catalytic wave will be equal to the
sum $i_c + i_d$, and if the diffusion current of A is appreciable, it
may be measured separately (in the absence of Z), and the catalytic
current obtained by difference. Equation (3.36) would predict a
linear dependence of i_c on concentration. However, the mechanism
under the specified conditions used to derive this equation is only
one of many possibilities for a catalytic process, and again as for
the CE mechanism both linear and nonlinear limiting current-vs.-
concentration relationships can be obtained [32]. Figure 3.11a and
3.11b show polarograms and calibration curves for the tungsten/
peroxide/oxalate system [33]. Note that limiting currents and wave
shapes are frequently far from ideal, and difficulties in obtaining
accurate polarographic parameters can be expected for this class of
electrode process. Discussion in Chap. 2 regarding specificity
and other difficulties should also be remembered when using a
catalytic wave.

Similar considerations to the above apply to disproportionation
and other partially regenerative mechanisms and to catalytic hydrogen waves [34]. Catalytic hydrogen waves are those in which the
normally irreversible and very negative reduction wave involving
hydrogen ions is shifted to more positive potentials by the presence

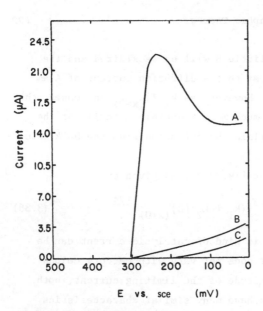

FIGURE 3.11a Kinetic wave for tungsten in peroxide/oxalate media:
(A) 3.0×10^{-5} M WO_4^{2-}, 8.0×10^{-2} M H_2O_2, 1×10^{-2} M $H_2C_2O_4$;
(B) 8.0×10^{-2} M H_2O_2, 1×10^{-2} M $H_2C_2O_4$; (C) 3.0×10^{-5} M WO_4^{2-},
1×10^{-2} M $H_2C_2O_4$.

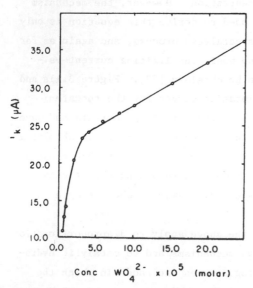

FIGURE 3.11b Kinetic current as a function of tungsten concentration:
8×10^{-2} M H_2O_2, 1×10^{-2} M $H_2C_2O_4$. [Reproduced from Anal. Chem. *44*,
184 (1972). © American Chemical Society.]

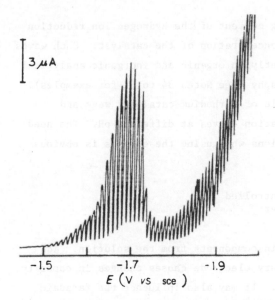

FIGURE 11c Catalytic wave for 2×10^{-8} M Rh(III) in ammonia buffer at pH 10, 10^{-7} M cysteine, and 0.002% Triton-X at 15°C. [Reproduced from J. Electroanal. Chem. *31*, App. 3 (1971).]

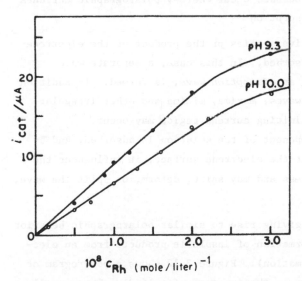

FIGURE 11d Catalytic wave height plotted as a function of Rh(III) concentration under the same conditions as Fig. 11c. [Reproduced from J. Electroanal. Chem. *31*, App. 3 (1971).]

of a catalyst. The limiting current of the hydrogen ion reduction wave is a function of the concentration of the catalyst. Such waves have been used quite frequently in organic and inorganic analytical applications of dc polarography (see Notes 34 to 37 for examples). Figure 3.11c shows an example of a rhodium-catalyzed wave and Fig. 3.11d shows the calibration curves at different pH. The need to carefully control conditions when using these waves is obvious from the data provided.

3.2.4 Limiting Currents Controlled by Adsorption or Other Surface Phenomena

Electrodes can adsorb certain components from the solution. Adsorption at a dropping mercury electrode causes changes in capacity current as shown in Chap. 2. It may also influence the faradaic current in many undesirable ways. This latter aspect is now considered in further detail.

Essentially two processes occur whereby polarographic currents may be influenced by adsorption:

1. The electroactive species or the product of the electrode reaction is adsorbed. In this case, a separate wave, usually called an *adsorption wave*, is formed. In addition to adsorption waves, maxima, minima and other irregularities on the limiting current region may occur.

2. Some other component of the solution is adsorbed, and by its presence at the electrode surface, it influences the electrode process and may shift, deform, or split the wave, for example.

Another phenomena giving rise to similar polarographic behavior to adsorption is the formation of insoluble products from an electrode process (film formation). Figure 3.12 shows a polarogram of the oxidation of mercury in the presence of sulfide. The overall electrode process is $Hg + S^{2-} \rightleftharpoons HgS + 2e$ and the observation of at least three waves is believed to be due to the formation of insoluble films of mercury sulfide.

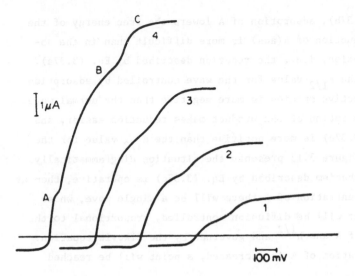

FIGURE 3.12 Polarograms of Na$_2$S in 0.1 M KCl. The waves were
recorded from -0.9 V (vs. nce) in the anodic direction. Na$_2$S concentration: (1) 2 × 10^{-4}; (2) 4 × 10^{-4}; (3) 6 × 10^{-4};
(4) 8 × 10^{-4} M. [Reproduced from Polarography 1964 *1*, 473 (1966).]

Theoretical treatment of adsorption and related surface phenomena are rather complex [38] because so many possibilities occur in
both the nature of the adsorption isotherm and the kinetics of adsorption and electron transfer coupled to the adsorption. For our
purposes, recognizing the presence of surface phenomena and finding
possible ways of minimizing deleterious effects are the important
ingredients to be gleaned from the theory, and only the simplest
theoretical interpretation is presented.

Adsorption is due to surface forces, the range of which usually
do not exceed molecular dimensions, so that theoretical descriptions
of adsorption phenomena consider what occurs when a monomolecular
layer of adsorbed material is formed. This is assumed in the following discussion.

Consider the electrode processes

$$A + ne \rightleftharpoons B \tag{3.37a}$$

$$A \rightleftharpoons A(ads) + ne \rightleftharpoons B \tag{3.37b}$$

$$A + ne \rightleftharpoons B \rightleftharpoons B(ads) \tag{3.37c}$$

In Eq. (3.37b), adsorption of A lowers the free energy of the system, and reduction of A(ads) is more difficult than in the absence of adsorption, i.e., the reaction described by Eq. (3.37a). Consequently, the $E_{1/2}$ value for the wave controlled by adsorption of the electroactive species is more negative than the normal wave. Conversely, adsorption of the product makes reduction easier, and $E_{1/2}$ for Eq. (3.37c) is more positive than the $E_{1/2}$ value for the normal wave. Figure 3.13 presents the situation diagrammatically.

If the mechanism described by Eq. (3.37c) is operative, then at a very low concentration of A there will be a single wave, whose limiting current will be diffusion controlled, proportional to the concentration of A and $h^{1/2}$ and governed by the Ilkovic equation. As the concentration of A is increased, a point will be reached

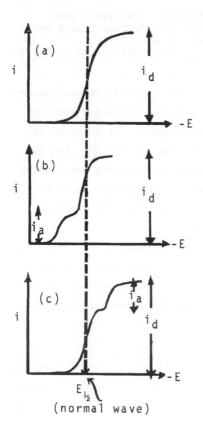

FIGURE 3.13(a) Waveforms in the absence of adsorption, (b) for the adsorption of the reduced form, and (c) for the adsorption of the oxidized form for dc polarographic reduction waves.

where enough B(ads) is formed during the life of the drop to cover
the entire surface. More than this amount of reduction can be
accommodated only via the "normal" process, where excess B is dissolved into the solution. Since it is more difficult to reduce A
to dissolved B than to reduce it to adsorbed B, the reduction of
excess A will produce a second wave at a more negative potential.
The original wave formed at low concentrations is now adsorption
rather than diffusion-controlled, and its limiting current i_a will
become independent of concentration of A at the same concentration
at which the second or normal wave is observed. The total height
of the two waves corresponds to reduction of all A reaching the
electrode via diffusion and is therefore diffusion-controlled and
proportional to both concentration of A and $h^{1/2}$. The limiting
currents of neither wave, however, is individually diffusion-
controlled. Analogous arguments hold for the case of reduction of
A(ads). However, in this case the single wave obtained at low concentrations of A is more negative than the normal wave which appears
only at high concentrations of A.

   If it is assumed that the oxidized form, which is subject to
a reversible reduction, is adsorbed, then equations based on the
Langmuir isotherm give the following results. The number of adsorbed moles per unit surface area of a dme, a, is

$$\frac{zw[0]_{(x=0)}}{1 + w[0]_{(x=0)}} \tag{3.38}$$

where

   z = maximum number of moles adsorbed per unit electrode
      surface

   w = adsorption coefficient

   $[0]_{(x=0)}$ = the equilibrium concentration of the oxidized form
      at the electrode surface

If the adsorption equilibrium is established rapidly and the electrode surface area [39] is $A = 0.85m^{2/3}t^{2/3}$, then from Faraday's law,

$$i = nFa\frac{dA}{dt} = nF\frac{dA}{dt}\frac{zw[0]_{(x=0)}}{1 + w[0]_{(x=0)}} \tag{3.39}$$

If the drop surface is fully covered as soon as the drop is formed (that is, $w[0]_{(x=0)} \gg 1$), then $i = i_a$, the adsorption current, and

$$i_a = nFz \frac{2}{3} 0.85 m^{2/3} t^{-1/3} \qquad (3.40)$$

The mean adsorption current is given by

$$(i_a)_{mean} = nFz 0.85 m^{2/3} t^{-1/3} \qquad (3.41)$$

It follows from Eq. (3.41) that the limiting adsorption current

$$i_a = \text{constant} \times h \qquad (3.42)$$

The simple dependence on the height of the mercury head and independence of concentration are two important criteria for assigning adsorption currents.

Similar results apply for the case where the reduced form is adsorbed and where film formation from an insoluble product occurs. In analytical work the total height of all waves is linearly dependent on concentration, and this height rather than just the adsorption wave can be used in the preparation of calibration curves.

Figures 3.14 and 3.15 summarize the characteristics of the different kinds of limiting current with respect to dependence on mercury column height and concentration. These two criteria are the simplest to apply in deciding the nature of the limiting current.

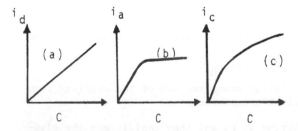

FIGURE 3.14 Types of plots obtained for dependence of limiting current on concentration: (a) diffusion- and most kinetically controlled limiting currents, (b) adsorption-controlled limiting currents, and (c) catalytic limiting currents.

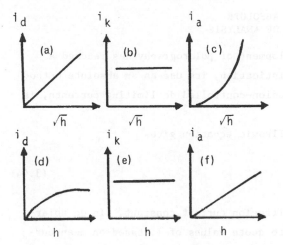

FIGURE 3.15 Plots of the dependence of limiting current on the mercury height: (a) to (c), dependence on $h^{1/2}$; (d) to (f), dependence on h. (a) and (d) are for diffusion-controlled limiting currents; (b) and (e) are for kinetically controlled limiting currents; (c) and (f) are for adsorption-controlled limiting currents.

3.2.5 Charging Current in DC Polarography

In Chap. 2 it was shown that the charging current is given by the expression

$$i_c = E' F(E) \frac{dA}{dt} (E_m - E)$$

Assuming a spherical drop, $A = 0.85 m^{2/3} t^{2/3}$ [39] and the result

$$i_c = C'_{F(E)} (\tfrac{2}{3} 0.85 m^{2/3} t^{-1/3}) (E_m - E) \tag{3.43}$$

is obtained for dc polarography.

It follows from this equation that the instantaneous charging current at constant potential reaches its maximum value at the beginning of the drop life and its minimum value at the end of the drop life. This is an important result, since one of the advances in dc polarography is based on the different time dependences of the faradaic ($t^{1/6}$) and charging currents ($t^{-1/3}$), as will be seen in Chap. 4.

3.3 DC POLAROGRAPHY AS AN ABSOLUTE
OR COMPARATIVE METHOD OF ANALYSIS

While the theoretical development of polarography has reached a
substantial degree of sophistication, its use as an absolute method
of analysis, even for diffusion-controlled dc limiting currents, is
not widespread.

Rearrangement of the Ilkovic equation gives

$$I = \frac{i_d}{Cm^{2/3}t^{1/6}} \qquad (3.44)$$

where I is the so-called diffusion current constant. In dc polarography some authors used to quote values of I, based on mean currents, which, in principle, could be used to form the basis of an
absolute method. Unfortunately, the value of I in fact has been
demonstrated to be dependent on capillary characteristics [3,4].
From previous discussion this is not unexpected, and the absolute
method may be used only to give analytical data at about the ±5%
confidence level. Rigorous restrictions on flow rate and drop time
accompanied the suggested use of polarography as an absolute method
of analysis. Theoretical uncertainties in non-dc polarographic
techniques, while frequently difficult to ascertain at the present
because of a paucity of data, are probably higher than this, and as
with many other instrumental methods of analysis, comparative determinations based on reference to standard solutions are almost always
used in practical situation. In the author's experience, this is certainly the recommended procedure and has some useful side advantages
in the systematic use of all polarographic methods. For example, a
useful procedure to check that interference-free polarographic data are
being obtained is to assume that the shape (for example, $E_{1/4} - E_{3/4}$)
in dc polarography) and $E_{1/2}$ value from the unknown solution are the
same as that found for the standards. Furthermore, subtle changes
in solution conditions, capillary angle, etc., can cause significant
changes in the value of i_d per unit concentration, and careful
matching of unknown and standards eliminates potential sources of

error likely to be found in using a value of I, particularly when
the value of I used is that reported from another laboratory, using
a capillary of unknown origin.

Finally, but not least importantly, the author has the suspicion
that the early endeavors to develop dc polarography as an absolute
method of analysis led to the adoption of an unnecessarily conservative approach. For example, during the early development stages
of dc polarography, it was found that the Ilkovic equation did not
apply at short drop times, and the practice of using drop times
below about 2 sec was generally discouraged for many years. Nowadays however, drop times down to 5 msec are being advantageously
exploited in analytical work, as is discussed in Chap. 4. Provided
a linear dependence of i_d on concentration is found, and one is prepared to use calibration curves, all restrictions on drop time, use
of vertical or horizontal capillaries, flow rate etc., are lifted,
and complete freedom to experiment is available. This is generally
a more attractive approach than trying to comply with the rigorous
restrictions necessarily placed on the experiment, if polarography
is to be used as an absolute method of analysis.

NOTES

1. A. Fick, Pogg. Ann. *94*, 59 (1855).

2. J. Heyrovský and J. Kůta, *Principles of Polarography*,
   Academic Press, New York, 1966, pp. 73–119.

3. L. Meites, *Polarographic Techniques*, 2nd ed., Interscience,
   New York, 1965, pp. 95–202.

4. I. M. Kolthoff and J. J. Lingane, *Polarography*, 2nd ed.,
   Interscience, New York, 1952.

5. D. Ilkovic, Collect. Czech. Chem. Commun. *6*, 498 (1934).

6. D. Ilkovic, J. Chim. Phys. *35*, 129 (1938).

7. J. Heyrovský and D. Ilkovic, Collect. Czech. Chem. Commun. *7*,
   198 (1935).

8. Note 2, pp. 205–266, and Note 3, pp. 203–266.

9. S. Vavricka and J. Koryta, Collect. Czech. Chem. Commun. *32*,
   2346 (1967).

10. Y. Matsui, R. Kawakado, and Y. Date, Bull. Chem. Soc. Jap. *41*, 2914 (1968).

11. Y. Matsui, Y. Kurosaki, and Y. Date, Bull. Chem. Soc. Jap. *43*, 1707, 2046 (1970).

12. Y. Matsui and Y. Date, Bull. Chem. Soc. Jap. *43*, 2052 (1970).

13. A. M. Bond, A. T. Casey, and J. R. Thackeray, J. Electroanal. Chem. *48*, 71 (1973).

14. A. M. Bond, A. T. Casey, and J. R. Thackeray, J. Electrochem. Soc. *120*, 1502 (1973).

15. J. Revenda, Collect. Czech. Chem. Commun. *6*, 453 (1934).

16. I. M. Kolthoff and C. S. Miller, J. Amer. Chem. Soc. *63*, 1405 (1941).

17. R. Haul and E. Scholz, Z. Elektrochem. *52*, 226 (1948).

18. A. A. Vlcek, Collect. Czech. Chem. Commun. *19*, 221 (1954).

19. R. Brdicka, V. Hamus, and J. Koutecký, in *Progress in Polarography* (P. Zuman and I. M. Kolthoff, eds.), Interscience, New York, 1962, vol. I, p. 145, and references therein.

20. R. Brdicka and K. Wiesner, Collect. Czech. Chem. Commun. *12*, 138 (1947).

21. J. Koutecký and R. Brdicka, Collect. Czech. Chem. Commun. *12*, 337 (1947); J. Amer. Chem. Soc. *76*, 907 (1954).

22. J. Koutecký, Collect. Czech. Chem. Commun. *18*, 11, 183, 311, 597 (1953); *19*, 857, 1045, 1093 (1954); *20*, 116 (1958).

23. R. Brdicka, Z. Elektrochem. *64*, 16 (1960).

24. W. Hans, Z. Elektrochem. *59*, 807 (1955).

25. J. Koryta and J. Koutecký, Collect. Czech. Chem. Commun. *20*, 423 (1955).

26. P. Rüetschi and G. Trümpler, Helv. Chim. Acta *35*, 1957 (1952).

27. P. Delahay, Ann. Rev. Phys. Chem. *8*, 229 (1957).

28. P. Delahay, J. Amer. Chem. Soc. *74*, 3506 (1952).

29. D. R. Crow, *Polarography of Metal Complexes*, Academic Press, New York, 1969, and references therein.

30. P. Delahay and G. L. Stiehl, J. Amer. Chem. Soc. *74*, 3500 (1952).

31. J. Koutecký, Collect. Czech. Chem. Commun. *18*, 311 (1953).

32. Note 2, pp. 380-393, and references therein.

33. I. M. Kolthoff and E. P. Parry, J. Amer. Chem. Soc. *73*, 5315 (1951); K. B. Yatsimirzki and L. I. Budarin, Zh. Neorg. Khim. *7*, 1824 (1962); T. A. O'Shea and G. A. Parker, Anal. Chem. *44*, 184 (1972).

34. Note 2, pp. 407-428, and references therein.

35. M. Brezina and P. Zuman, *Polarography in Medicine, Biochemistry and Pharmacy*, Interscience, New York, 1956.

36. S. G. Mairanovskii, J. Electroanal. Chem. *6*, 277 (1963), and references therein.

37. See, for example, R. A. F. Bullerwell, J. Polarogr. Soc. *12*, 12 (1966); P. W. Alexander and G. L. Orth, J. Electroanal. Chem. *31*, App. 3 (1971).

38. Note 2, pp. 287-335.

39. Note 2, pp. 39-40.

Chapter 4

ADVANCES IN DC POLAROGRAPHY

During the history of polarography, many advances in dc techniques
have been made. Rather than discuss the techniques in chronological
order of development, they are presented here in an order aimed at
systematically highlighting how limitations found in conventional
dc polarography (listed in Chap. 3) have been overcome. However, it
is also important to note that with certain methods, advantages
gained in one direction are at the expense of a loss elsewhere, and
this may need to be taken into account in assessing the overall
usefulness of each technique.

4.1 RAPID DC POLAROGRAPHY
   WITH SHORT CONTROLLED DROP TIMES

One of the disadvantages of conventional polarography compared with
other analytical methods has always been that the natural drop times
of the dropping mercury electrode (dme)-- usually between 2 and
8 sec-- necessitate reasonably slow scan rates of potential, and the
time required for recording a polarogram is correspondingly long.
Scan rates of potential must be slow for two major reasons: to
avoid violating the constant potential-current conditions assumed
for theoretical purposes, and to provide a high degree of precision
of measurement. Each drop represents one datum point on the current-
voltage curve, and especially over the steeply rising portion of a

dc polarogram, a large number of data points are necessary. Obviously, the faster the scan rate, the fewer data points on the graph, and this results in a decrease in precision. Thus, even if the theoretical constant potential-current requirement did not need to be met, there would still be an excellent practical reason why the scan rate cannot be increased beyond well-defined limits when natural drop times are used. Figure 3.4 and the discussion in Chap. 3 on the limitations of dc polarography also emphasize this difficulty.

In electroanalytical methods with stationary electrodes, such as the hanging drop mercury electrode, platinum electrode, glassy carbon electrode, etc. (to be discussed in Chap. 5), fast scan rates of potential are used routinely. The theory for stationary electrode voltammetry includes terms for the scan rate of potential, and so the first restriction given above for polarography does not apply, nor the second, because a continuous current-voltage curve is obtained. Similarly with streaming (mercury) electrodes, no restriction is placed on scan rate. However, although voltammetric methods may be used with considerable time saving, they have several well-established difficulties and disadvantages compared with the polarographic method, and fast scan rates with a dme method are still worth considering.

If, for the polarographic method, time saving is required through increasing of the scan rate, then the only real possibility is shortening the drop time. In principle, this could be achieved by using suitable glass capillaries and/or increased mercury column height, but a much more convenient method is to mechanically knock or remove the mercury drop at selected time intervals to produce a short controlled drop time. This has been the preferred technique and is the one discussed here. Most commercially available instruments now include capabilities for short controlled drop times (see Chap. 2).

In the theoretical sense the drop time could be shortened indefinitely until the mercury drops come so close together that a streaming mercury electrode, rather than a dropping mercury electrode, is obtained. Thus, at some drop time or range of drop times,

a transition between a dme and a streaming electrode should be encountered. The nature of electrode processes and the physical
characteristics of a short drop time dme itself could also be expected to undergo certain transitions as the drop time is decreased.

Various features of short controlled drop time dc polarography
have been examined in the literature (see Notes 1 to 16, for instance).

The use of short controlled drop times permits the application
of scan rates of potential up to several hundred millivolts per
second. In routine analytical applications of polarography, the
resultant shorter recording times possible with this rapid technique
may be of considerable advantage if a large number of analyses are
to be performed.

However, a fast scan rate of potential is by no means the only
advantage of the rapid technique, as pointed out in recent reviews
[9,13]. Cover and Connery [5], using a dropping mercury electrode
with drop times as short as 5 msec, which they refer to as a vibrating dropping mercury electrode (vdme), observed that maxima
were suppressed without the addition of surfacants, and that catalytic and kinetic waves could be minimized or eliminated. This is
a general feature of rapid dc polarograms, and ill-defined curves
found at natural drop times often are considerably simplified [13].

Inhibition of dme response by adsorption of species on the
electrode surface may prevent collection of useful analytical data
with polarography. Adverse effects of such inhibition often include
nonlinear response to the concentration of electroactive species,
erratic drop behavior, or polarographic waves so distorted that
meaningful current measurement is prevented. Elimination or at
least minimization of these effects is obviously desirable.
Connery and Cover [6] also examined a number of systems where adsorption phenomena were present at the dme. They considered several cases of adsorption of electroinactive species and two examples
where the product of the electrode reaction was adsorbed, and
showed the superiority of the vdme over the dme as an analytical
tool for these systems.

Another phenomenon which inhibits dme response, and which is often associated with analytically undesirable behavior similar to that described above, is the formation of insoluble reaction products on the electrode surface-- a situation which is particularly relevant to anodic polarographic waves corresponding to the formation of mercury compounds. Abnormal behavior associated with this important class of electrode process was examined by Canterford et al. [16], and again much of the anomolous behavior disappears or is minimized at short drop time.

Figures 4.1 to 4.4 show examples of rapid polarograms and other relevant data. Many of the advantages are clearly illustrated. Considering the obvious advantages in decreasing the time scale of analysis and simplifying electrode processes, it is surprising that the rapid polarographic method has not been used more widely. However, early opcculations and recommendations on the use of short drop time techniques were not encouraging, to say the least, and presumably this provided the inertia preventing their exploitation.

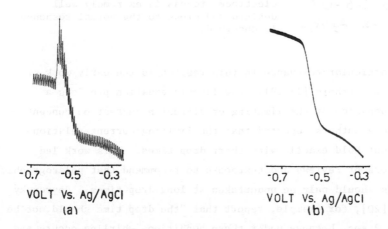

FIGURE 4.1 The use of short drop times can often eliminate maxima. Solution is 1.5×10^{-3} M tin(IV) in 5 M HCl and the electrode process is Sn(II) + 2e $\rightleftharpoons$ Sn(Hg). (a) Conventional dc polarogram with maximum; (b) maximum is eliminated at a drop time of 0.16 sec.

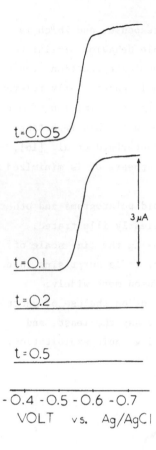

t=0.05

3 μA

t=0.1

t=0.2

t=0.5

-0.4 -0.5 -0.6 -0.7

VOLT v s. Ag/AgCl

FIGURE 4.2 The use of short drop times can sometimes eliminate inhibition of the electrode process caused by surface-active species. Solution is 5.5×10^{-4} M Cd(II) in 2.5 M HCl and also contains 2.5×10^{-4} M tribenzylamine (TBA). The electrode process is inhibited by the TBA and no Cd(II) reduction is observed under conventional conditions. However, with very short drop times the $Cd(II) + 2e \rightleftharpoons Cd(Hg)$ electrode process is extremely well defined and close to the normal response is observed.

Of particular relevance in this respect is the early work of Maas [17] and others [18,19]. The Ilkovic equation predicts a linear dependence of the limiting or diffusion current on concentration. These authors reported that the limiting current relationships do not hold exactly with short drop times. This work led many writers of reviews and textbooks to recommend that polarographic techniques should only be undertaken at long drop times. Heyrovský and Kůta [20], for example, report that "the drop time should not be less than 2 sec, because under these conditions whirling occurs and destroys the diffusion layer; the currents are increased. The most suitable drop time is from 3 to 5 sec." Meites in his

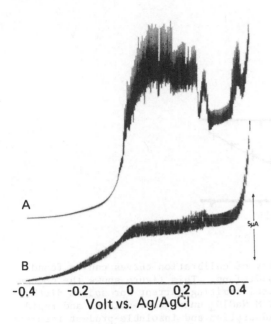

FIGURE 4.3 The use of short drop times frequently improves the definition of the limiting current region when sparingly soluble products are formed. Electrode process is simplistically written as
$$2Hg + 2I^- \overset{ads}{\rightleftharpoons} Hg_2I_2\!\!\downarrow + 2e,$$ where $[I^-]$ is 1.2×10^{-3} M, and supporting electrolyte is 1 M $NaClO_4$: (A) t = 2.9 sec; (B) t = 0.16 sec. [Reproduced from Anal. Chem. *45*, 1327 (1973). © American Chemical Society.]

textbook [21] also reports in the same vein, as do Kolthoff and Lingane [22] in their standard reference book.

Recently, however, Cover and Connery [7], using drop times down to the millisecond region, have found close correlations with much of the existing theory, and they specifically comment that their data do not show the sharp increase in diffusion current constant found by other workers with drop times below 2 sec. They suggest that this may result in part from their lack of data in the 0.01- to 1.0-sec drop time range, although they note that other data [3] in this range also conform to existing theory, as has now been verified elsewhere [9,14].

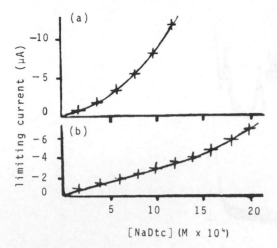

FIGURE 4.4 Improved linearity of calibration curves can be found under rapid polarographic conditions. This figure shows the concentration dependence of the total limiting current for sodium diethydithiocarbonate (NaDtc) in 1 M $NaClO_4$ under conventional and rapid polarographic conditions. Adsorption and insoluble-product formation are involved. Electrode process is simplistically written as

$$Hg + 2Dtc^- \overset{ads}{\rightleftharpoons} Hg(Dtc)_2 + 2e:$$ (a) t = 2.9 sec; (b) t = 0.16 sec. (Data taken from Note 16.)

4.1.1 Rapidly Dropping Mercury
       Electrodes and Their Theory

Different workers have utilized many different approaches to achieve short drop times. Much of the early work varied the drop time by altering the mercury column height or constructing capillaries of varying geometry. Nowadays the favored approach for obtaining short drop times is through mechanically knocking the dme at selected time intervals, to give a controlled drop time shorter than that which would occur if the mercury drop were allowed to fall naturally. Thus, the apparent anomalies could well arise from the different approaches to obtaining short drop times and from the fact that the experiments being conducted are not identical. Fisher et al. [23] made the following comment:

The convention has somehow become established that dme's should be operated at a drop time between 2 and 6 sec; this is a misleading oversimplification. It is not the drop time alone but rather the combination of drop time and mercury flow rate that is important in the case of dme's operated with a freely forming and falling drop, and in the case of dme's where drop time is properly controlled by mechanical hammering the magnitude of the flow rate is more important than the fact that the controlled drop time is <2 sec.

This idea seems to have been partially substantiated by Bond and O'Halloran [14], who deliberately set out to study the apparent anomaly in the literature. Provided the short drop times were achieved by mechanical control of the drop time, the Ilkovic (or extended versions of it) and other related dc polarographic theory appear to remain valid at least phenomenologically. Importantly, under such conditions, linear i_d-vs.-concentration curves are definitely obtained, and for the analytical chemist this is probably the result of greatest significance. There now seems no valid theoretical reason why short drop times should not be used provided they are produced by mechanical control. Dependence on drop knocker design can arise, but this will be neglected here.

4.1.2 Recording, Damping, and
   Maximum Scan Rates of Potential

With conventional dc polarography, scan rates of potential are usually in the range 5 to 15 min V^{-1}, depending on the drop time and accuracy of measurement required. Because of the large current fluctuations associated with each drop during the considerable period of growth, X-Y recorder damping is frequently necessary, particularly with low concentrations.

In the case of rapid polarography, with an extremely short drop time, very little time is available for growth of the mercury drop. Hence, measured current fluctuations are small and damping is completely unnecessary even with the lowest detectable concentrations. The removal of the need for damping is an extremely convenient feature of rapid polarography and facilitates ready measurements with minimum possibility of recorder distortion. This implies, of course, that maximum currents will be used in rapid polarography, since mean currents are not directly measurable from undamped recorders. Provided the response time of the recorder is sufficiently fast [24], the only restriction on the usable scan rate should be that a maximum of about 5 mV should be covered per drop.

4.1.3 Reproducibility

With any technique, reproducibility and precision of results are always of great importance. Results show that the considerable time saving obtained with the rapid polarographic method is not gained at the expense of these factors, as is often the case when an attempt is made to increase the rate of measurement. In fact, an improvement in reproducibility is sometimes observed [9], making the technique highly attractive.

4.1.4 Dependence of i_d on Various Parameters

Dependence upon Mercury Column Height and Drop Time. With careful drop knocker design and use, the Ilkovic equation or extensions have been shown to be valid, if not quantitatively, at least phenomenologically, over drop times ranging from those used with the conventional dme down to the millisecond region, as previously noted.

Assuming the Ilkovic equation is obeyed, it follows that increasing the mercury column height at constant drop time increases the flow rate of mercury, and i_d increases accordingly.

The dependence upon $t^{1/6}$ predicted by the Ilkovic equation would indicate that the decrease in i_d, going from a drop time of 3 sec, say, in the conventional method, to $t = 0.16$ sec in the rapid method would not be particularly large, because, for example, with $t = 3$ sec, $t^{1/6} = 1.2$ $sec^{1/6}$, and with $t = 0.16$ sec, $t^{1/6} = 0.74$ $sec^{1/6}$. Considerations such as mercury flow rate variations are neglected in these calculations.

Dependence on Concentration and Detection Limits. Figure 4.5 shows plots of i_d vs. concentration for rapid and conventional polarography. In general, linear dc plots are observed with the rapid polarographic method when the corresponding conventional method gives linear plots.

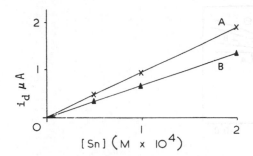

FIGURE 4.5 Comparison of analytically used calibration curves in
rapid and conventional dc polarography for the electrode process
Sn(II) + 2e $\rightleftharpoons$ Sn(0) in 5 M HCl: (A) i_d vs. concentration of tin
using conventional dc polarography; (B) i_d vs. concentration of tin
using rapid dc polarography, with t = 0.16 sec.

However, in certain circumstances where the short drop time of the
rapid polarographic method causes alterations to the electrode
process, it can be noted that this analogy between the rapid and
conventional method does not follow necessarily. For example,
where adsorption or other surface phenomena are eliminated (Figs.
4.2-4.4) by the use of short drop times, the rapid method may give
a linear concentration plot, where conventional dc polarography
would give a nonlinear calibration curve. The lower faradaic currents obtained with the rapid dc method are clearly illustrated in
Fig. 4.5.

The analytically usable limit of concentration detection of
rapid dc polarography in most cases has been found to be about 10^{-5} M,
a level slightly higher than that found in conventional dc polarography. The limit of detection is necessarily less favorable than
for conventional dc polarography because the ratio of charging
current to faradaic current increases, the shorter the drop time.
Figure 4.6 shows the charging current contribution as a function of
drop time, and this can be seen to increase as the drop time decreases. It has already been shown in Fig. 4.5 that the faradaic
current decreases with decreasing drop time. To a first-order
approximation, the faradaic current i_f is proportional to $Ct^{1/6}$

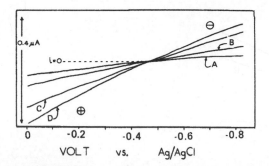

FIGURE 4.6   Charging current in 0.25 M $K_2C_2O_4$. Note magnitude of
charging current is larger than the shorter drop time except at the
point of zero charge, where it is zero for all drop times; see
Chap. 2:   (A) t = 2 sec; (B) t = 0.4 sec; (C) t = 0.1 sec;
(D) t = 0.05 sec.   [Reproduced from J. Electroanal. Chem. *68*, 257
(1976).]

for a diffusion-controlled process.   The charging current i_c is
proportional to $t^{-1/3}$ and is independent of C.   Hence,

$$\frac{i_f}{i_c} = k \frac{Ct^{1/6}}{t^{-1/3}} = kCt^{1/2} \tag{4.1}$$

where

    k = a constant for a particular potential

    C = concentration

    t = drop time

Thus, for normal drop times of, say, 4 sec,

$$\left(\frac{i_f}{i_c}\right)_{conv} = 2kC \tag{4.2}$$

For rapid controlled drop times of, say, 0.16 sec,

$$\left(\frac{i_f}{i_c}\right)_{rapid} = 0.4\ kC \tag{4.3}$$

Therefore, at any given concentration, the faradaic-to-charging
current ratio is less favorable under rapid polarographic conditions.

In summary, rapid polarography is akin to using the current-time
curve early in the drop life, and equations derived on this basis
give an acceptable description of the technique. The elimination of
adsorption and kinetic influences can be understood in this manner,
since current-time curves, while extremely complex late in the drop
life, tend to be normal at the early stages, i.e., when rapid polarograms are effectively recorded.

Since drop times down to the millisecond region are available,
a time scale of four or more orders of magnitude can now be exploited
in dc polarography, and kinetic complications arising from the kinetic
$B \xrightarrow{k} C$ step in the electrode process $A \xrightleftharpoons{e} B \xrightarrow{k} C$ can be eliminated
(Notes 5, 9, and 13, for example). Since catalytic and kinetic
processes give rise to limiting currents that are frequently complex,
nonlinear functions of concentration, and since such processes are
generally analytically undesirable, the ability to eliminate these
steps while retaining the useful level of response associated with
a diffusion-controlled process is most advantageous. Figure 4.7
summarizes the limiting current-drop time response for three systems [5]. The kinetic process (formaldehyde wave) and the catalytic
wave (uranium-catalyzed reduction of nitrate) decrease much more
rapidly than that of the diffusion-controlled wave of cadmium as
the kinetic contribution to the electrode process is varied.

For systems inhibited by adsorption on the surface of the dme,
the improved polarographic behavior under rapid conditions has been
attributed to the decreased drop time and the increased rate of
surface area formation, which together minimize the extent of surface
coverage during detector life [6,13]. All evidence obtained to date
indicates that adsorption of electroinactive substances is sufficiently slow [13] so that the use of very short drop times can
effectively eliminate the inhibiting effects of even strongly surface active materials. Where products of electrode processes are
adsorbed, the short controlled drop time (i_1-vs.-concentration)
response is linear over a much wider range. This has been found to
be true for the methylene blue, arsenic(III), quinine, quinoline,
and 3-amino quinoline systems [6,15].

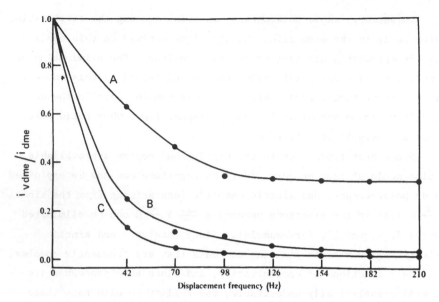

FIGURE 4.7 Limiting current–frequency (drop time) response of vdme
for kinetic waves. The response for all three systems decreases
with vibrational frequency (drop time). However, the kinetic process decrease is much more rapid than that for the diffusion-controlled
cadmium.   (A) 4.0 mM Cd(II), 0.1 M KNO_3; (B) 4.0 mM formaldehyde,
0.102 M NaOH; (C) 0.02 mM U(VI), 0.100 M KCl, 0.01 M HCl,
1.00 mM KNO_3. [Reproduced from Rev. Anal. Chem. *1*, 14 (1972).]

Similar arguments hold for film formation. The appearance of
a second dc wave indicates inhibition of the electrode process when
the surface of the electrode is completely covered by a film of the
reaction product. Because of the higher rate of surface area formation at short drop times, the concentration at which total coverage of the electrode surface occurs would be expected to be higher
under rapid conditions. This prediction was confirmed by observing
the concentration at which the second wave appeared as a function of
drop time in a number of systems [16]. For example, for sulfide the
second wave appeared under conventional conditions (t = 2.9 sec) at
3×10^{-4} M but did not become evident at a drop time of 0.16 sec
until the concentration was increased to 7×10^{-4} M [compare
Fig. 5(a) and 5(b) of Note 16].

Of course, if short enough drop times were used, only one wave would be observed over the entire concentration range. Figure 4.8 shows some current-time curves measured on the sulfide system. When the overall electrode process is diffusion-controlled, the current is proportional to $t^{1/6}$ (Fig. 4.8), but in potential regions where films of the reaction product inhibit the electrode process, the curves deviate markedly from this shape. However, it can be seen clearly that very early in the drop life, where rapid polaro-graphic measurements are effectively made, almost normal shape is observed (that is, $i \propto t^{1/6}$), irrespective of the behavior later in the drop life. This example illustrates the earlier discussion.

4.1.5 Determination of Numbers of Electrons in Consecutive Electrode Processes

When a number of consecutive electrode processes are observed from the one complex at different potentials, and all limiting currents are diffusion-controlled, then according to the Ilkovic equation,

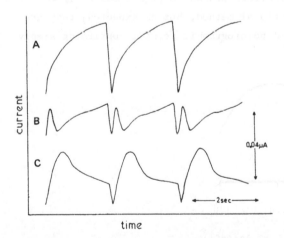

FIGURE 4.8 Current-time curves for 1×10^{-3} M sulfide in 1 M $NaClO_4$ at various potentials. Drop time is 2.0 sec: (A) −0.250 V; (B) −0.625 V; (C) −0.675 V (vs. Ag/AgCl). [Reproduced from Anal. Chem. *45*, 1327 (1973). © American Chemical Society.]

the ratios of the various i_d values should be close to the ratio of
the numbers of electrons involved in each electrode process. However, with the conventional method, the drop time is well known to
be markedly potential dependent, as may be evidenced by the so-called
electrocapillary curves commonly used to study electrode phenomena.
Consequently, $m^{2/3} t^{1/6}$ in the Ilkovic equation is potential dependent, and ratios of i_d values of two consecutive electrode processes,
separated by a considerable potential, may give a completely false
idea of the relative numbers of electrons in the respective electrode processes. By contrast, in the rapid polarographic method,
the drop time is controlled and therefore independent of the
potential. This eliminates a variable usually encountered in
polarography.

Figure 4.9 shows a plot of drop time vs. potential using the
conventional method. Figure 4.10 shows a comparison of rapid and
conventional dc polarograms [26] for reduction of $Pt(SacSac)_2$ in
acetone ($SacSac^- =$ dithioacetylacetonate). A considerable difference
in i_d values of the two electrode processes $Pt(II) \overset{e}{\rightleftharpoons} Pt(I) \overset{e}{\rightleftharpoons} Pt(0)$
is obvious with the conventional method, but as expected, they are
almost equal with the rapid polarographic method, confirming simply

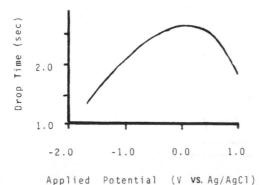

Applied Potential (V vs. Ag/AgCl)

FIGURE 4.9 Drop time-vs.-potential plot in acetone. Supporting
electrolyte is 0.1 M tetraethylammonium perchlorate. (Data taken
from Note 9.)

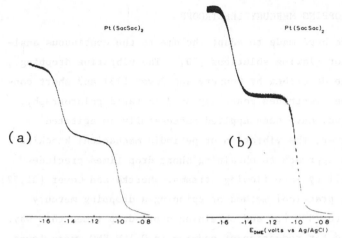

FIGURE 4.10 Comparison of (a) rapid and (b) conventional dc polarograms of Pt(SacSac)$_2$ in acetone [9,26]. Supporting electrolyte is
0.1 M tetraethylammonium perchlorate. Note that both waves are of
equal height at short controlled drop times.

and conveniently that equal numbers of electrons are involved in
both electrode processes.

That this use of the rapid polarographic method is possible,
confirms again the earlier finding that the Ilkovic equation can
still be used successfully in applications of polarography with
short drop times.

4.1.6 Other Applications

The fast scan rates permitted by the rapid polarographic method have
allowed other applications of the technique in glass-corrosive hydro-
fluoric acid media [27], in the study of complexes [28], and in the
extension of the potential where the final current rise occurs in
acidic solutions owing to reduction of hydrogen ions [29]. The use
in flowing or agitated solutions also appears to be attractive [13]
(see Sec. 4.2 as well).

In summary, the use of short drop time is most useful, bringing the time for recording a polarogram into the seconds time domain.
The only disadvantage is an increase in the limit of detection
because of a less favorable faradaic-to-charging current ratio.
Overall, the author believes the technique to be a most attractive
advance in dc polarography [9,13].

4.2 SPINNING DROPPING MERCURY ELECTRODES

Many efforts have been made to adapt the dme to the continuous analysis of stirred or flowing solutions [30]. The vibrating dropping
mercury electrode described by Connery and Cover [13] and short controlled drop time electrodes generally used in rapid polarography,
as discussed above, have been applied successfully in agitated
solutions. However, the vibration or periodic mechanical knocking
of the electrode approach to obtaining short drop times precludes
sealing the capillary into flowing streams. Mortko and Cover [31,32]
have developed a practical method of spinning a dropping mercury
electrode at up to 7000 rpm and obtaining short drop times this way.
Limiting currents for reduction of cadmium in 0.1 M KNO_3 were found
to be independent of flow rate of solution over the range 0.001 to
0.5 liter min^{-1} at a rotation rate of 7000 rpm using one particular
cell. Under conventional dc polarographic conditions convection
currents contribute to the response in flowing solutions to give an
erratic response, and this technique is therefore not appropriate
to continuous monitoring of solutions which may range from quiescence
to extreme turbulence at various points in time. Elimination or
minimization of kinetic and adsorption nuances also occurs with the
short drop time spinning dme as described in Sec. 4.1 on rapid
polarography, and this, combined with the insensitivity to erratic
convection currents introduced by solution movement, makes this
approach ideal for continuous analysis of flowing systems.

In addition to taking into account the short drop time behavior
associated with the spinning dme, the influence of the rotation of
the electrode on the mass transport process needs to be considered.
This influence is much more severe than any vibration of an electrode
caused by mechanical knocking of the capillary in rapid dc polarography. Kolthoff and Okinoka have described a rotating dropping
mercury electrode [33-36] and enhanced sensitivity is found compared
with a nonrotating capillary because of an improved faradaic-to-
charging current ratio. In addition to the above-mentioned advantages accruing from decreasing the drop time, it is clear that the

rotation of the electrode in its own right discriminates even further against several undesirable phenomena [31,32]. The combination of short drop times and high rotational speeds, therefore, makes this electrode arrangement well worth considering for analytical work.

4.3 CURRENT-SAMPLED DC POLAROGRAPHY

Rapid dc polarography utilizes advantages gained in moving to short drop times. However, one obvious disadvantage noted in this approach is that the larger the charging-to-faradaic current ratios, the shorter the drop time. It follows conversely, assuming the current at the end of the drop life is being measured, that the longer the drop time, the more favorable the faradaic-to-charging current ratio, and the lower the concentration limit that can be detected.

Hence, by recording only the current flowing through the cell near the end of the drop life, e.g., the last 5 to 20 msec of a 5-sec drop, instead of the entire current-time curve for each drop, optimization of sensitivity is obtained in dc polarography. With the aid of "sample and hold" circuitry and modern electronics, this kind of measurement presents little difficulty. Figure 4.11 shows the readout obtained with current sampled dc polarography, and the derivation from the i-E curve in conventional dc polarography is obvious.

Many commercially available instruments provide this kind of readout, and it has become an integral part of the modern circuitry employed in polarography. This technique is also known by the names *Strobe* and *Tast* polarography. However, the term *current-sampled polarography* is preferred and used in this text because it explicitly defines the principle behind the technique. Furthermore, current sampling near the end of the drop life now plays a predominant role in many modern polarographic methods, and in the future it is likely to be accepted as the normal method of recording polarographic i-E curves, even when short drop times are used, as in rapid polarography.

Despite the increased sensitivity this method of measurement provides in comparison to conventional dc polarography, the increase

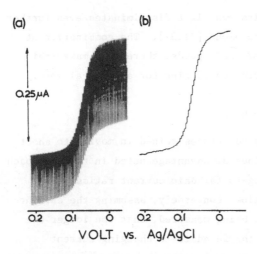

FIGURE 4.11 Comparison of dc and current-sampled dc polarography. Solution is 1×10^{-4} M Cu in 1 M $NaNO_3$, and electrode process is $Cu(II) + 2e \rightleftharpoons Cu(Hg)$: (a) dc polarogram; (b) current-sampled dc polarogram with same drop time (2 sec) as (a). [Reproduced from Anal. Chem. *44*, 721 (1972). © American Chemical Society.]

is only marginal (particularly relative to the undamped dc polarograms recorded with fast-response X-Y recorders), and several far superior methods for discriminating against the charging current are available, as will be seen later. The real improvement lies in the form of readout obtained. The virtual elimination of the serrations due to the continuous mercury dropfall-dropgrowth sequence provides smooth i-E curves. Damping of component apparatus is unnecessary; as with all usual drop times, area change during the current-sampling period is small and oscillations are negligible. Polarograms obtained by this method are simpler and can be evaluated more exactly than in normal dc polarography, where excessive serrations cause considerable difficulty in measurement. The readout form is also well suited to conversion to digital format and therefore for use with computer technology and for capacitance current compensation [37]. Perusal of the literature [37-48] shows that most of the claimed advantages of current-sampled readout have been derived from the "smoothed" readout form rather than the increased sensitivity. Figure 4.12 shows a useful advantage derived from this area and

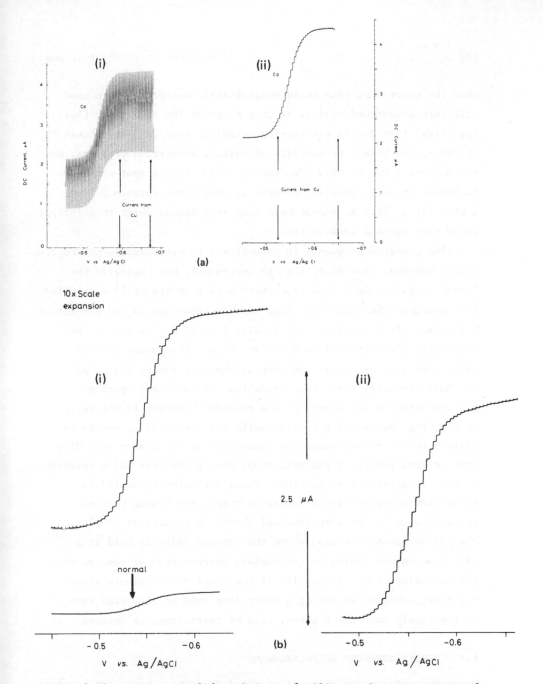

FIGURE 4.12 Current-sampled techniques facilitate the measurement of an electrode process in the presence of another more positively reduced one. (a) Measurement of the cadmium(II) electrode process in the presence of the more positively reduced copper (II). (i) Conventional dc polarography; natural drop time used. (ii) Current-sampled dc polarography; Medium is 0.5 M NaClO$_4$. Controlled drop time used. [Cd] = 1.3 × 10^{-3} M and [Cu = 1.65 × 10^{-3} M. (b) Use of scale expansion and current sampling to facilitate measurement of an electrode process in the presence of a more positively reduced species. [Cd] = 1.3 × 10^{-3} M. (i) [Cu] = 6.35 × 10^{-3} M; (ii) 4.00 × 10^{-2} M. Medium is 0.5 M NaClO$_4$. [Reproduced from Anal. Chim. Acta *62*, 415 (1972).]

even the short drop time rapid polarographic method has been used
with current-sampled readout at drop times as low as 50 msec [46].
The slight increase in sensitivity gained at very long drop times is
offset by the higher probability of maxima, adsorption problems, and
the necessary use of slow scan rates. Thus, the current-sampled
technique is still usually employed at drop times between 0.5 and
5 sec, rather than at excessively long drop times, which in principle
would give optimum sensitivity.

The sequence of events in recording a current-sampled polarogram
varies somewhat from instrument to instrument, but basically the
format is as follows: The experiment uses a mechanically controlled
drop time, so that the exact drop time is known and is independent of
potential. At a fixed period, usually 5 to 20 msec prior to the
drop fall, the current is sampled and read. This point is then
plotted on the i-E curve. As soon as the drop falls, the sample
and hold circuitry comes into operation and holds the apparent
current value on the Y-axis of the recorder constant at the value
read during the sampling period, while the potential or X-axis is
shifted in the normal manner as determined by the scan rate. With
the sampling period of the next drop, the new current value relevant
to the new potential is plotted. Thus, the current-sampled dc
polarogram is built from a series of steps, the rising portion
corresponding to the time interval where the current is sampled,
and the horizontal section where the current value is held at a
set value corresponding to the maximum current read at the end of
the preceding drop. Naturally, if the steps are extremely close
together, achieved by having a short drop time or slow scan rate,
an apparently smooth i-E curve, void of serrations, is observed.

4.4 CURRENT-AVERAGED DC POLAROGRAMS

Current-sampled polarograms of the kind described above effectively
use maximum currents and eliminate from the readout the large
current fluctuations resulting from the growth and fall of the
drops in the dme. To obtain a signal that is directly proportional

to the average, rather than maximum value of the cell current, a low-pass filter that provides linear damping can be used. This can be achieved with a parallel-T filter [49-54]. It is important that nonlinear transformations by the recorder be avoided [53,55]. For dme's having drop times of up to 5 sec, a quadruple parallel-T, RC filter terminated by a low time-constant RC network has been designed [49]. For dme's having shorter drop times of around 0.5 sec, a dual parallel-T, RC filter has been shown to be most suitable. By means of the current-averaging, low-pass filter systems, a voltage signal is obtained which is highly reproducible, directly proportional to the average current, and suitable as the input signal to the X-Y recorder or to a computer to perform derivatives, as will be described in Sec. 4.5. The top right-hand side of Fig. 4.13 shows the readout of a current-averaged polarogram, and the advantages of this form of i-E curve are immediately apparent. It can be noted that, as with many other techniques, the use of short controlled drop times is generally preferred in this mode of current readout [50].

The use of parallel-T filters to record average current dc polarograms is superior in most respects to the use of ordinary RC damping because the final form of the data is readily amenable to further electronic manipulation if required (e.g., in taking derivatives) and is well suited to computational procedures as the analog-to-digital conversion of the signal is readily achieved. Furthermore, this method of obtaining average currents has less potential for introducing wave distortion into the readout because of the method of data acquisition. Polarograms with solutions of about the same composition as Fig. 4.13 have been recorded with the current-sampled, T-filter, and RC damping approaches [41,42,47, 53]. The thallium and cadmium waves are not resolved with ordinary RC damping, but they are resolved with both the other methods of recording dc polarograms.

Current-sampled or T-filter techniques can be considered essentially equivalent in performance in most respects. One method

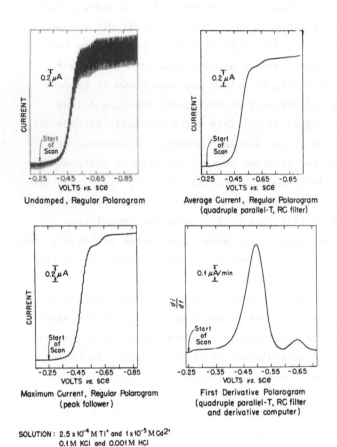

Undamped, Regular Polarogram

Average Current, Regular Polarogram
(quadruple parallel-T, RC filter)

Maximum Current, Regular Polarogram
(peak follower)

First Derivative Polarogram
(quadruple parallel-T, RC filter
and derivative computer)

SOLUTION : 2.5 x 10⁻⁴ M Tl⁺ and 1 x 10⁻⁵ M Cd²⁺
0.1 M KCl and 0.001 M HCl

FIGURE 4.13 Examples of different readout forms in dc polarography.
[Reproduced from Polarography 1964 *1*, 89 (1966).]

uses sample and hold circuitry to record maximum currents, the other
parallel-T filters to record average currents, so that excellent
electronic approaches are now available to record the two types of
current. Preference for the former method may come about in the
future because of the need, together with several other important
polarographic techniques, such as pulse polarography, to utilize
maximum currents. Present trends toward construction of multifunctional instrumentation may therefore tend to provide this form of
readout as the inherent part of polarographic instrumentation, and
the introduction of new circuitry for average currents may not be
considered as frequently. However, the author envisages that either
of these approaches should become the "normal" readout form in analytical dc polarographic work and the use of heavily damped RC networks to produce average currents should in time be completely replaced by these far superior forms of readout.

4.5 DERIVATIVE DC POLAROGRAPHY

As the name indicates, this technique should consist of a plot of
di/dE or $(\Delta i/\Delta E)_{\Delta E \to 0}$ versus E and a peak-shaped rather than sigmoidal
curve should result. Figure 4.14 shows an example of a derivative
polarogram recorded with an RC circuit. The simultaneous determination of phenazine-1-carboxylic acid and 2-hydroxyphenazine in the
evaporated chloroform extract of *Pseudomonas aureofaciens* in
alkaline phosphate buffer has been described with this kind of
circuitry [57]. However, the history of derivative dc polarography
has been plagued by approaches that introduce undesirable instrumental artifacts and give a nontheoretical response [58]. Thus,
despite the potentially significant advantages provided by a plot
of di/dE versus E, reviewers of this technique of dc polarography
have until recently suggested that because of instrumental performance, it is unsuitable for quantitative measurements [58]. The
original methods based on the use of two synchronized dme's, RC
differentiating circuits [59-64] and other approaches have been
critically reviewed by Schmidt and von Stackelberg [58] and need

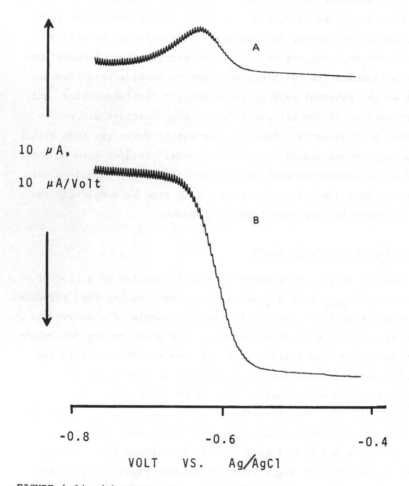

10 μA,

10 μA/Volt

-0.8   -0.6   -0.4

VOLT   VS.   Ag/AgCl

FIGURE 4.14   (A) Derivative polarogram recorded with an RC circuit.
1×10^{-3} M cadmium in 1 M HCl.   (B) Conventional dc polarogram of
same solution.

not be discussed further, since they are unlikely to find a place
in modern analytical procedures.

However, with advanced instrumentation and attention to experimental design, excellent derivative dc polarographic results can
now be obtained and the technique challenges many of the more sophisticated methods such as pulse, ac and square-wave polarography in
simplicity of instrumentation and application [54]. Satisfactory
first- or second-derivative dc polarograms can now be recorded at
concentrations as low as 10^{-7} M for reversible electrode processes
[56].

Likely procedures for obtaining successful types of derivative
plot can be understood by considering the nature of the polarographic
experiment. If the normal dc polarographic technique is considered,
it can be appreciated that any differentiation will produce a derivative of the i-t behavior of each drop, whereby extremely large
serrations would be produced. To avoid this undesirable feature,
very heavy damping of the detecting system must be employed to
generate a curve as shown in Fig. 4.14, but this damping can cause
severe instrumental distortion. Thus, it is the i-t behavior associated with each drop of mercury in conventional dc polarography
that mitigates against successful derivative dc polarography, and
just taking the straightforward derivative of the complete cell
response can never be completely successful.

The obvious approach to obtaining successful derivative dc
polarograms and to eliminate the above problem is to take the derivative after the data are in either the current-sampled or time-
averaged (parallel-T-filter) forms; that is, apply the derivative
circuitry to the two previously described dc polarographic methods,
both of which virtually eliminate from the readout oscillations
resulting from growth and fall of the mercury drop.

Figure 4.15 shows a block diagram of a derivative polarograph
[54]. This particular model uses a three-electrode potentiostat,
an electronic scan generator to vary the cell potential linearly
with time, efficient filtering of the signal fluctuations

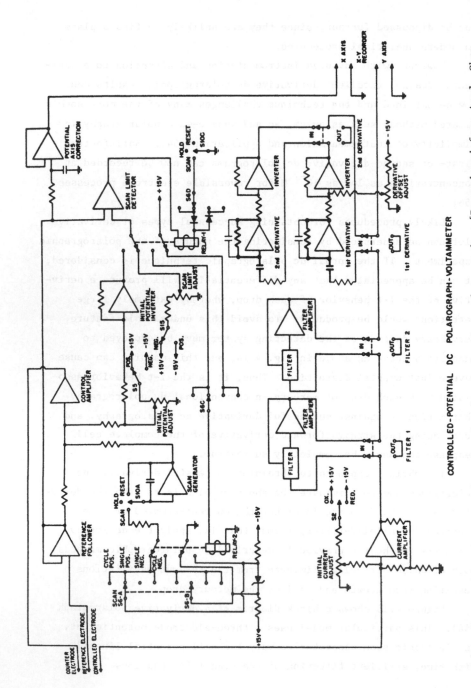

CONTROLLED-POTENTIAL DC POLAROGRAPH-VOLTAMMETER

FIGURE 4.15 Block diagram of a circuit of a derivative polarograph. [Reproduced from Anal. Chem. 41, 772 (1969).] © American Chemical Society.]

(parallel-T, RC filters), time-averaging circuitry, and time-
derivative computer networks to obtain both first- and second-
derivative polarograms. This instrumental approach [49-54, 56] has
been highly developed over a period of time at the Oak Ridge National
Laboratories, and from data available in the literature the performance is extremely good (particularly with short controlled drop
times around 0.5 sec), giving close to the theoretical response,
high reproducibility, and sensitivity significantly superior to dc
polarography [54,56]. Figure 4.16 [54] shows that the reproducibility
of triplicate polarograms is essentially the thickness of the recorder
pen at moderate sensitivities. Figure 4.17 shows that the operation
at high sensitivity is also excellent. Residual or charging current
contributions with this technique are generally less than in normal
dc polarography [46]. Naturally, the resolution is also improved
when taking a derivative and using a peak- rather than sigmoidal-
shaped curve, and Fig. 4.18 demonstrates this qualitatively.
Applications of this technique have appeared in the literature
[53,65] supporting arguments that derivative dc polarography in

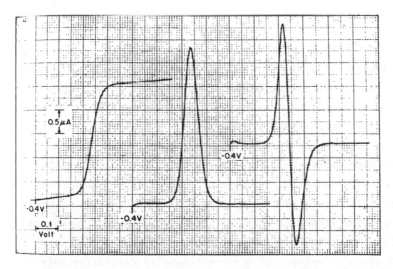

FIGURE 4.16 Reproducibility of triplicate polarograms 0.5-sec
controlled drop time; dc, first derivative and second derivative
responses shown. Cell solution: 5×10^{-4} M Cd, 1 M KCl, 0.001 M
HCl. [Reproduced from Anal. Chem. *41*, 772 (1969). © American
Chemical Society.]

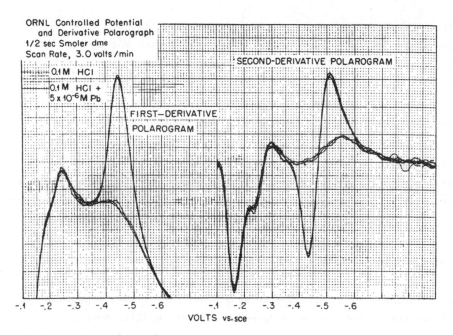

FIGURE 4.17 Derivative techniques enable lower concentrations to be
determined than conventional dc polarography. [Reproduced from
Polarography 1964 1, 89 (1966).]

this form has now "come of age" and can be considered to be a significant advance in dc polarography.

The use of the current-sampled derivative method [66,67] has
received considerably less attention, although it seems to be
equally as attractive in principle. In current-sampled dc polarography the potential is applied in increments of say 5 mV per drop.
The current is measured discontinuously, but each time shortly
before the drop falls. A current-sampled derivative polarogram
records the difference (Δi) between two consecutive current measurements (or $\Delta i/\Delta E$) as a function of potential. Since ΔE is a constant, a plot of either Δi or $\Delta i/\Delta E$ will give a derivative-shaped
curve. Figure 4.19 shows the principle behind this method, and
Fig. 4.20 gives an example of a current-sampled derivative polarogram. Of course, the smaller ΔE is, the closer the curve approaches
a mathematical derivative di/dE. Recently, this method has been

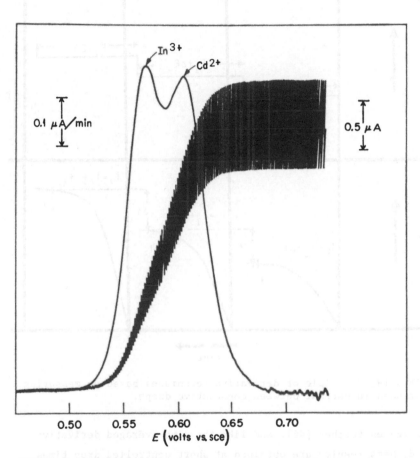

FIGURE 4.18 Qualitative comparison of resolution by conventional and derivative dc polarography. Conditions: 1×10^{-4} M In^{3+} and 2×10^{-4} M Cd^{2+} in 0.1 M KCl. Regular: undamped; derivative: quad. parallel-T, RC current avg filter. [Reproduced from Polarography 1964 *1*, 89 (1966).]

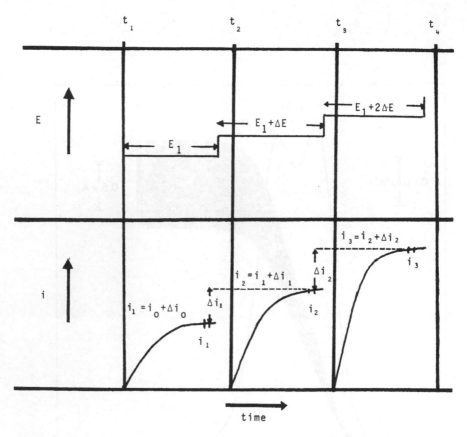

FIGURE 4.19 Principle of derivative techniques based on measuring differences in current between consecutive drops.

investigated further [46], and like the time-averaged derivative method, best results are obtained at short controlled drop times and with moderately fast scan rates. These conditions correspond to what might be called *rapid derivative* dc polarography and, as well as providing optimum instrumental conditions and fast scan rates, the elimination of maxima and adsorption phenomena would be anticipated. It should be noted that unless $\Delta E \to 0$, the measured peak height I_p, while being a linear function of concentration, will be a function of scan rate and drop time. Accordingly, both the unknown and the standard solutions must be recorded at the same

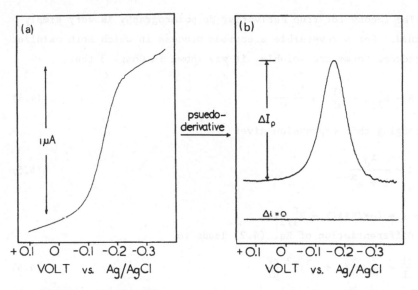

FIGURE 4.20 A current-sampled derivative polarogram: 1×10^{-3} M
Fe(III) in 0.25 M $K_2C_2O_4$; t = 0.5 sec. (a) Current-sampled dc
polarogram. (b) Pseudoderivative polarogram. [Reproduced from
J. Electroanal. Chem. *68*, 257 (1976).]

scan rate. Even if this restriction did not need to be invoked on
theoretical grounds, this procedure would still be recommended,
since the risk of instrumental artifacts entering into derivative
curves is considerable. Most commercially available instruments
having capabilities for performing derivative polarograms introduce
at least some minor form of nontheoretical scan rate dependence.
These arise in the main from inadequacies in the electronics used
to perform derivatives.

This conclusion provides a convenient point to state specifically what should be generally obvious to most users of instrumental
analysis. That is, as a general precaution, it is strongly advised
that calibration and unknown solutions for all polarographic methods
should always be recorded under conditions as close to identical as
possible, regardless of whether the theory indicates that this is
essential.

The theory for true derivative dc polarography is very simply attained. For a reversible electrode process in which both oxidized and reduced forms are soluble, it was shown in Chap. 3 that

$$E = E^r_{1/2} + \frac{RT}{nF} \ln \frac{i_d - i}{i} \qquad (4.4)$$

Rearranging this expression gives

$$i = \frac{i_d}{1 + e^x} \qquad (4.5)$$

where $x = (nF/RT)(E - E^r_{1/2})$.

Differentiation of Eq. (4.2) leads to

$$\frac{di}{dE} = -i_d e^x (1 + e^x)^{-2} \frac{nF}{RT} \qquad (4.6)$$

The maximum value of di/dE is the peak current I_p, and this is defined by the condition $d^2 i/dE^2 = 0$. Since

$$\frac{d^2 i}{dE^2} = i_d \left(\frac{nF}{RT}\right)^2 \frac{e^x (e^x - 1)}{(1 + e^x)^3} \qquad (4.7)$$

this condition is reached with $e^x = 1$; that is, di/dE = I_p when $x = 0$ or when $E = E^r_{1/2}$, so that

$$I_p = -\frac{i_d}{4} \left(\frac{nF}{RT}\right) \qquad (4.8)$$

Since i_d is a linear function of concentration, so is I_p for a *reversible* electrode process. Substituting Eq. (4.8) into (4.5) and putting di/dE equal to I gives

$$I = 4I_p \frac{\exp (nF/RT)(E - E^r_{1/2})}{[1 + \exp (nF/RT)(E - E^r_{1/2})]^2} \qquad (4.9)$$

or

$$E = E^r_{1/2} + \frac{2RT}{nF} \ln \left(\frac{I_p}{I}\right)^{1/2} \pm \left(\frac{I_p - I}{I}\right)^{1/2} \qquad (4.10)$$

Thus, a plot of E vs. $\log (I_p/I)^{1/2} \pm [(I_p - I)/I]^{1/2}$ should be a straight line of slope $2(2.303RT/nF)$ and this could be used to define a reversible derivative dc polarogram. In actual fact the half width of a derivative polarogram is used more commonly to define the reversibility or otherwise.

Examination of Eq. (4.10) shows that E at half the wave height is given by

$$E = E^r_{1/2} + \frac{2RT}{nF} \ln (\sqrt{2} - 1) \tag{4.11a}$$

or

$$E = E^r_{1/2} + \frac{2RT}{nF} \ln (\sqrt{2} + 1) \tag{4.11b}$$

Subtraction of Eq. (4.11a) from (4.11b) to obtain the half width $\Delta E_{p/2}$ gives Eq. (4.12)

$$\Delta E_{p/2} = \frac{2RT}{nF} \ln \frac{\sqrt{2} + 1}{\sqrt{2} - 1} = \frac{4RT}{nF} \ln (\sqrt{2} + 1)$$

$$= (1.52)(2.303) \left(\frac{RT}{nF} \right) \tag{4.12}$$

At 25°C, $\Delta E_{p/2}$ has a value close to 90/n mV, and so this simple two-point analysis is a convenient way of assigning an electrode process as reversible from its derivative dc polarogram. Alternatively, values of I/I_p may be computed readily as a function of $E - E^r_{1/2}$ from Eq. (4.6), and the reversibility may be assessed in this manner over the entire wave.

For the second derivative a maximum $(\phi_p)_{max}$ and minimum $(\phi_p)_{min}$ are observed as shown in Fig. 4.22. From similar calculations for obtaining the first derivative

$$(\phi_p)_{max} = - (\phi_p)_{min} = \frac{\sqrt{3}}{18} i_d \left(\frac{nF}{RT} \right)^2 \tag{4.13}$$

and

$$\phi = \frac{d^2 i}{dE^2} = 6\sqrt{3} \, (\phi_p)_{max} \frac{e^x (e^x - 1)}{(1 + e^x)^3} \tag{4.14}$$

Thus, a plot of $(\phi_p)_{max}$ or $(\phi_p)_{min}$ versus concentration would be linear. More conveniently, the peak-to-peak current ϕ_{p-p}, where

$$\phi_{p-p} = \frac{2\sqrt{3}}{18} \, i_d \left(\frac{nF}{RT}\right)^2 \tag{4.15}$$

is used in analytical work. The use of a peak-to-peak current avoids the need for a zero reference mark and provides a precisely measurable parameter. This would appear to be probably the most attractive feature of the second-derivative method.

Derivatives higher than the second appear to be without merit and even improvement in resolution, if any, is too marginal to justify the use of such higher derivatives.

The preceding theory applies only to reversible electrode processes. It should be noted that in using I_p or ϕ_p peak values in preparing calibration curves, one is now dealing with a current that flows on the rising part of the i-E curve and not on the plateau. This means that only in the reversible case will I_p be diffusion-controlled, and often a kinetically controlled derivative current will be used where i_d would be diffusion-controlled. For example, for a totally irreversible electrode process with an α value of 0.5, I_p will be kinetically controlled and only attain a value of about half the diffusion-controlled value. The sensitivity (I_p per unit concentration) is therefore slightly decreased. By way of comparison, in normal dc polarography, both the reversible and irreversible electrode processes give equivalent diffusion-controlled limiting currents, and the sensitivities are essentially the same for both classes of electrode processes. Thus, the concept of the nature of the electrode mechanism influencing the sensitivity is actually first encountered in derivative dc polarography. However, in ac polarography and other techniques such phenomena will be seen to be far more important, and factors of 100 or greater in change in sensitivity may originate in going from a reversible to an irreversible electrode process. Differences in derivative polarography are therefore usually relatively small.

Equations for the pseudo-derivative, current-sampled $\Delta i/\Delta E$-vs.-potential curves also can be derived readily for the reversible case from the Heyrovský-Ilkovic equation [46]. Results are similar to those for a true derivative, provided ΔE is small, and they need not be considered further, since no new concepts are involved. Similarly, derivative curves of catalytic and other classes of electrode processes need not be considered, since the normal rules of differential calculus apply. If the dc waves are drawn out, the derivative curve will not be as suitable as for steeply rising waves. If the dc wave exhibits kinks and inflections or other irregularities, then the derivative may have more than one peak and be unsuitable for analytical work. However, for systems giving well-defined sigmoidal-shaped dc waves, derivative dc polarography is likely to show substantial advantages in terms of limit of detection, e.g., ease of using and interpreting $f(i)$-vs.-E curves and resolution. Presently available data suggest that 10^{-6} to 10^{-7} M concentrations can be determined by derivative dc polarography [46, 54,56] which is about one order of magnitude improvement over that offered by normal dc polarography.

4.6 SUBTRACTIVE DC POLAROGRAPHY

The application of subtractive techniques is widespread in instrumental methods. In uv-visible spectroscopy the absorbance of a cell containing the solvent and one containing the sample in the same solvent can be recorded. The contribution of the solvent at each wavelength is then electronically subtracted, and the absorbance of the sample plotted out at each wavelength is automatically corrected for solvent contribution.

In dc polarography, subtractive methods, utilizing twin dropping mercury electrodes placed in two cells, have been suggested frequently over the years [68-72]. In spite of indisputable successes, any attempted widespread application to routine analytical work generally founders on the intrinsic difficulties associated with maintaining uniformly dropping mercury electrodes for long periods of time.

This technique, in principle, has considerable advantages. The sensitivity can be increased substantially, because not only is the charging current compensated, but the residual current from impurities present in the solvent or introduced from addition of the supporting electrolyte is also subtracted. The resolution also can be made superior, especially if the appropriate amount of interfering species is added to the second cell containing the blank solution. However, probably only an extremely experienced polarographer will be able to utilize these advantages with conventional analog instrumentation, and even then probably for a "one off" experiment rather than in routine analysis.

There are several possible approaches to undertaking dual-cell subtractive polarography. Twin electrodes dipping into two cells connected via a salt bridge containing a reference electrode which is common to both cells, provides one possibility. Each cell has its own auxiliary electrode, which may be a mercury pool electrode, for example. Figure 4.21 shows the dual-cell arrangement for this kind of experiment. The two dme's must be matched as closely as possible, and this is a nontrivial task. Use of controlled drop time methods, which allow independent variation of flow rate of mercury at each electrode, alleviates some of the difficulties. With controlled drop time the height of the mercury columns can be varied independently until the flow rates of both capillaries are equal. The drop times are already fixed; so when this latter condition is reached, the capillaries should provide drops of equal area, and therefore, the capillaries behave identically. Additionally, the two cells should be physically the same, the two auxiliary electrodes should be identical, and completely symmetrical positioning of all electrodes should be undertaken. Thus, the reference electrode should be equidistant from both dme's, and the relative positions of the dme and auxiliary electrode in both cells should be identical. Considerable care should also be exercised in the oxygen removal procedure. Although dissolved oxygen may not interfere with the reaction being considered, differences in the oxygen concentration of each cell can cause severe problems.

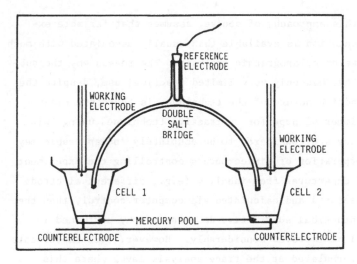

FIGURE 4.21 Dual-cell polarography using a double salt bridge.
(By courtesy of Princeton Applied Research Corporation.)

Another approach to using subtractive techniques is to have two
completely separate networks, i.e., two identical potentiostats,
each controlling identical sets of dme, reference electrode and
auxiliary electrode. Undoubtedly, this is a theoretically ideal
approach, but naturally it is more expensive (since two identical
potentiostats are required) and is probably even more difficult to
implement on a routine basis. Undoubtedly, the only really practical
approach to subtractive polarography is to undertake two separate
experiments with the same instrumentation and record the solvent
and sample solutions separately from two different runs. The data
from the solvent can be stored electronically and subtracted from
subsequent scans on solutions containing the electroactive species
of interest. When performed with the aid of a digital computer
(see Chap. 10), the latter technique is in fact attractive because
the need for dual cells is eliminated. Indeed, in the author's
opinion, only with a computerized polarographic system using a
single cell in which the background is stored in memory and subtracted
from the result on the test solution would subtractive polarography
be worth considering. However, variable oxygen levels still cause

difficulties. This approach, of course, assumes that far more expensive instrumentation is available than usually associated with much
of the literature on polarographic analysis. The reason why the subtractive method has had only very limited practical use, despite the
apparently attractive nature of the technique, is therefore clear.

The one glimmer of hope for subtractive techniques using twin
electrodes if they are considered to be absolutely indispensable may
lie in the incorporation of minicomputers controlling the experiment
(Chap. 10). If departures from ideality (e.g., different electrode
areas) can be measured and calculated via computer control, then the
possibility of practical subtractive dc polarography performed in
this manner will be improved considerably. However, such experiments
have yet to be formulated at the trace analysis level where this
technique should be useful, and the author is unable to see how this
approach can be superior to the one where the background and test
solutions are recorded in separate experiments using the same capillary. At the very least, the use of two synchronized dme's
and associated circuitry to perform subtractive techniques will
introduce considerable additional instrumental and computational
complication, and this in itself will mitigate against their widespread use, even if the present operating difficulties are ultimately
overcome. However, regardless of the exact methodology, it is
again stressed that the real use of subtractive dc polarography is
likely to remain in the domain of the analytical chemist having
access to a computerized polarograph.

4.7 LINEAR CHARGING CURRENT COMPENSATION

When polarograms are run at high sensitivities, the current required
to charge the double layer introduces considerable slope into the
i-E curve, as discussed previously. With a sloping base line,
evaluation of i_d and $E_{1/2}$ must be made via geometric constructions.
Charging current compensator circuits, which are a feature of the
majority of commercially available polarographs, apply a compensating current to the current measuring amplifier, the magnitude of

which increases in a linear fashion as the potential progresses.
[73]. This is done by using the linear dc potential ramp output
as the source of compensating current. Since the charging current
is not exactly a linear function of potential (Chap. 2), considerable
caution must be exercised in applying this kind of compensation.
Figure 4.22 shows the charging current in 0.1 M KCl before and after
the application of linear compensation. At either side of the point
of zero charge or electrocapillary maximum (ecm), the linear approximation is an excellent one; however, different slopes apply in
different potential ranges, and so the charging current compensation
must be carefully applied only with respect to a particular potential
range. Around the ecm and, indeed, near the solvent limit or other
distinctly nonlinear regions, application of this technique of compensation will give incorrect results. Figure 4.23 shows that the
determination of low cadmium concentration would be rendered invalid
by the use of linear capacitance compensation, because cadmium is

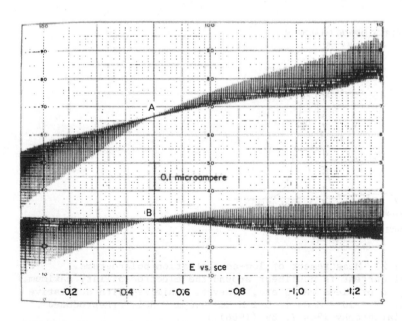

FIGURE 4.22 Residual or charging current curves, 0.1 M KCl:
(A) normal; (B) with linear charging current compensation.
[Reproduced from C. G. Enke and R. A. Baxter, J. Chem. Educ. *41*,
202 (1964).]

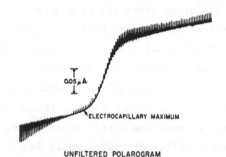

ELECTROCAPILLARY MAXIMUM

UNFILTERED POLAROGRAM
|—0.1 V—|

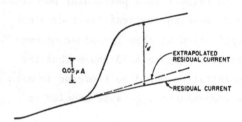

AVERAGE CURRENT POLAROGRAM
|—0.1 V—|

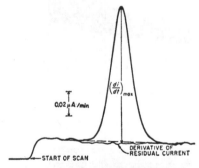

DERIVATIVE OF AVERAGE CURRENT POLAROGRAM
|—0.1 V—|

FIGURE 4.23 Measurement of low-concentration polarograms.
Residual current is nonlinear around the electrocapillary maximum,
leading to problems which are, however, minimized with derivative
methods. 2×10^{-5} M Cd in 0.1 M KCl/0.001 M HCl. [Reproduced
from Polarography 1964 *1*, 89 (1966).]

reduced near the ecm in 0.1 M KCl and a linear extrapolation of the
residual current is not valid. Note also that the derivative plot
obviates most of the difficulties encountered with the charging
current. Modern polarographic techniques which actually discriminate against the charging current on a general basis are undoubtedly
preferable to methods which try to compensate for the charging
current in some restricted fashion. These approaches, rather than
the more marginal methods like linear charging current compensation
are the ones which should receive most attention from the analytical
chemist.

4.8 CHARGING CURRENT COMPENSATION WITH SUPERIMPOSED AC SIGNAL

The complete elimination of charging current is obviously not feasible
with a linear form of compensation, as noted above. To achieve this,
knowledge of the double-layer parameters at each potential is required.
Barker [74] more than 20 years ago suggested the use of ac superimposition for charging current compensation. Recently this approach
has been perfected by Poojary and co-workers [75,76].

From Chap. 2, the charging current in polarography i_c can
be expressed as

$$i_c = A \left(\frac{\partial q}{\partial E} \right)_A \frac{dE}{dt} + q \frac{dA}{dt} \tag{4.16}$$

or

$$i_c = \left(\frac{\partial q'}{\partial E} \right)_t \frac{dE}{dt} + \left(\frac{\partial q'}{\partial t} \right)_E \tag{4.17}$$

In Eq. (4.17) the total charge q' is expressed as a product of the
charge density q and area A.

In dc polarography dE/dt is assumed to be 0, so that

$$i_c^{dc} = \left(\frac{\partial q'}{\partial t} \right)_E \tag{4.18}$$

From Chap. 2, in the presence of a small-amplitude sine wave of amplitude ΔE and frequency ω, the ac charging current i_c^{ac} is given by

$$i_c^{ac} = \left(\frac{\partial q'}{\partial E}\right)_t \Delta E \, \omega \cos \omega t \qquad (4.19)$$

Comparison of Eqs. (4.18) and (4.19) shows that the ac charging current provides information on $(\partial q'/\partial E)_t$, while what is required for dc charging current compensation is a knowledge of $(\partial q'/\partial t)_E$.

At least three options are open to obtain the required information. First, integrate $(\partial q'/\partial E)_t$ with respect to E to give $(q')_t$. Then, differentiate the output q' with respect to time and obtain $(\partial q'/\partial t)_E$. This may be referred to as the *differentiator-integrator* (DI) *method* and it is the most rigorous approach. As an alternative, the approximations

$$\frac{\partial q'}{\partial t} \simeq \delta_1 \frac{\bar{q}'}{t} \qquad (4.20)$$

where the average charge

$$\bar{q}' = \frac{1}{t} \int_0^t q' \, dt \qquad (4.21)$$

or

$$\frac{\partial q'}{\partial t} \simeq \delta_2 \left(\frac{q'}{t}\right) \qquad (4.22)$$

can be used to obtain $(\partial q'/\partial t)_E$.

For a dropping mercury electrode, the time variation in q' is due to A or $t^{2/3}$, so that $\delta_1 = 10/9$ and $\delta_2 = 2/3$, or in general, if

$$A = pt^n \qquad (4.23)$$

then

$$\delta_1 = n(n + 1) \qquad (4.24)$$

and

$$\delta_2 = n \qquad (4.25)$$

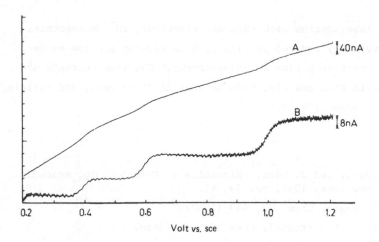

FIGURE 4.24 DC polarogram of 3×10^{-6} M Pb(II), 3×10^{-6} M Cd(II),
and 3×10^{-6} M Zn(II) in 0.1 M KCl: (A) uncompensated dc polarogram;
(B) dc polarogram compensated for charging current by the DI method.
$\Delta E = 10$ mV (p - p); f = 220 Hz; drop time = 0.17 sec. [Reproduced
from J. Electroanal. Chem. *75*, 135 (1977).]

Figure 2.24 shows a polarogram obtained by the DI method near
the 10^{-6} M level, and it can be seen that almost perfect charging
current compensation can be obtained over a wide range of potential.
While approaches of this kind still need to be assessed in greater
detail, there is no doubt that they offer substantial scope for
improving dc polarography and for taking detection limits below
10^{-6} M in aqueous media.

4.9 STATIC MERCURY DROP ELECTRODES
AND MINIMIZATION OF CHARGING CURRENT

Since this book went to press, Princeton Applied Research Corporation
has produced an electrode which radically departs from the dme used
since the inception of the polarographic method by Heyrovský. The new
design enables all of the area growth to occur over the first 50 to
200 msec of drop life, and when current measurements are made at,
say 0.5 sec, the electrode area is constant. This revolutionary
approach, using a so-called static mercury drop electrode, means
dA/dt at current measurement time is zero.

In our laboratories with this new electrode, 10^{-7} M concentrations of reversibly reduced species such as cadmium can now be determined by short drop time dc polarography. Charging currents are very small with this new electrode because dA/dt is zero, and exciting developments appear likely.

NOTES

1. J. Heyrovský and J. Kůta, *Principles of Polarography*, Academic Press, New York, 1966, pp. 34, 41.

2. S. Wolf, Angew. Chem. *72*, 449 (1960).

3. D. Wolf, J. Electroanal. Chem. *5*, 186 (1963).

4. V. G. Mairanovskii, Zavod. Lab. *31*, 1187 (1965).

5. R. E. Cover and J. G. Connery, Anal. Chem. *41*, 918 (1969).

6. J. G. Connery and R. E. Cover, Anal. Chem. *41*, 1191 (1969).

7. R. E. Cover and J. G. Connery, Anal. Chem. *41*, 1797 (1969).

8. A. M. Bond, J. Electroanal. Chem. *23*, 277 (1969).

9. A. M. Bond, J. Electrochem. Soc. *118*, 1588 (1971).

10. D. A. Berman, P. R. Sanders, and R. J. Winzler, Anal. Chem. *23*, 1040 (1951).

11. J. G. Connery and R. E. Cover, Anal. Chem. *40*, 87 (1968).

12. P. W. Carr and L. Meites, J. Electroanal. Chem. *12*, 373 (1966).

13. R. E. Cover, Rev. Anal. Chem. *1*, 141 (1972).

14. A. M. Bond and R. J. O'Halloran, J. Phys. Chem. *77*, 915 (1972).

15. R. E. Cover and J. T. Folliard, J. Electroanal. Chem. *30*, 143 (1971).

16. D. R. Canterford, A. S. Buchanan, and A. M. Bond, Anal. Chem. *45*, 1327 (1973).

17. J. Maas, Collect. Czech. Chem. Commun. *10*, 42 (1938).

18. J. J. Lingane and B. A. Leveridge, J. Amer. Chem. Soc. *66*, 1425 (1941).

19. F. Buckley and J. K. Taylor, J. Res. Nat. Bur. Stand. *34*, 97 (1945).

20. See Ref. 1, p. 85.

21. L. Meites, *Polarographic Techniques* 2nd ed., Interscience, New York, 1965, pp. 111-125.

22. I. M. Kolthoff and J. J. Lingane, *Polarography*, 2nd ed., Interscience, New York, 1952, pp. 63-98.

23. D. J. Fisher, W. L. Belew, and M. T. Kelley, in *Polarography 1964* (G. J. Hills, ed.), Macmillan, London, 1966, vol. 1, p. 101.

24. A. M. Bond and J. H. Canterford, Anal. Chim. Acta *70*, 177 (1974).

25. A. M. Bond and D. R. Canterford, Anal. Chem. *44*, 721 (1972).

26. A. M. Bond, G. A. Heath, and R. L. Marin, Inorg. Chem. *10*, 2026 (1971).

27. A. M. Bond and T. A. O'Donnell, Anal. Chem. *41*, 1801 (1969).

28. A. M. Bond, Anal. Chim. Acta *53*, 159 (1971), and references cited therein.

29. J. T. Folliard and R. E. Cover. J. Electroanal. Chem. *33*, 463 (1971).

30. J. V. A. Novak, in *Progress in Polarography* (P. Zuman and I. M. Kolthoff, eds.), Interscience, New York, 1962, vol. 2, p. 569, and references cited therein.

31. H. J. Mortko and R. E. Cover, Anal. Chem. *45*, 1983 (1973).

32. H. J. Mortko and R. E. Cover, Amer. Lab. *7*, 51 (1975).

33. Y. Okinaka and I. M. Kolthoff, J. Amer. Chem. Soc. *74*, 3326 (1957).

34. I. M. Kolthoff, Y. Okinaka, and T. Fujinaga, Anal. Chim. Acta *18*, 295 (1958).

35. I. M. Kolthoff and Y. Okinaka, Anal. Chim. Acta *18*, 83 (1958).

36. I. M. Kolthoff and Y. Okinaka, in *Progress in Polarography* (P. Zuman and I. M. Kolthoff, eds.), Interscience, New York, 1962, vol. 2, p. 357, and references cited therein.

37. H. Schmidt and M. von Stackelberg, *Modern Polarographic Methods*, Academic Press, New York, 1963, pp. 13-16.

38. E. Wahlin, Radiometer Polarog, *1*, 113 (1952).

39. E. Wahlin and A. Besle, Acta Chem. Scand. *10*, 935 (1956).

40. K. Kronenberger, H. Strehlow, and A. W. Elbel, Polarogr. Ber. *5*, 62 (1957).

41. A. W. Elbel, Z. Anal. Chem. *173*, 70 (1960).

42. A. Besle, Acta Chem. Scand. *10*, 943, 947, 951 (1956).

43. P. O. Kane, J. Polarogr. Soc. *8*, 10 (1962).

44. H. B. Mark, Jr., and C. N. Reilley, J. Electroanal. Chem. *3*, 54 (1962).

45. A. M. Bond, Anal. Chim. Acta *62*, 415 (1972).

46. A. M. Bond and R. J. O'Halloran, J. Electroanal. Chem. *68*, 257 (1976).

47. K. Kronenberger and W. Nickels, Z. Anal. Chem. *186*, 79 (1962).

48. Y. Yasumori, Bunseki Kagaku *7*, 354 (1958).

49. M. T. Kelley and D. J. Fisher, Anal. Chem. *28*, 1130 (1956).

50. M. T. Kelley and D. J. Fisher, Anal. Chem. *30*, 929 (1958).

51. M. T. Kelley, H. C. Jones, and D. J. Fisher, Anal. Chem. *31*, 1475 (1959).

52. M. T. Kelley, D. J. Fisher, and H. C. Jones, Anal. Chem. *32*, 1262 (1960).

53. D. J. Fisher, W. L. Belew, and M. T. Kelly, in *Polarography 1964* (G. J. Hills, ed.), Macmillan, London, 1966, vol. 1, pp. 89-134.

54. H. C. Jones, W. L. Belew, R. W. Stelzner, T. R. Mueller, and D. J. Fisher, Anal. Chem. *41*, 772 (1969).

55. L. Nemec, Collect. Czech. Chem. Commun. *25*, 2085 (1960).

56. W. L. Belew, D. J. Fisher, H. C. Jones, and M. T. Kelley, Anal. Chem. *41*, 779 (1969).

57. S. Mann, Arch. Mikrobiol. *71*, 304 (1970).

58. Note 7, pp. 8-13, and references cited therein.

59. J. Heyrovský, Analyst *72*, 229 (1947); Anal. Chim Acta *2*, 537 (1948).

60. J. Heyrovsky, Chem. Listy *43*, 149 (1949).

61. J. Vogel and J. Riha, J. Chim. Phys. *47*, 5 (1950).

62. J. Riha, Collect. Czech. Chem. Commun. *16*, 479 (1951).

63. J. J. Lingane and R. Williams, J. Amer. Chem. Soc. *74*, 790 (1952), and references cited therein.

64. M. P. Leveque and F. Roth, J. Chim. Phys. *46*, 480 (1949).

65. W. L. Belew, D. J. Fisher, M. T. Kelley, and J. A. Dean, Microchem. J. *10*, 301 (1966).

66. J. Glickstein, S. Rankouitz, C. Auerbach, and H. O. Finston, Advan. Polarog. *1*, 183 (1960).

67. C. Auerbach, H. L. Finston, G. Kissel, and J. Glickstein, Anal. Chem. *33*, 1480 (1961).

68. G. Semerano and L. Riccoboni, Gazz. Chim. Ital. *72*, 297 (1942).

69. Kanevskii, Zh. Prikl. Khim., Mosk. *17*, 514 (1944).

70. K. G. Powell and G. F. Reynolds, in *Polarography 1964* (G. J. Hills, ed.), Macmillan, London, 1966, vol. 1, pp. 249-259.

71. L. Airey and A. A. Smales, Analyst, *75*, 287 (1950).

72. M. T. Kelley and H. H. Miller, Anal. Chem. *24*, 1895 (1952).

73. C. G. Enke and R. A. Baxter, J. Chem. Educ. *41*, 202 (1964).

74. G. C. Barker, *Proc. Cong. Modem. Anal. Chem. Ind.*, St. Andrews, Heffer, Cambridge, 1957, p. 215.

75. A. Poojary and S. R. Rajagopalan, J. Electroanal. Chem. *62*, 51 (1975).

76. S. R. Rajagopalan, A. Poojary, and S. K. Rangarajan, J. Electroanal. Chem. *75*, 135 (1977).

Chapter 5

LINEAR SWEEP VOLTAMMETRY
AND RELATED TECHNIQUES

5.1 NOMENCLATURE

In this chapter polarographic, or more strictly speaking voltammetric,
methods based on utilizing a rapid linear (or approximately linear)
sweep of the potential range are considered. Unless otherwise stated,
it is assumed that the solution is unstirred (mass transfer does not
occur by forced convection) and migration currents are eliminated
by the presence of an electrolyte as in polarography. When applied
to the dropping mercury electrode (dme), this technique enables the
entire potential range to be covered on one drop. That is, the
voltage sweep time is short compared with the drop time. In dc
polarography, as practiced with modern instrumentation, the potential
is also applied to the electrodes in the form of a linear ramp.
However, in contrast to the linear sweep method, the voltage sweep
time in polarography is much greater than the drop time, and the
potential covered per drop is only on the order of millivolts or
less. Furthermore, another important distinction arises from the
fact that in polarography it is assumed for theoretical purposes that
constant potential-current curves are being recorded. In the linear
sweep method, the constant potential assumption is lifted and the
theory is solved under conditions of a continuously changing potential. Thus, when the scan direction is negative,

$$E = E_i - vt \qquad\qquad\qquad\qquad\qquad (5.1)$$

where

E_i = initial potential

v = scan rate

t = time after scan has commenced

Because the recording of the i-E curve at fast scan rates originally required the use of an oscilloscope, the technique of linear
sweep voltammetry (LSV) has been referred to historically as *oscillographic polarography* or *cathode-ray polarography*. Nowadays, fast-
response X-Y recorders or digital display (see Fig. 2.18, for example) may be used as alternative forms of readout in many applications.
Since the mode of measurements is really only incidental to the technique, the use of nomenclature associated with the readout device is
best avoided. Similarly, the term *fast sweep polarography*, frequently
used as a synonym for LSV at a dme, could be misleading because the
rapid polarographic method utilizing very short drop times and discussed in Chap. 4 also employs fast sweep rates. It should be noted
however, with this latter method, that the voltage sweep time is
still long compared with the drop time and this technique is still
correctly referred to as a polarographic method. Another synonym,
stationary electrode polarography, indicates that the method designated as LSV at a dme in this book is one in which the scan rate
occurs over only a fraction of the life of the mercury drop. For
theoretical purposes, the potential is considered to have been
applied to a stationary electrode because during the recording of
the i-E curve, the growth of the drop is negligible. Theory presented
for LSV at a dme, therefore, can be applied to any stationary electrode (using modified diffusion equations applicable to the electrode
in question if necessary) and, as such, is relevant to the whole
array of solid electrodes, e.g., platinum, carbon, gold, which are
used in analytical voltammetry when mercury electrodes are not suitable. Other names for LSV, such as *peak polarography*, are rejected
on grounds of ambiguity, as will become obvious in subsequent discussion.

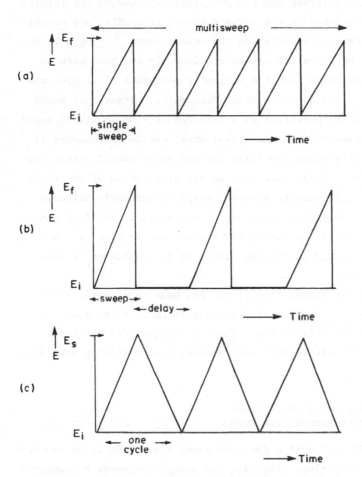

FIGURE 5.1 Some applied waveforms used in linear sweep voltammetry and related techniques. (a) Single sweep achieved by linear potential ramp. E_i = initial potential; E_f = final potential. If more than one scan is required, a saw tooth waveform can be used during which the potential is returned to E_i immediately on reaching E_f and the scan repeated. (b) A delay period between sweeps is frequently used in LSV at a dme and sometimes at stationary electrodes. (c) Cyclic (triangular wave) voltammetry. E_s = switching potential corresponding to E_f in LSV. As in LSV, a delay between cycles could also be required, particularly for work at a dme [see (b)].

If at some point on the linear potential ramp, the potential
scan direction is reversed and the potential returned to its initial
value, material reduced in the forward sweep, if stable, may be oxidized back to starting material on the return sweep. Using a triangular voltage, the potential may be continuously switched between
the initial potential E_i and the switching potential E_s to give the
electrochemical technique of cyclic voltammetry. Figure 5.1 shows
diagrammatically the potential as a function of time in linear sweep
and *cyclic voltammetry*. In analytical work, cyclic voltammetry is
not used frequently since the first forward sweep usually gives the
required analytical data. However, in the elucidation of electrode
mechanisms, the extension to reverse sweeps is extremely valuable
[1], and in the context of being useful as an aid in deciding the
analytical utility of the normal LSV method, rather than as a widely
used analytical method in its own right, it is considered in some
detail in this book.

When the conventional form of LSV has been considered, it will
be seen that many of the "tricks" applied in Chap. 4 can also be
employed to improve this method. Thus, derivative and staircase
(current-sampled) voltammetry, for example, form part of the related
methodology.

5.2 SYNCHRONIZATION OF SWEEP TIME
AT A DROPPING MERCURY ELECTRODE

To apply the LSV method at a dme, the sweep time needs to be synchronized with the drop time. That is, the sweep time needs to commence
at a selected and accurately known time, t_0 sec after the drop growth
has commenced. Additionally, the drop must not fall off before completion of the sweep. Ideally, the growth of the mercury drop during
the sweep duration should be negligible, so that constant area conditions can be assumed. Since

$$A = 0.85m^{2/3}t^{2/3} \tag{5.2}$$

where A = area, m = flow rate of mercury, and t = time, then

$$\frac{dA}{dt} = \frac{2}{3} \, 0.85m^{2/3}t^{-1/3} \tag{5.3}$$

From Eq. (5.3) the rate of change of area is least toward the end of
the drop. Ideally, therefore, the sweep should be initiated late
in the drop life, e.g., 2 sec or more after growth commences. Thus
any timing circuitry developed for use with LSV at a dme needs to
(a) be initiated by the fall of the drop or else "know" when the
drop life commenced; (b) introduce a delay period, t_0 sec; and then
(c) initiate the potential scan, with adequate time left to complete
the scan before the drop growth-drop fall cycle is completed.

Since the instant of drop detachment is characterized by an
extremely rapid change in circuit resistance (impedance) at most potentials, the circuit properties of the cell readily permit electronic
detection of the instant of drop fall. The exception to this is at
or near the potential of the electrocapillary maximum or point of
zero charge, where difficulties may arise because of extremely low
current levels (see Chap. 3). Many variations for the detection of
the drop fall based on electronic methods have been reported [2-9].
The falling mercury drop has also been used to interrupt a light beam
and optically detect the drop fall instantly [10-14]. Figure 5.2
provides an example of a "reflecting type" drop fall detector which
monitors the amount of light reflected by the drop as it grows [14].
When the drop detaches, the amount of light reflected by the drop
falls to zero. Advocates of the optical approach point out that many
electronic methods require an external electrical connection to the
cell and state that this is undesirable. This seems an unwarranted
criticism in analytical work, since linear calibration curves and
extremely high accuracy are obtained with electronic detection of
the drop fall. Furthermore, no experimental evidence relevant to
analytical use of these detection methods has been put forward; so
this author concludes that the ease and convenience of electronic
detection methods highly recommend them over optical and mechanical
methods. Obvious objections to optical methods are based on their
limited use in colored and turbid solutions or solvents with

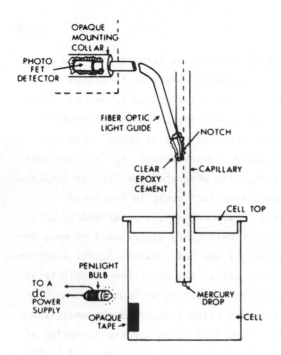

FIGURE 5.2 An optical drop fall detector. The drop is illuminated
from the side by a focused lamp mounted outside the cell. Part of
the light reflected up the capillary is diverted out of the capillary
by a fiber optic light guide. The drop fall monitor measures the
light from the fiber optic input, and the instant of drop fall can
be monitored. [Reproduced from Anal. Chem. 46, 802 (1974).
© American Chemical Society.]

unsuitable refractive indexes, and experimental difficulties in
alignment of the optics. These seem well founded in comparison to
the unsubstantiated arguments raised against electronic methods.

The modern practice of having controlled drop times simplifies
apparatus even further, because the system "knows" when a drop is
about to be mechanically dislodged. Hence, the timing circuitry for
the complete operation can be triggered by the same impulse that
detaches the mercury drop. Critics of this approach usually suggest
that stirring resulting from displacement of the capillary by the
mechanical blow delivered to dislodge the drop will cause problems.
Such claims are totally unsupported by experimental evidence,

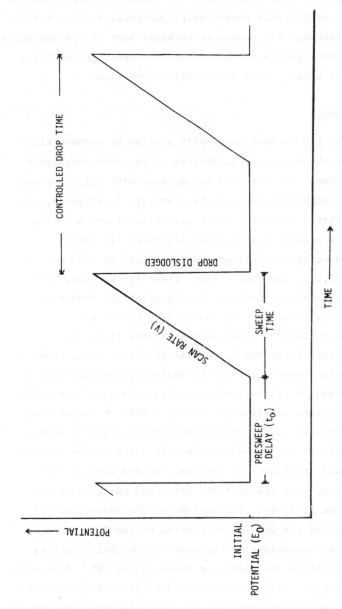

FIGURE 5.3 Sequence of operations for recording a linear sweep voltammogram at a dme with controlled drop time. [Reproduced from Anal. Chem. 46, 1934 (1974). © American Chemical Society.]

and extremely reliable analytical data are obtained by this method.
In the overall context, this is probably the simplest approach and it
is most compatible with trends toward multifunctional instrumentation
in which a "drop knocker" has become an integral part of the apparatus.
Figure 5.3 shows the sequence of operations involved in recording a
LSV voltammogram at a dme, using a controlled drop time.

5.3 SOLID ELECTRODES

In analytical work, LSV is most frequently applied at mercury electrodes, because of the high reproducibility of the area obtainable,
particularly at a dme. The ease and convenience with which the dme
may be used, once synchronization of the scan rate is achieved, also
highly recommend this electrode. Other varieties of mercury electrodes, such as the hanging mercury drop electrode [11,15-19],
frequently used in stripping voltammetry (see Chap. 9), and mercury
pool electrodes [20,21], have been used. These latter kinds of electrodes, while achieving genuinely stationary electrode conditions,
do not offer the convenience or reproducibility of the dme.

A host of solid electrodes are also available [1,22-24].
Almost all of the more inert metals such as platinum, gold, rhodium,
iridium, silver, etc., have been used in analytical applications of
LSV. Electrodes based on carbon materials [1] have also been widely
used, e.g., wax-impregnated graphite, glassy carbon, boron carbide,
carbon paste. Such electrodes, while permitting analytical applications not amenable to mercury electrodes, do not have the asset of
a constantly renewed, reproducible surface, nor are they as convenient to use, since much preparation, care, and maintenance are
required to use them reliably. Discussions of the advantages and
disadvantages of solid electrodes are covered by the literature
[1,22-23] and can be summarized by the observation that a mercury
electrode is likely to be the preferred material for LSV. Generally,
specific reasons for not using mercury would be given as precursors
when venturing to an alternative surface: "the electroactive species
oxidizes or interacts in some way with mercury," "mercury electrodes

cannot be used in this 'on stream' application," or "unsatisfactory waves are obtained at mercury because of strong adsorption of material," etc. These comments should not be taken to denigrate the importance of solid stationary electrodes in analytical voltammetry, because the exceptions to using mercury electrodes incorporate some of the most important applications. Thus, the electroanalytical chemist certainly needs to be aware of the potentialities of solid electrodes in addition to recognizing the problems relative to mercury electrodes.

It is probably as well to emphasize at this juncture that despite the fact that much of the data and discussion presented in this section are derived from mercury electrodes, the basic theory for stationary mercury electrodes is usually very similar to that of stationary solid electrodes (differences may arise from geometrical considerations). Usually, therefore, the theoretical equations presented are more or less applicable to stationary electrodes in general. However, one needs to recognize the important practical aspect that the heterogeneous pathway in the electrode process can be modified substantially by changing the electrode surface. Thus, k_s values at mercury electrodes can sometimes be much faster than at solid electrodes, and what is presented as an example of a reversible electrode process at a dme may only pertain to mercury surfaces. The reversible class of electrode processes is generally the most suitable for analytical work with many modern polarographic methods, and changes in reversibility in changing electrode material may be important. Similarly, surface phenomena such as adsorption are frequently extremely dependent on the nature of the electrode material, and strong adsorption which may cause considerable complexity in an example cited at mercury may be absent at, say, platinum, or vice versa.

Finally, it should be noted that if the solution is stirred or the electrode rotated, forced convection occurs. The theory presented at stationary electrodes is no longer even approximately correct under these conditions. Analytical applications using solid electrodes, in fact, are carried out frequently under conditions of forced convection, and only the theory given at the end of this chapter rather than that immediately following, is valid in these circumstances.

5.4 THEORY FOR FARADAIC PROCESSES

The time scale of the LSV method in unstirred solutions is governed by the scan rate. Intuitively, therefore, one can understand phenomenological events by analogy with dc polarography, and by knowing what happens in this previously considered technique when the drop time is changed. Thus, the shorter the drop time in polarography, the more irreversible the electrode process, i.e., the wave becomes more drawn out and further removed from $E°$; in LSV, the faster the scan rate, the more irreversible the measured response, and analogous changes in wave shape and position occur. Under the section on rapid polarography (Chap. 4), it was shown that chemical steps in EC mechanisms can sometimes be eliminated at short drop time. Fast scan rates achieve the same effect in LSV. Finally, the charging current-to-faradaic current ratio, and therefore, the limit of detection, was seen to be less favorable the shorter the drop time; in LSV, the faster the scan rate, the higher the charging current-to-faradaic current ratio, and a concomitant decrease in limit of detection also occurs, the higher the scan rate. Figure 5.4 shows a LSV voltammogram and polarogram recorded on the same solution and using the same capillary. The peak rather than sigmoidal shape of the LSV curves is explained by depletion terms occurring in the mass transfer. The peak height is a function of concentration and is the parameter generally used in analytical work.

On application of a rapidly increasing potential a current starts to flow as the reduction potential of the electroactive species is reached. As in polarography, the current flow increases as the potential becomes more negative (reduction assumed), in response to the increasing concentration gradient at the electrode surface caused by electrolysis. However, unlike polarography,

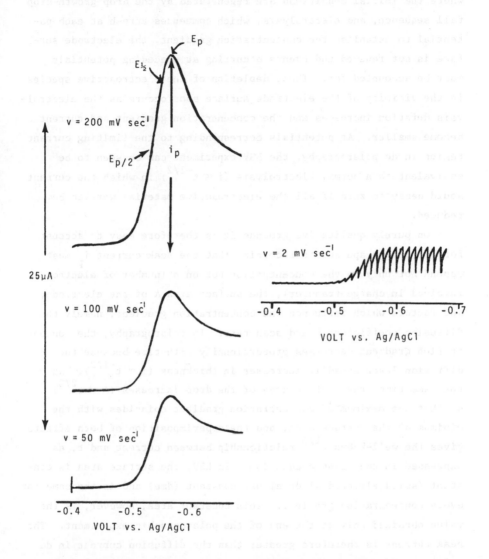

FIGURE 5.4 Comparison of a dc polarogram and linear sweep voltammo-grams for 5×10^{-4} M Cd(II) in 1 M KNO_3. DC polarogram was recorded with a drop time of 5 sec. Linear sweep voltammograms were recorded with the same capillary using a presweep delay period of 3 sec.

where the initial conditions are regenerated by the drop growth–drop fall sequence, and electrolysis, which commences afresh at each potential to establish the concentration gradient, the electrode surface is not renewed and events occurring at preceding potentials must be accounted for. Thus, depletion of the electroactive species in the vicinity of the electrode surface soon occurs as the electrolysis duration increases and the concentration gradient and current become smaller. At potentials corresponding to the limiting current region in dc polarography, the LSV experiment can be seen to be equivalent to a normal electrolysis ($i \propto t^{-1/2}$) in which the current would decay to zero if all the electroactive material were to be reduced.

On purely qualitative grounds it is therefore easy to account for the peak shape and to recognize that the peak current i_p must depend not only on the concentration but on n (number of electrons involved in charge transfer), the surface area A of the electrode, and factors which influence the concentration gradient, namely the diffusion coefficient D and scan rate. In polarography, the concentration gradient decreases proportionally with time because the diffusion layer steadily increases in thickness ($i \propto t^{-1/2}$). At the same time, the surface area of the drop increases ($A \propto t^{2/3}$) so that the maximum of concentration gradient coincides with the minimum of the surface area, and the superimposition of both effects gives the well-known $t^{1/6}$ relationship between current and t, as expressed in the Ilkovic equation. In LSV, the surface area is constant (solid electrodes) or almost constant (dme) and is the same for every concentration gradient. This constant area, however, is the value obtained only at the end of the polarographic experiment. The peak current is therefore greater than the diffusion current in dc polarography. Naturally, the faster the scan rate, the larger the current because the diffusion layer has less time to increase in thickness and the dependence of i_p on v is explained.

If the product of the electrode process *is stable*, then on reversing the scan direction this material can be oxidized back to

starting material. Figure 5.5 shows a cyclic voltammogram for a
chemically reversible system. Not surprisingly, a chemically reversible system is characterized by equal peak heights for the reduction
and oxidation processes. If the product is unstable and reacts
before the reverse scan takes place, then no wave will be seen on
the reverse scan. Similarly, adsorption of product or reactant will
perturb the system compared with the simple electron transfer case.
The ease with which cyclic voltammetry can be used to diagnose mechanistic complications can now be appreciated. Figure 5.6(a) shows
the forward sweep of an i-E curve for an oxidation when the rest
potential t_0 is set in an electroactive region. Figure 5.6(b)
shows an example of an oxidation as it normally would be recorded.

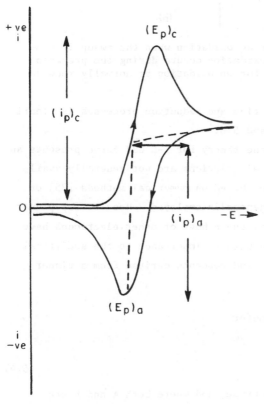

FIGURE 5.5 Reversible cyclic voltammogram for a reduction process.

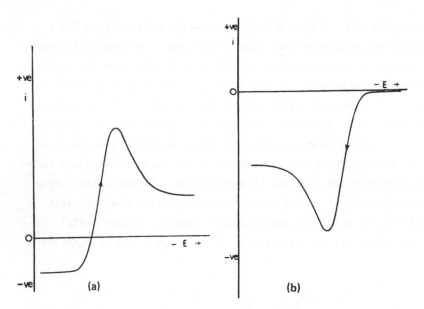

FIGURE 5.6   (a) LSV curve for an oxidation when the sweep is commenced at a potential where oxidation occurs during the presweep
delay period.   (b) LSV curve for an oxidation as normally recorded.

The distinction between oxidation and reduction processes is clearly
seen by comparing Figs. 5.5 and 5.6.

A rigorous solution to the theory for i–E LSV curve presents an
arduous task because analytical solutions are not generally available.   Mathematical procedures based on numerical methods [25] or
digital simulation [26] are generally employed.   Results for planar,
cylindrical, spherical, wedge, and a host of other electrodes have
been considered.   Only the results of importance to the analytical
chemist are considered below, and concepts derived from a planar
electrode are used.

5.4.1   Reversible Charge Transfer

For a reversible reduction

$$A + ne \rightleftharpoons B \tag{5.4}$$

taking place at a planar electrode, and where both A and B are

solution soluble, the boundary value problem for LSV, using the
same symbols defined previously, is

$$\frac{\partial C_A}{\partial t} = D_A \frac{\partial^2 C_A}{\partial x^2} \tag{5.5}$$

$$\frac{\partial C_B}{\partial t} = D_B \frac{\partial^2 C_B}{\partial x^2} \tag{5.6}$$

For $t = 0$, $x \geq 0$,

$$(C_{x,t})_A = C_A \qquad (C_{x,t})_B = C_B (\approx 0) \tag{5.7}$$

for $t \geq 0$, $x \rightarrow \infty$,

$$(C_{x,t})_A \rightarrow C_A \qquad (C_{x,t})_B \rightarrow 0 \tag{5.8}$$

for $t > 0$, $x = 0$,

$$D_A \left(\frac{\partial C_A}{\partial x}\right) = -D_B \left(\frac{\partial C_B}{\partial x}\right) = \frac{i}{nFA} \tag{5.9}$$

$$\frac{C_A}{C_B} = \exp[\frac{nF}{RT} (E - E°)] \tag{5.10}$$

[Note: From Eq. (5.10), results are only applicable to the
$A + ne \rightleftharpoons B$ case and not all reversible systems. Thus,
$A + ne \rightleftharpoons B(Hg)$ or $A + B + ne \rightleftharpoons C$ would yield different results.
This must not be overlooked.] Other equations required are

$$E = E_i - vt \qquad 0 < t \leq t_s \tag{5.11a}$$

$$E = E_i - 2vt_s + vt \qquad t_s \leq t \tag{5.11b}$$

where t_s = time when potential direction is reversed, and where
Eq. (5.11a) is relevant to LSV or the first forward sweep in cyclic
voltammetry, and Eq. (5.11b) extends the calculation into the time
region of cyclic voltammetry. Notes 25 and 26 should be consulted
for further details of the solution to these equations. In general,

$$(i)_{rev} = nFAC_A \sqrt{\pi a D_A} \; f(\chi) \tag{5.12}$$

where $a = nFv/RT$, and $f(\chi)$ is a function that can be calculated. At the peak potential E_p where $f(\chi)$ attains its maximum value and $(i)_{rev} = (i_p)_{rev}$,

$$(i_p)_{rev} = k'n^{3/2} A D_A^{1/2} C_A v^{1/2} \quad (k' = \text{constant at constant } t) \tag{5.13}$$

Furthermore, E_p is independent of scan rate. The polarographic $E_{1/2}$ value ($\approx E°$) can be estimated from a reversible LSV curve from the fact that it occurs at a point 85.2% of the way up the wave. The separation of E_p and $E°$ at 25°C is 28.5/n mV or

$$(E_p - E°)n + \frac{RT}{F} \ln \left(\frac{D_A}{D_B}\right)^{1/2} = 0.285 \tag{5.14}$$

or

$$E_p = E_{1/2} - 1.1 \frac{RT}{nF} \tag{5.15}$$

It is sometimes convenient to use the half-peak potential ($E_{p/2}$) as a reference point, even though this has no direct thermodynamic significance. $E_{p/2}$ precedes $E_{1/2}$ by 28.0/n mV at 25°C or

$$E_{p/2} = E_{1/2} + 1.09 \frac{RT}{nF} \tag{5.16}$$

or

$$E_p - E_{p/2} = \frac{0.057}{n} \tag{5.17}$$

at 25°C. In the case of an oxidation, these equations are still valid, except that negative signs are usually associated with i (anodic currents are negative) and the term 1.1 RT/nF in Eq. (5.15) should be preceded by a plus sign, etc. With cyclic voltammetry, the separation in peak potential between cathodic $(E_p)_c$ and anodic scans $(E_p)_a$ will be

$$(E_p)_c - (E_p)_a = 2(1.11 \frac{RT}{nF}) \tag{5.18}$$

or 57/n mV at 25°C, which provides a convenient test of reversibility, in addition to the expected observation of equivalent peak heights in the forward and reverse scan directions. Figure 5.5 shows some of the results diagrammatically. All equations and discussion on cyclic voltammetry assume that the switching potential is substantially more negative than $(E_p)_c$ for reduction or substantially more positive than $(E_p)_a$ for oxidation.

5.4.2 Nonreversible Charge Transfer

Reversibility, in terms of LSV, means that the rate of electron transfer (k_s) is sufficiently fast, with respect to the scan rate, so that Eq. (5.10) is applicable within the limit of experimental error. k_s values of about 0.1 cm sec^{-1} can be considered reversible with scan rates usually used in analytical work, for example, 50 mV/sec to 1 V/sec.

If the k_s value is sufficiently slow compared with the potential sweep rate, so that the surface concentrations of A and B cannot maintain the Nernstian values, terms involving α and k_s must be introduced into the expressions, as is the case for quasi-reversible and irreversible electrode processes in dc polarography. For the totally irreversible case

$$(i)_{irr} = n(\alpha n_\alpha)^{1/2} A D_A^{1/2} v^{1/2} C_A f(\chi') \tag{5.19}$$

where $f(\chi')$ is the function for the irreversible case, which is analogous to $f(\chi)$ in Eq. (5.12), and n_α is the number of electrons transferred up to and including the rate-determining step, and

$$(i_p)_{irr} = k''n(\alpha n_\alpha)^{1/2} A D_A^{1/2} v^{1/2} C_A \tag{5.20}$$

where k'' is a constant. For an irreversible one-electron transfer, where $n = n_\alpha = 1$ necessarily, and all other experimental factors remain constant, it can be shown [27] that

$$\frac{(i_p)_{irr}}{(i_p)_{rev}} = 1.1\alpha^{1/2} \tag{5.21}$$

Thus, if one assumes a value of α of 0.5, then

$$\frac{(i_p)_{irr}}{(i_p)_{rev}} = 0.77 \tag{5.22}$$

Consequently, unlike dc polarography where the limiting current is independent of the kinetics, in LSV, i_p is a function of α for the irreversible case. Therefore, in analytical work utilizing non-reversible electrode processes, it is essential that the solution to be determined should be similar to the standards from which the calibration curves are prepared. That is k_s and α should be similar in both cases, which means that E_p for both standards and unknowns as well as some property of the wave shape, e.g., $(E_p - E_{1/2})$ or $(E_p - E_{p/2})$, should be shown to be identical. In dc polarography, surfactants are frequently used to eliminate maxima. The same in-gredients, if present in a solution on which a LSV scan is to be used for quantitative analysis, could alter k_s and/or α, and reference to a calibration curve prepared in the absence of the surface-active ingredient would give erroneous results.

For the totally irreversible case at 25°C,

$$E_p - E_{p/2} = \frac{0.048}{\alpha n_\alpha} \tag{5.23}$$

and

$$E_p = E° - \frac{RT}{\alpha n_\alpha F} (0.780 + 0.5 \ln \frac{\alpha n_\alpha D_A Fv}{RT} - \ln k_s) \tag{5.24}$$

Thus, in contrast to the reversible case, E_p is dependent on α, k_s, and v. The distinction between reversible and irreversible electrode processes is therefore readily made.

In cyclic voltammetry, for the quasi-reversible electrode proc-ess, the separation in cathodic and anodic peak potentials, $(E_p)_c - (E_p)_a$, is a function of k_s and scan rate [28]. For a given

scan rate, the smaller the k_s, the larger the separation in peak
potentials, and vice versa; so cyclic voltammetry readily enables
the reversibility or otherwise of an electrode process to be evaluated. Figure 5.7 shows the situation diagrammatically. Numerical
solution of the appropriate integral equation defines $(E_p)_c - (E_p)_a$
in terms of a function ψ [28], where

$$\psi = \gamma^\alpha \frac{k_s}{\pi^{1/2} D_A^{1/2} (nF/RT)^{1/2} v^{1/2}} \qquad (5.25)$$

Since γ is the square root of the ratio of the diffusion coefficients
γ^α is unity for most practical applications. Using a value of
$D_A = 1 \times 10^{-5}$ cm^2 sec^{-1} and $nF/RT = 39.2$ V^{-1}, gives

$$\psi \simeq 28.8 \frac{k_s}{v^{1/2}} \qquad (5.26)$$

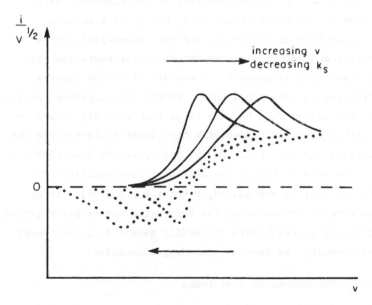

FIGURE 5.7 Dependence of linear sweep and cyclic voltammograms for
quasi-reversible electrode process on scan rate or k_s.

The variation of ΔE_p with ψ, in tabular form or as a working curve, is available in Note 28, and after experimentally measuring ΔE_p k_s can be calculated from these data.

Despite the dependence of $(i_p)_{irr}$ on the electrode kinetics, Eq. (5.22) shows that the sensitivity of the LSV method is almost independent of the degree of reversibility. By contrast, ac and some other modern polarographic methods virtually cannot be used for irreversible electrode processes, since the current per unit concentration is very low. The relative insensitivity of the current per unit concentration in LSV on k_s and α is therefore worth recalling when reading later chapters.

In summary, it can be seen that species exhibiting irreversible processes can be readily determined by LSV, provided the precautionary check that E_p and $E_p - E_{p/2}$ are identical to standard solutions is undertaken, for example. Alternatively, linear i_p vs. concentration plots frequently permit the method of standard additions to be used, which avoids the need for careful matching of the standards matrix. However, in view of the independence of i_p and E_p on electrode kinetics for reversible processes and because irreversible waves can be considerably drawn out (i.e., peak-height measurements are less accurate and resolution is poorer), reversible electrode processes are more attractive in analytical work. Except in exceptional circumstances, this conclusion will be seen to be true with all modern polarographic techniques, and undoubtedly the closer to reversible the electrode process, the more reliable the polarographic determination. The ability to be able to report the degree of reversibility of an electrode process must be emphasized as being a prime skill to be gained by the worker contemplating the systematic use of polarography. The electrochemical reversibility is readily gauged in linear sweep and cyclic voltammetry, as seen in preceding discussion.

5.4.3 Effect of Uncompensated Resistance

Qualitatively, the influence of uncompensated resistance is similar to decreasing k_s in LSV, and indeed, this analogy holds with almost all modern polarographic techniques. Uncompensated resistance terms therefore assume considerable importance in LSV, as wave shape and position are influenced by iR drop. In dc polarography, while ohmic iR drop also leads to a shift in $E_{1/2}$ and change in wave shape, the limiting current i_d is essentially unaltered and uncompensated resistance is unlikely to lead to the reporting of erroneous analytical data. The use of a two-electrode system, while often acceptable in dc polarography, is fraught with danger in LSV and most other modern polarographic methods, under even mildly severe operating conditions with respect to cell resistance. The necessity for using a three-electrode potentiostat in modern polarographic methods, as recommended in Chap. 2, can now be further appreciated.

When a faradaic current flowing through the working electrode is considered, the empirically observed effects of the iR drop during the linear scan of the potential are as follows [29]: A shift of E_p toward more negative potentials in the case of reduction, or a shift toward more positive potentials for oxidation (the combined effect of these observations is an increase in separation of peak potentials in cyclic voltammetry), a decrease of i_p, and an enlargement of the peak width. The magnitude of these effects increases with concentration of the electroactive species, the cell resistance, and scan rate of potential. In the presence of uncompensated resistance, nonlinear plots of i_p vs. C can be obtained for even the simple charge transfer process.

The use of a potentiostat and, if necessary, positive feedback circuitry (Chap. 2), minimize problems associated with uncompensated resistance in LSV, and few difficulties should be encountered under conditions normally present in analytical voltammetry with this kind of instrumentation. However, in high-resistance nonaqueous solvents, iR drop problems could still be evident and the usually recommended procedure of ensuring that E_p and $E_{p/2}$ on the unknown are the same as for standard solutions, should ensure recognition of potential problems having their origin in ohmic loss. Further, if the absence of chemical or adsorption pathways is indicated, linear plots of i_p vs. concentration should be obtained over wide concentration levels. Curvature at the higher concentration levels, coupled

with a shift in the E_p in the expected direction and broadening of
the wave, would be indicative of the presence of uncompensated resistance so that simple diagnostic criteria for detecting iR drop problems are available.

5.4.4 Influence of Coupled Chemical Reactions

Having considered the influence of coupled chemical reactions in
some detail with dc polarography, the extension of the ideas to LSV
is readily accomplished.

For example, with the EC mechanism,

$$A + ne \rightleftharpoons B \xrightarrow{k_f} C$$

E_p and i_p are governed by the magnitude of the rate constant k_f at
slow scan rates. However, assuming reversible charge transfer is
retained, the follow-up chemical reaction may be outrun at fast scan
rates, as is the case when using short drop times in dc polarography,
and the LSV curve can revert to the reversible charge transfer case
at sufficiently fast scan rate. In dc polarography, while k_f influences $E_{1/2}$, it has no effect on i_d; however, in LSV, both E_p and
i_p are dependent on k_f. The analytically used parameter i_p is now
under kinetic control in LSV, and cautions recommended previously
for this situation apply. The actual value of i_p in the presence
of a follow-up chemical reaction exceeds that in a reversible electron transfer step; however, provided it is a first-order reaction,
linear i_p-vs.-concentration plots are obtained.

If the follow-up chemical reaction is reversible, then k_f, k_b,
and the equilibrium constant of the chemical reaction enter into the
description of the i-E curve as in dc polarography, and analogous
arguments to those presented with dc polarography also follow with
other nuances of the EC mechanism.

In cyclic voltammetry the EC mechanism is characterized by
unequal cathodic-to-anodic peak heights, since all or part of B has
reacted to give C, rather than being available for oxidation back to
starting material.

Quantitative treatments of linear sweep and cyclic voltammograms with the different kinds of coupled chemical reactions are available in the literature [25-28,30-36], and details can be consulted in the original work.

Qualitatively, from the analytical chemist's point of view, all that needs to be recognized is that the results presented for dc polarography can be readily extrapolated to LSV. For example, values of i_p/C are low for CE mechanisms (undesirable for analytical work) but i_p/C is large for catalytic or regenerative mechanisms (may be useful in analytical work); linear plots of i_p versus C are obtained for straight charge transfer, whereas in the presence of coupled chemical reactions, i_p versus C is frequently nonlinear, e.g., catalytic processes. Figure 5.8 gives an example for the catalytic system

$$A + ne \rightleftharpoons B \qquad B + Z \xrightarrow{k} A$$

The expected dependence of i_p on the rate constant k for the chemical process is found. At very high scan rates, the chemical step is

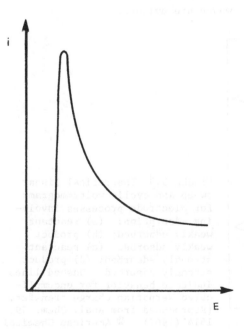

FIGURE 5.8 Linear sweep voltammogram for the catalytic mechanism.

unimportant and the peak characteristics are those for the charge
transfer. As the scan rate decreases, i_p reduces less rapidly than
for a noncatalytic process because A is regenerated by the chemical
reaction; that is, $i_p/v^{1/2}$ is not independent of v, and E_p varies
with v.

5.4.5 Influence of Adsorption

As in dc polarography, adsorption can cause splitting of waves and
other undesirable phenomena. In the presence of weakly adsorbed
material, linear sweep voltammograms may exhibit enhancement of the
peak currents [37-39]. On the other hand, if the product or reactant is strongly adsorbed, a separate adsorption peak may occur prior
to or after the normal peak [39-42], as in dc polarography (Chap. 3).
Figure 5.9 summarizes the various possibilities, and the obvious
analogy with dc polarography is evident. Figure 5.10 shows an example of methylene blue, in which the product of the electrode process
is strongly adsorbed [40]. At low concentrations, virtually only
one wave is observed, while at higher concentrations, both the
normal and adsorption-controlled waves are evident.

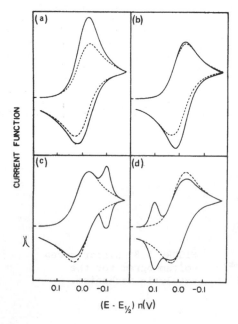

FIGURE 5.9 Theoretical linear
sweep and cyclic voltammograms
for electrode processes involving adsorption: (a) reactant
weakly adsorbed; (b) product
weakly adsorbed; (c) reactant
strongly adsorbed; (d) product
strongly adsorbed. Dashed lines
indicate behavior for uncompli-
cated Nernstian charge transfer.
[Reproduced from Anal. Chem. *39*,
1514 (1967). © American Chemical
Society.]

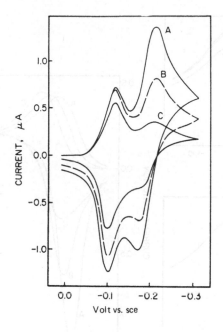

FIGURE 5.10 Cyclic voltammograms for methylene blue in buffered solution (pH 6.5; v = 44.5 mV/sec): A, 1.00×10^{-4} M; B, 0.70×10^{-4} M; C, 0.40×10^{-4} M. [Reproduced from Anal. Chem. *39*, 1527 (1967). © American Chemical Society.]

Figure 5.11 shows the various influences of adsorption. Three relatively easy tests can be applied to test for adsorption:

1. Adsorption-controlled waves are frequently symmetrical about i_p, unlike normal waves.

2. $i_p/C\ v^{1/2}$ generally increases rapidly with increasing scan rate; however, i_p/Cv may remain nearly constant.

3. While i_p/C is constant for many processes, in the presence of adsorption an increase of i_p/C with decreasing concentration is usually observed, sometimes leveling off to a constant value for low concentrations.

Based on discussion in Chaps. 3 and 4 for dc polarography, the majority of observations are intuitively obvious with respect to adsorption. The same applies to film formation and related inhibition phenomena. In many respects, however, adsorption is even more deleterious in LSV than in dc polarography, as evidenced by Figs. 5.9 to 5.11. Generally, the use of LSV methods in the presence of adsorption is fraught with danger and must be undertaken only on those occasions

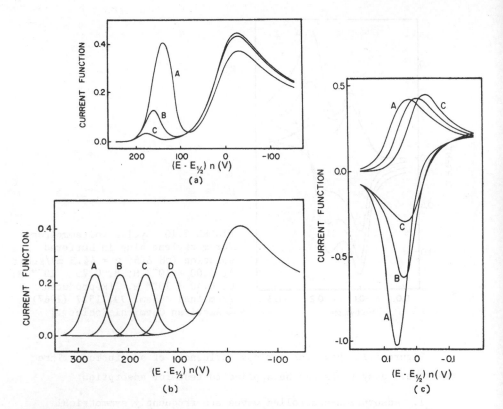

FIGURE 5.11 Some influences of adsorption in linear sweep voltammetry. (a) Effect of variation of concentration for a reaction in
which the product is adsorbed. Relative concentrations C:B:A are
1:4:16. (b) Variation of LSV curves with product strongly adsorbed
as a function of free energy of adsorption. Free energies of adsorption are 29.4, 24.8, 20.2, and 15.6 $kcal/mol^{-1}$, respectively,
in curves A to D. (c) Influence of scan rate on cyclic voltammograms
with product weakly adsorbed. Relative scan rates A:B:C are
$4 \times 10^4 : 2.5 \times 10^3 : 1$. [Reproduced from Anal. Chem. *39*, 1514 (1967).
© American Chemical Society.]

where it can be ascertained that the adsorption characteristics of
both unknown solutions and standards are identical.

5.4.6 Diagnostic Criteria for
   Various Electrode Mechanisms

Table 5.1 summarizes simple diagnostic criteria for recognizing some
of the common classes of electrode processes in linear sweep and
cyclic voltammetry.

TABLE 5.1 Dependence of i_p on Scan Rate and Concentration and of E_p on Scan Rate for Some Common Classes of Electrode Process in Linear Sweep and Cyclic Voltammetry (Reduction Assumed)

Mechanism	LSV			Cyclic voltammetry
	Dependence of i_p on v	Dependence of i_p on C	Dependence of E_p on v	Dependence of $(i_p)_c$ and $(i_p)_a$ on v
(I) Reversible $A + ne \rightleftarrows B$	$i_p \propto v^{1/2}$	Linear	E_p independent of v	$(i_p)_c$ and $(i_p)_a \propto v^{1/2}$; thus $(i_p)_a/(i_p)_c = 1$ for all v
(II) Irreversible $A + ne \rightarrow B$	$i_p \propto v^{1/2}$	Linear	E_p becomes more negative with increasing v	$(i_p)_a = 0$, $(i_p)_c \propto v^{1/2}$; thus $(i_p)_a/(i_p)_c = 0$ for all v
(III) EC $A + ne \rightleftarrows B \rightarrow C$	Approaches reversible case (I) with increasing v	Generally linear; i_p/C slightly greater than reversible case	E_p more positive than reversible case; E_p becomes more negative with increasing v	$(i_p)_c > (i_p)_a$ at low scan rates; $(i_p)_a \rightarrow (i_p)_c$ and $(i_p)_a/(i_p)_a \rightarrow 1$ with increasing v
(IV) CE $Y \rightarrow A + ne \rightleftarrows B$	Plot of $i_p/v^{1/2}$ versus v decreases with increasing v	Frequently nonlinear; $i_p/C <$ reversible case	E_p becomes more positive with increasing v	$(i_p)_a/(i_p)_c$ increases with increasing v

TABLE 5.1 (Continued)

| Mechanism | LSV | | | Cyclic voltammetry |
	Dependence of i_p on v	Dependence of i_p on C	Dependence of E_p on v	Dependence of $(i_p)_c$ and $(i_p)_a$ on v
(V) Regenerative $A + ne \rightleftharpoons B \rightarrow A$	Approaches reversible case (I) with increasing v	Frequently nonlinear; $i_p/C >$ reversible case	E_p becomes more positive with increasing v	$(i_p)_a/(i_p)_c \approx 1$ for all v
(VI) Reactant adsorbed $A \rightleftharpoons A(ads) + ne \rightleftharpoons B$	i_p increases with increasing v	Nonlinear	E_p becomes more negative with increasing v	$(i_p)_a/(i_p)_c \lesssim 1$; approaches 1 at low scan rates
(VII) Product adsorbed $A + ne \rightleftharpoons B \rightleftharpoons B(ads)$	i_p decreases slightly with increasing v	Nonlinear	E_p becomes more positive with increasing v	$(i_p)_a/(i_p)_c \gtrsim 1$; approaches 1 at low scan rates

5.5 LINEAR SWEEP VOLTAMMETRY
IN ANALYTICAL APPLICATIONS

Immediately after considering electrode processes and prior to discussing advances in the methodology, it is useful to consider several
practical applications of LSV. The simple electron transfer case
provides asymmetric peak-shaped curves whose peak current is proportional to concentration over wide concentration ranges. This class
of electrode process is eminently suitable for analytical work, and
most discussions in review articles are usually directed toward it.
However, the introduction of mechanistic complications is extremely
common and creates considerable difficulties, reminiscent of problems
of nonideality in dc polarography. Certainly it should not be assumed
that in LSV a peak-shaped curve will always be obtained, nor will
linear peak current-vs.-concentration plots necessarily be found, as
seen from Table 5.1. Since discussion of advances in LSV and almost
everything else in the following pages will be based essentially on
the reversible case and to ensure that a sense of reality is retained,
it is as well to provide examples of analytical applications utilizing
complex electrode processes.

In dc polarography, the reduction of tellurium is complex, and
in many supporting electrolytes, irregular, poorly shaped waves are
obtained [43-46]. In many media, the same observations are made
with LSV. The occurrence of nonlinear i_p-vs.-C plots in LSV is
therefore not unexpected. Figure 5.12 gives the calibration curves
for both forward and reverse sweeps for tellurium at a Hg electrode.
In 1.5 M phosphoric acid, the peak height of tellurium for the
forward scan is linear between about 0.05 and 0.7 ppm [47]. Above
this range, nonlinear peak currents are obtained, but by the use of
the reverse scan, the linear region may be extended to about 5 ppm.
The use of the reverse scan, rather than the more usual forward scan,
is sometimes advantageous in LSV, particularly for the more mechanistically complicated electrode processes.

Visually, linear sweep curves for tellurium have the asymmetric
peak-shaped appearance of an i-E curve expected with this voltammetric

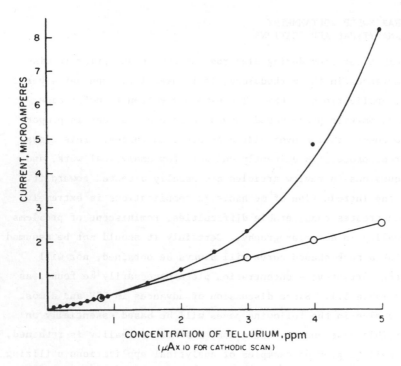

FIGURE 5.12 Comparison of the forward (cathodic) and reverse (anodic)
scans for the determination of tellurium by linear sweep voltammetry
in 1.5 M phosphoric acid. [Reproduced from Anal. Chem. *37*, 1516
(1965). © American Chemical Society.]

method. However, application of standard diagnostic criteria would
show that mechanistic complications are operative, and the specificity
of the analytical method may be questioned. Table 5.2 shows some
relevant data. From 0.5 to 0.1 mg of potentially interfering ions
was added to 20-μg portions of tellurium in 100 ml of 1.5 M phosphoric
acid. Positive and negative interferences can be obtained, vanadium
and chromium providing extremities.

The determination of boron may also be cited as an example of
the use of a complex electrode process [48]. Figure 5.13(a) shows
the linear sweep curve. Rather than a peak-shaped curve, a sigmoidal
i-E curve, reminiscent of examples in dc polarography, is found. A
plot of the limiting current versus square root of concentration
[Figure 5.13(b)] is linear on this occasion. Perusal of data in the

TABLE 5.2   Interferences in Tellurium
Determinations[a]

Ion added	Te Recovered %
Bismuth	101.0
Cadmium	98.8
Chromium	84.9
Manganese	98.8
Cobalt	97.6
Nickel	100.0
Vanadium	109.0
Copper	95.4
Lead	97.6
Tin	95.4

[a]As reported in Note 47.

original paper [48] reveals that the current is kinetically controlled and influenced by a considerable number of variables. The
determination of boron, in complex solutions, would therefore be
expected to present considerable difficulty.

Finally, the determination of iron via two methods is considered
[49]. Figure 5.14(a) shows the reverse sweep curve for the direct
determination of Fe(III) in citrate media. Fe(III) is reversibly
reduced to Fe(II), and a linear plot of i_p versus concentration is
found over virtually all concentrations. Figure 5.14(b) is a forward
sweep curve for the reduction of Fe(III) in the presence of the dye,
solochrome violet RS (SVRS). The reduction peak of this dye splits
up in the presence of Fe(III). The peak height of the wave as
a result of the reduction of the iron complex [49,50] is proportional
to iron concentration, for iron concentrations below 3×10^{-5} M.
Examination of the wave shape shows it to be almost symmetrical about
E_p and very different from the reversible one-electron reduction
[Figure 5.14(a)]. These observations are indicative of the use of
an adsorption-controlled process. Larger i_p-vs.-C values are

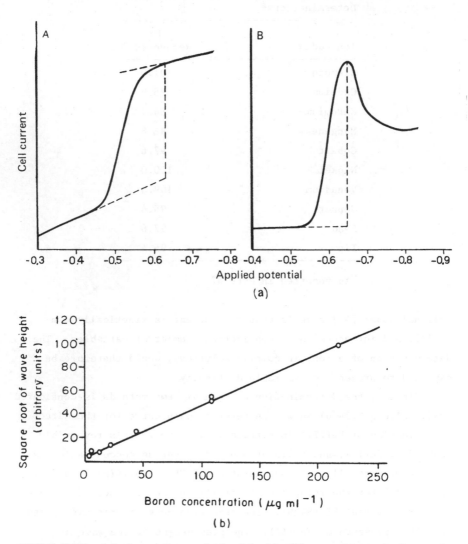

FIGURE 5.13   (a) Comparison of the kinetically complicated electrode
process for determining boron by LSV and the reversible cadmium
electrode process: A, boron 43.28 µg ml^{-1}) in 0.12 M sodium sulfite
and 0.2 M mannitol; B, 10^{-3} M cadmium in 0.1 M KCl.   (b) Plot of
square root of wave height vs. concentration is linear in the determination of boron by LSV.   [Reproduced from J. Polarogr. Soc. 7, 2
(1961).]

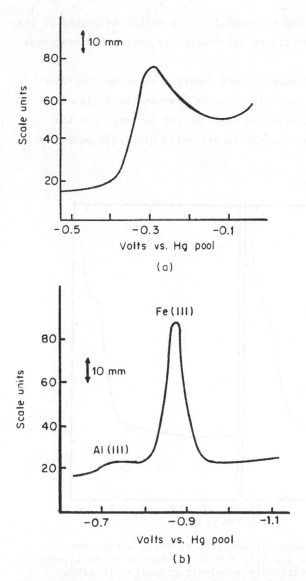

FIGURE 5.14 (a) Linear sweep voltammogram of 9×10^{-6} M Fe(III) in 0.5 M citrate buffer at pH 7 (reverse sweep). (b) Linear sweep voltammogram of 3×10^{-6} M Fe(III) in 0.003% SVRS/0.1 M HClO$_4$/0.2 M NaOAc at pH 5 (forward sweep). [Reproduced from Z. Anal. Chem. *269*, 349 (1974).]

obtained and therefore higher sensitivity, but this is gained at the expense of advantages mentioned previously for reversible electrode processes.

All the preceding examples were reported using mercury electrodes. Figures 5.15a and 5.15b provide examples at solid electrodes. The same considerations apply as for mercury, but the additional difficulties inherent in all solid electrode work must be taken into account.

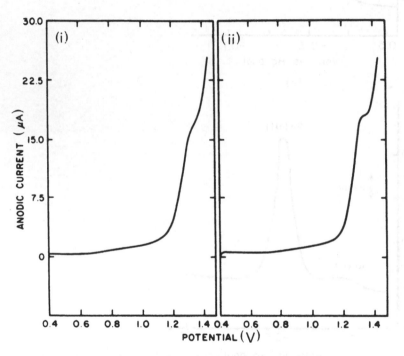

FIGURE 5.15a   Linear sweep voltammogram at solid electrodes: (i) 5×10^{-4} M caffeine and (ii) 5×10^{-4} M *theo*-bromine in acetate buffer (pH 4.7) at a stationary pyrolytic graphite electrode. Scan rate is 3.3 mV/sec; electrode area, 12.5 mm^2. [Reproduced from J. Electroanal. Chem. *30*, 407 (1971).

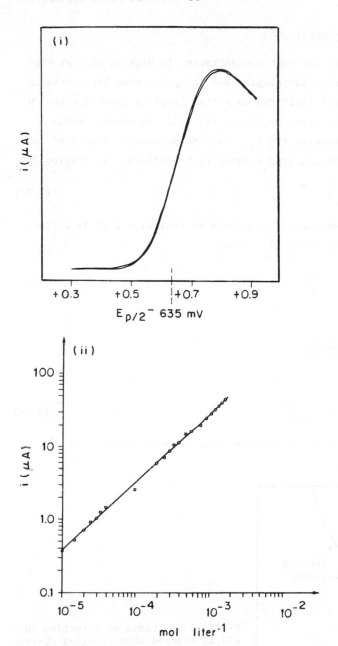

FIGURE 5.15b Oxidation of adrenaline in Britton–Robinson buffer
pH = 3.6.   Scan rate is 1.5 v/min at silicone rubber–based graphite
electrodes: (i) concentration of adrenaline = 9.09 × 10^{-4} M
(3 scans); (ii) calibration curve.   (Reproduced by courtesy of
Hungarian Scientific Instruments.)

5.6 THE CHARGING CURRENT

Although the current per unit concentration is high in LSV, at high
instrumental sensitivities sloping and irregular base lines arising
from charging current limit the detection level to about 1×10^{-7} M
for a reversible electrode process (Fig. 5.16) in aqueous media.

Using the expression $E = E_i - vt$, and procedures described in
Chap. 2, enables the charging current to be derived. As previously,

$$i_c = - \frac{d}{dt} (Aq')$$ (5.27)

where q' = charge density of electrode at potential E (E is defined
relative to E_m as in Chap. 2). Thus,

$$i_c = -q' \frac{dA}{dt} - A \frac{dq'}{dt}$$ (5.28)

or since

$$C_{dl} = \frac{dq'}{dE} = \frac{dq'}{dt} \frac{dt}{dE}$$

as in Chap. 2,

$$i_c = -q' \frac{dA}{dt} - A C_{dl} \frac{dE}{dt}$$ (5.29)

From Eq. (5.1) $dE/dt = -v$, and the final expression is

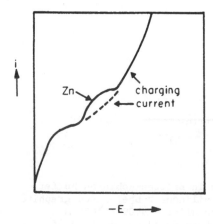

FIGURE 5.16 Limit of detection in
LSV is reached when charging current
masks faradaic current. Example
given is 3×10^{-7} M Zn in 0.25 M
NH_3/NH_4Cl.

$$i_c = -q' \frac{dA}{dt} + AC_{dl}v \tag{5.30}$$

At a solid electrode $dA/dt = 0$, while in LSV at a dme, provided the scan rate is fast and the i-E curve is recorded near the end of the drop life, $dA/dt \rightarrow 0$. With the latter approximation applicable to the dme case, one can write

$$i_c = AC_{dl}v \tag{5.31}$$

C_{dl} is frequently referred to as the *differential capacity*, and LSV curves for the charging current therefore have a similar shape to those for the dependence of differential capacity on potential obtained by other methods [51,52]. The adsorption or desorption of surface-active species can therefore give rise to large capacity currents in LSV. Figure 5.17 provides an example of the charging current for 0.5 M Na_2SO_4 in the presence of octyl alcohol [53]. The potential at which the substance adsorbs or desorbs is characterized by a sharp maximum resulting from the large change in differential capacity accompanying the rearrangement of the electrode double layer. It can be appreciated, therefore, that the magnitude of i_c in LSV can vary markedly [53]. The scan rate determines the magnitude of i_c as well as the differential capacity, and the latter is extremely dependent on the presence or absence of surface-active species.

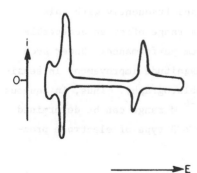

FIGURE 5.17 Charging current for 0.5 M Na_2SO_4 in the presence of octyl alcohol. Peaks indicate adsorption and desorption of the surface film. In the potential region between the peaks, octyl alcohol is adsorbed [53].

5.7 SENSITIVITY

The limit of detection in LSV, as in other polarographic methods,
is governed by the ratio of faradaic to charging current. A general
description covering all possibilities is prohibitive, since the
ratio depends on the nature of the electrode process. Assuming the
absence of adsorption phenomena influencing either the double layer
or faradaic processes, the results for the $A + ne \rightleftarrows B$ class of electrode process are as follows:

$$i_f = k' \, v^{1/2} \, C \tag{5.32a}$$

$$i_c = k'' \, v \tag{5.32b}$$

where k' and k'' are constants. Thus,

$$\frac{i_f}{i_c} = k''' \, C \, v^{-1/2} \tag{5.33}$$

where k''' is a constant, and the faster the scan rate or the lower
the concentration, the less favorable the faradaic-to-charging current
ratio. Typical scan rates used in analytical work are in the range
20 mV sec^{-1} to the volt-per-second range. The lower range of scan
rates enables the use of X-Y recorders as the readout device, introduces fewest problems from iR drop, and generally has the most favorable faradaic-to-charging current ratio. However, at 20 mV sec^{-1}
with a dme, dA/dt may not be considered zero, and scan rates of
between about 100 to 200 mV sec^{-1} are used frequently with this
important electrode. Scan rates in this range offer an acceptable
compromise for achieving close to optimum performance. Under such
conditions, approximately an order of magnitude improvement in sensitivity is gained over conventional dc polarography. Thus, in aqueous
media, concentrations in the 10^{-7} to 10^{-6} M range can be determined
by LSV at mercury electrodes for $A + ne \rightleftarrows B$ type of electrode processes, as indicated previously.

5.8 COMPARISONS WITH DC POLAROGRAPHY

Before discussing the advances in LSV, it is essential that the inherent advantages and disadvantages of this method relative to conventional dc polarography be understood. This approach is necessary to achieve a correct perspective on the relative merits of the two polarographic methods. In Chap. 4, it was seen that advances in dc polarography can bring the performance of dc polarography, in many instances, up to or beyond that of the LSV method as considered so far. Obviously, therefore, stating that the sensitivity of LSV is an order of magnitude better than in dc polarography is meaningless unless the conditions of the comparison are explicitly stated. Clearly, anomalies and/or inaccuracies in providing comparisons are easily introduced if one is not careful. For example, in the literature it is frequently stated that the accuracy and reproducibility of LSV at a dme are lower than that of dc polarography, the reason given being that the voltammogram produced in the readout is relatively small. In this kind of comparison, data from an X-Y recorder (polarography) are being compared usually with data obtained from an oscilloscope, and all that is being said is that the X-Y recorder used was a better readout device than the oscilloscope. Had the dc polarograms also been recorded on the same oscilloscope, or the LSV curves on the same X-Y recorder (assuming response time fast enough), then probably it would be found that the reproducibility of both methods is almost equivalent. In the author's laboratories, better than 1% reproducibility is achieved routinely with either method at a dme. Indeed, on an X-Y recorder, the reproducibility obtainable with both methods is frequently governed by the thickness of the recorder pen.

In routine analytical work, the real advantages of LSV lie in the increased sensitivity, the speed of recording of the i-E curve, the specificity obtainable via the use of a larger time scale available with respect to scan rate, and improved resolution of the peak-shaped curve. The last advantage must be qualified somewhat because

corrections for the presence of preceding electrode processes from
an asymmetric peak-shaped curve are difficult, frequently requiring
somewhat arbitrary extrapolation procedures. Figure 5.18 illustrates
this difficulty. In dc polarography, the limiting current of the
preceding electrode process is often almost independent of potential
and corrections are more easily undertaken. Similarly, the correction for charging current is frequently easier in dc polarography,
being closer to a linear function of potential over a small range of
potential, than in LSV:

$$(i_c)^{dc} \propto \left(\frac{\partial q'}{\partial t}\right)_E \qquad (i_c)^{lsv} \propto \left(\frac{\partial q'}{\partial E}\right)$$

As a general rule, however, the advantages of LSV are considerable, and this technique can be used in an extremely wide variety
of analytical applications, particularly when the advances considered
in Sec. 5.9 are incorporated into the methodology. The extensive use

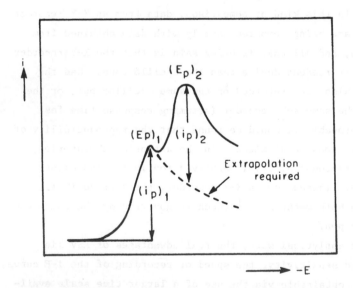

FIGURE 5.18 The determination of a species in the presence of
another more positively reduced species presents considerable
difficulty in LSV because the reference current arising from the
first process must be extrapolated.

of this polarographic method made by the U.S. National Bureau of
Standards (NBS) illustrates how favorably this method of polarography
can compete with spectroscopic and other widely used analytical
methods [54]. Matrices analyzed in the NBS laboratories by LSV
include lunar rock and soil from "Apollo" flights, botanicals, metal
organics, air particulates, effluents, and other environmental
samples. Endeavors to analyze the same samples by equivalent versions
of dc polarography would not have been competitive with existing
spectroscopic analytical methods.

5.9 ADVANCES IN VOLTAMMETRY

5.9.1 Staircase Voltammetry

As in dc polarography, many of the advances involve the use of techniques for discriminating against charging current. The use of
staircase voltage waveforms to reduce the charging current interference was suggested by Barker [55]. Figure 5.19 shows the form
of the applied potential in staircase voltammetry. Instead of using
a linear potential ramp, the potential is changed in a stepwise manner. Each step is of magnitude ΔE and follows the previous step by
τ sec. An applied voltage form of N steps produces an instantaneous
current with N discontinuities. A voltammogram consists of N current
samples, one during each of the N steps measured t_m sec after the
potential increment.

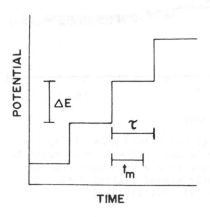

FIGURE 5.19 Waveform used in
staircase voltammetry. Symbols
are explained in the text.
[Reproduced from J. Electroanal.
Chem. *45*, 361 (1973).]

This technique achieves a result similar to that of current sampling in dc polarography, except that the additional complication of the dependence on drop time does not enter the argument. The technique differs from LSV in that the time scale of the experiment is no longer necessarily governed by scan rate, but by τ, within certain specified limits. Obviously, if τ were extremely long (equal to the drop time) and ΔE small, the experiment would simulate the conditions of dc polarography performed at a dme. However, if τ were infinitely small the ramp would revert to linear and the time scale would be governed by the rate of change of ΔE or the scan rate. In actual practice, the experiment is conducted under conditions where τ is small and so the shape of the i-E curves approaches that for LSV. Figure 5.20 provides examples of staircase voltammograms.

On application of the potential ΔE, large faradaic and charging currents initially flow through the cell. In any controlled-potential electrolysis experiment, the faradaic current decays toward

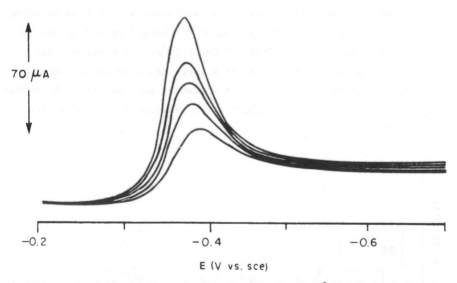

FIGURE 5.20 Staircase voltammograms of 2.7×10^{-3} M Pb(II) in 1 M HClO$_4$ with varied sampling time. τ = 50 msec; ΔE = 10 mV. Multiple exposure photograph of the analog output from a transient recorder after a D/A conversion of the internally stored digital data. [Reproduced from J. Electroanal. Chem. *45*, 361 (1973).]

zero as the electrolysis time proceeds. The capacitance current
also decays as a function of time, but at a faster rate than the
faradaic current. Hence, the larger τ is, the more favorable is
the faradaic-to-charging current ratio. The LSV experiment, corre-
sponding to $\Delta E \to 0$, $\tau \to 0$, $\Delta E/\tau$ = v, obviously has the least favorable charging-to-faradaic current ratio despite the fact that the
absolute magnitude of the faradaic current is larger. Figure 5.21
describes the above phenomena diagrammatically. The mode of discrimination against the charging current is common to many modern polarographic methods, as will be seen in subsequent chapters.

Despite the considerable advantage to be gained by using a
staircase voltage form, this form of voltammetry has been used by
few analytical chemists. Initial reservations based on the difficulties associated with more complex instrumentation should have
disappeared by now with the advent of digital electronics. Digital
signal sources produce staircase waveforms naturally, and digital
data gathering devices provide the kind of current sampling needed.
The ready availability of digital components and circuits has virtually erased the differences in complexity between staircase and
linear scan instrumentation. Indeed, when one remembers that in
the limit $\Delta E \to 0$, $\tau \to 0$, $\Delta E/\tau \simeq$ v, the two techniques are equivalent,
the differences themselves are tending to disappear. Future trends,
particularly those associated with the use of laboratory computers
(Chap. 10) may well see the digital staircase ramp becoming an

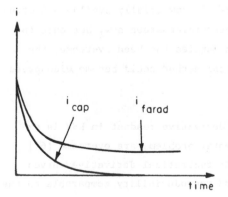

FIGURE 5.21 The charging current
decays faster than the faradaic
current as a function of time.
This provides the means for discriminating against charging
current in staircase voltammetry
and many other polarographic/
voltammetric methods.

integral part of polarographic instrumentation, and in this case
LSV may come to be regarded by future generations as a special case
of staircase voltammetry.

In light of the above arguments, it is the author's belief
that staircase voltammetry should be widely employed. However, practical analytical applications are virtually nonexistant. Mann [56]
and Nigmatullin and Vyaselev [57] have demonstrated the improved
sensitivity over the linear scan method resulting from the substantially decreased double-layer charging current problem, and have
shown the close conformity of results to the LSV theory under limiting situations. Christie and Lingane [58] have explained theoretically the differences between staircase and linear sweep voltammetry.
Finally, Ferrier and co-workers [59,60] have implemented a digital
version of the instrumentation and undertaken substantial theoretical
work, as have Zipper and Perone [61]. All work to date shows the
similarities to LSV in terms of response to electrode processes,
and little additional information with respect to the response to the
various classes of electrode processes needs to be given. Theoretical
investigations [59] show that as expected, the time scale is governed
by τ and electrode processes appear reversible when $k_s \tau^{1/2} > 10^{-2}$
cm sec$^{-1/2}$, quasi-reversible for $10^{-2} > k_s \tau^{1/2} < 10^{-4}$ cm sec$^{-1/2}$,
and irreversible for $k_s \tau^{1/2} < 10^{-4}$ cm sec$^{-1/2}$.

According to all expectations associated with the preceding
investigations, analytical chemists would do well to seriously consider using staircase voltammetry in preference to LSV. The present
paucity of analytical data and lack of commercially available instrumentation appear to be the rate-determining steps now, but once the
inertia problem generated by these hurdles has been overcome, the
application of this electroanalytical method could become widespread.

5.9.2 Derivative Voltammetry

As in dc polarography, the use of derivative readout in LSV is extremely advantageous once instrumental problems are overcome [62].
Modern electronic approaches enable theoretical derivative curves
to be obtained with an accuracy and reproducibility comparable to the

normal LSV curves [63]. However, at least an order of magnitude
increase in sensitivity can be gained, and the ability of the method
to resolve the reduction wave of a small amount of one species in
the presence of an excess of a more easily reducible species is significant [62,64].

The equations for derivative LSV curves are derived readily from
the theory of Nicholson and Shain [34]. Figure 5.22 shows a comparison of the normal and first-derivative LSV curves for reduction of
cadmium. The extremely sharp nature of the derivative plot suggests
that improved revolution is likely to be a feature of the method.
This figure also demonstrates the high reproducibility which is
retained in taking the derivative. Stephens and Harrar [65] have
examined the analytical use of the second derivative. Using a digital
readout system, a short-term reproducibility approaching 0.1% can
be achieved for concentrations above 10^{-4} M. Figure 5.23 shows the
considerably improved resolution offered by second-derivative methods.
In the context of resolution, the use of reverse sweep derivative
LSV is also useful because it is much easier to determine a small
peak in the face of a large wave than to see a small residual peak
in the negative part of an earlier wave. Comparison of Fig. 5.24(a)
and (b) illustrates this clearly. The determination of 5 ppm of

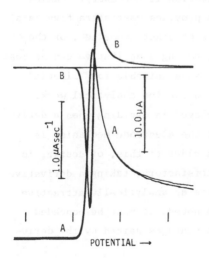

FIGURE 5.22 Normal and first derivative linear sweep curves for reduction of cadmium(II) at the hanging
mercury drop electrode: (A) normal
linear sweep curve; (B) first derivative linear sweep curve. Solution
is 2×10^{-3} M Cd(II) in 0.2 M KCl.
$v = 0.0327$ V sec^{-1}. Curves are
recordings of three replicate determinations superimposed. Distance
between vertical markings is 0.50 V.
First marking is at 0.00 V vs. sce.
[Reproduced from Anal. Chem. *37*, 2
(1965). © American Chemical
Society.]

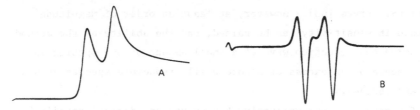

FIGURE 5.23   Normal (A) and second derivative (B) linear sweep
voltammograms of 10^{-3} M Pb(II) and 10^{-3} M Cd(II) in 2 M acetic
acid, 2 M ammonium acetate.   v = 1.0 V sec^{-1}.   [Reproduced from
Chem. Instr. 1, 169 (1968).]

cadmium in the presence of 10 ppm indium is being examined. A
cadmium peak is obtained which is so poorly resolved as to be im-
measurable if the forward sweep is used, but which is readily measured if the reverse scan is used so that the peak positions are
reversed [62].

Considerable effort has been expended on derivative LSV, with
respect to improving both sensitivity and resolution [66-69].
Figure 5.25 shows a computer-averaged second-derivative LSV curve
for 2×10^{-7} M Cd(II) in 1.0 M KCl. Reliable analytical data over
three to five orders of magnitude have been obtained with derivative
LSV [64], and the evidence on the systems chosen indicates considerable success. However, a word of caution is necessary. Taking
the derivative of a reversible system provides most attractive data;
but if the normal curve exhibits kinks or other anomalies, or the
electrode is kinetically complicated, taking a derivative can be most
deleterious, giving rise to complex shapes suitable for "kinetic"
art exhibitions but totally unusable in routine analytical work.
Considerable discretion should be employed in deciding when a derivative plot will prove advantageous if the electrode mechanism is
complex and not well understood. The rider to this, of course, is
that if the normal LSV method is unsatisfactory, taking a derivative
in its own right is unlikely to provide an analytically attractive
curve. The above caution having been noted, it must be conceded
that the demonstrated and potential advantages gained by the derivative LSV method are considerable.

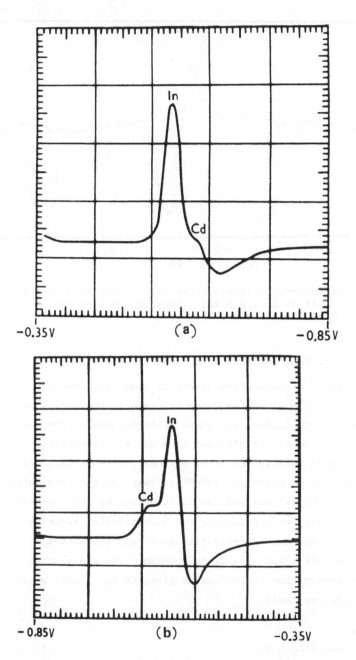

FIGURE 5.24 Derivative linear sweep voltammograms using (a) forward and (b) reverse sweep directions for the determination of indium and cadmium. [Reproduced from J. Polarogr. Soc. 9, 45 (1963).]

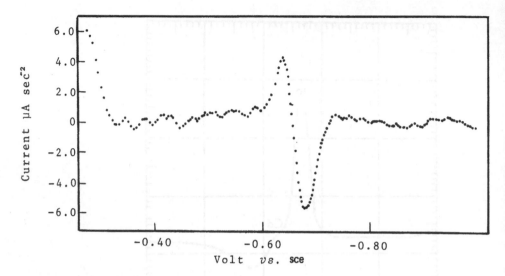

FIGURE 5.25 Computer-averaged second derivative linear sweep voltammogram of 2×10^{-7} M Cd(II) in 1.0 M KCl. [Reproduced from Anal. Chem.
44, 899 (1968). © American Chemical Society].

5.9.3 Convolution or Semi-Integral Techniques

The essential feature of convolution potential sweep voltammetry is
that quantities proportional to the concentrations of the diffusing
species at the electrode surface are immediately derivable from the
convoluted current which may be obtained from Laplace transforms,
from the semi-integral function or other mathematical function, and
recorded as a function of potential [70-85]. That is, the convolution
procedure, or semi-integral analysis as it is termed by some workers,
is a mathematical means for eliminating the mass transfer terms, or
$t^{-1/2}$ response, from the current-potential data, and the curves
revert to the sigmoidal shape of a dc polarogram. The convolution
function used by Saveant and co-workers is given in Eq. (5.34) where
χ is the integration variable:

$$I = \frac{1}{\pi^{1/2}} \int_0^t \frac{i(\chi)}{(t-\chi)^{1/2}} \, \partial\chi \qquad (5.34)$$

Figure 5.26 provides an example of a convoluted linear sweep
voltammogram and Fig. 5.27 a calculated reversible cyclic voltammogram
with the semi-integral plotted as a function of potential. The
equivalence to dc polarograms in terms of shape is clearly seen. The
present disadvantage of having two names (convolution or semi-integral
voltammetry) for techniques founded on the same principle is almost
inevitable when different research groups work independently on the
same new problem, and only time will resolve this undesirable nomenclature difficulty. However, as Nicholson points out [86], whether
the solution is obtained analytically by means of Laplace transform,
semi-integral transforms, finite difference simulations, numerical
integration, or other methods, the "solution" process is still one of
convolution; so the general term *convolution voltammetry* seems to be
more appropriate.

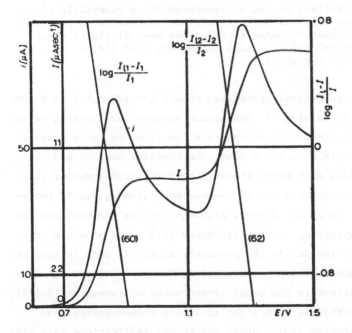

FIGURE 5.26 Convoluted and linear sweep voltammograms for
the two waves of *meta*-dinitrobenzene in dmf. Logarithmic
analysis using convoluted current I demonstrates the analogy
with a dc polarogram. [Reproduced from J. Electroanal. Chem.
44, 169 (1973).]

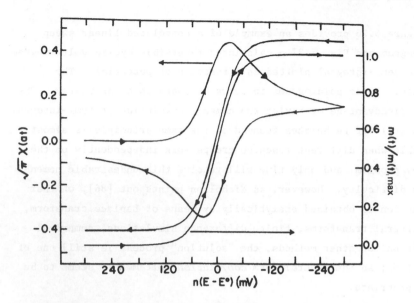

FIGURE 5.27 Theoretical cyclic voltammogram for a reversible electrode reaction and the semi-integral m(t) as a function of potential.
The return of the semi-integral function has been displaced upward
for clarity. [Reproduced from Anal. Chem. 45, 1298 (1973).
© American Chemical Society.]

    Since the mathematical procedures reduce the i-E curve to a form
equivalent to a dc polarogram, the author assumes the charging current
affects the semi-integral or convolution function, as is the case in
dc polarography. In dc polarography, the charging current and, indeed, the sigmoidal wave shape itself, limit the usefulness of the
method, and the analytical use of convoluted voltammograms is therefore also likely to be restricted, although "tricks" to minimize the
effect are now appearing in the literature [82]. One possible analytical advantage implied in the presently available data is that the
limiting magnitude of the semi-integral is less sensitive to iR drop
than the peak current in the usual linear sweep voltammogram [76,77].
The use of the method as a detector in liquid chromatography has
also been demonstrated [83]. Other analytical implications have also
been considered [82,83]. However, disadvantages of the relatively
large charging current contribution and a sigmoidal-shaped curve are

likely to be restrictive in most instances, and the predominant use of the method would appear to be confined to the area of electrode kinetics.

The semi-integral calculation approach is, of course, completely general and is not restricted to LSV methods. However, since most information is available with respect to this technique, it is discussed under this section.

5.9.4 Deconvolution or Semidifferential Techniques

Convolution or semi-integral voltammetry is the convolution of the observed current to give the sigmoidal-shaped response characteristic of the dc polarogram. An analytically more attractive approach would surely be to deconvolute the observed current to give a curve resembling a derivative of the dc polarogram. However, despite the obviously superior approach of deconvolution (semiderivative or semidifferential) voltammetry as compared with convolution (semi-integral) voltammetry, this mathematical approach has received relatively little attention.

Oldam and Spanier [87] suggested the use of semidifferentiation and Goto and Ishii [88] performed some experiments with this technique. Figure 5.28 shows a comparison of the curves obtained for a 10^{-5} M solution of Cd(II) in 0.11 M KCl using cyclic (linear sweep), semi-integral (convolution), and semidifferential (deconvolution) voltammetry. There is no doubt that, as expected, the sensitivity of detection and resolution in semidifferential voltammetry is better than in either cyclic voltammetry or semi-integral analysis.

Recently, Surprenant et al. [89] and Smith [90] have suggested that the fast Fourier transform technique with a digital computer is ideally suited for deconvolution of staircase voltammograms. Figure 5.29 shows the Fourier transform deconvolution of a steady-state staircase voltammogram. Since the Fourier transform and semidifferentiation can perform exactly the same task of deconvolution, the general term *deconvolution voltammetry* seems more appropriate than semidifferential voltammetry, as does *convolution voltammetry* in preference to semi-integral voltammetry.

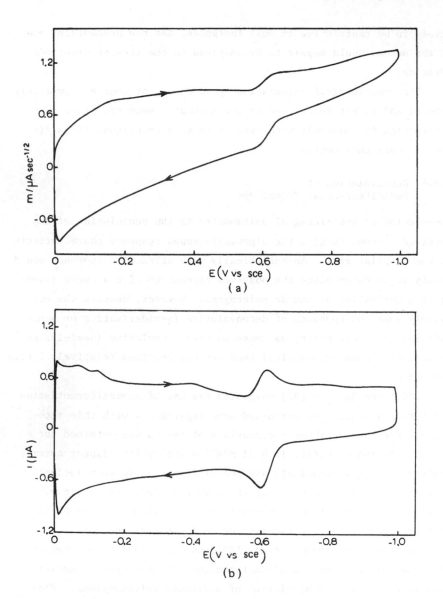

FIGURE 5.28   (a) i-E curve in cyclic (linear sweep) voltammetry at a hanging mercury drop electrode for 1.0×10^{-5} M Cd(II) in 0.11 M KCl.   Electrode area = 0.047 cm^2; scan rate = 200 mV sec^{-1}.
(b) Semi-integral of current versus potential in cyclic (linear sweep) voltammetry.   Conditions as in (a).

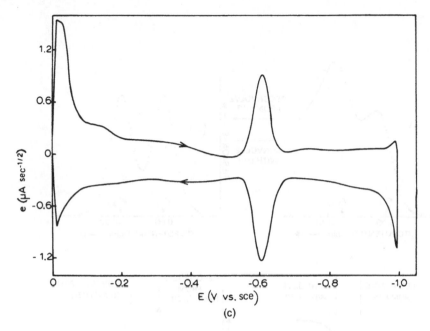

FIGURE 5.28, continued (c) Semidifferential of current versus potential in cyclic (linear sweep) voltammetry. Conditions as in (a).
[Reproduced from J. Electroanal. Chem. *61*, 361 (1975).]

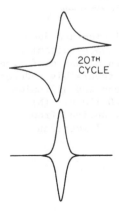

FIGURE 5.29 Fourier transform deconvolution of a
steady-state staircase voltammogram. [Reproduced
from J. Electroanal. Chem. *75*, 125 (1977).]

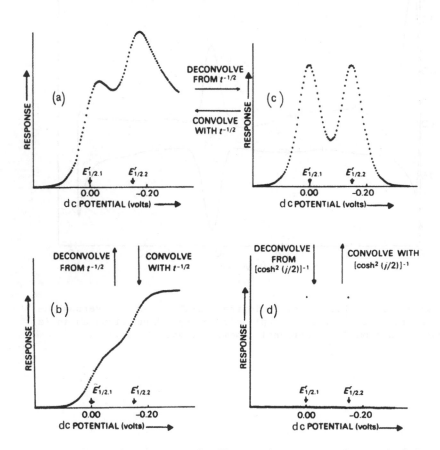

FIGURE 5.30 Illustration of effects of some convolution and deconvolution operations on stationary electrode linear sweep voltammetric response of two-component reversible system with planar diffusion.
(a) represents original voltammogram, assuming two reversibly reduced
components characterized by equal bulk concentrations and diffusion
coefficients with $(E_{1/2}^r)_1 = 0.000$ V and $(E_{1/2}^r)_2 = -0.150$ V. (b),
(c), and (d) are convolution or deconvoluted curves derived from
(a). [Reproduced from Anal. Chem. *48*, 517A (1976). © American
Chemical Society.]

Deconvolution procedures actually suggest a range of tricks which may improve voltammetric measurements. Ensuing discussion on this aspect is taken from Note 90. The reversible or diffusion-controlled LSV curve may be considered to be distorted by a broadening function which has the $t^{-1/2}$ form with planar diffusion. Deconvolution of $t^{-1/2}$ from a reversible LSV curve produces the dc derivative or $1/\cosh^2 (j/2)$ function, where $j = (nF/RT) (E - E_{1/2}^r)$. The effect of this operation is shown in Fig. 5.30. The advantage relative to the inverse operation of convolution, as far as peak resolution is concerned, is also evident in Fig. 5.30. Pursuing to its extremity the argument of what can be theoretically achieved, the deconvolution voltammogram could then in turn be deconvolved with the $1/\cosh^2 (j/2)$ function to obtain two impulse functions located at the $E_{1/2}^r$ values of each electrode process, with a concentration-dependent magnitude as shown in Fig. 5.30(d). Realistically, of course, noise and computer round-off and quantization errors, etc., would prohibit such a procedure for obtaining the complete ideal in terms of resolution, but one can envisage deconvolution steps which provide a compromise between the ideal of Fig. 5.30(d) and the function in Fig. 5.39(c). The crucial experiments in this area are still to be done. However, the effect of measurement delay time in staircase voltammetry at planar electrodes, for example, has in fact been shown to be removable by deconvolution voltammetry [89], and if pursued with vigor in the future, it is obvious that data enhancement concepts using deconvolution techniques could well achieve substantial success. Certainly the author can see a much brighter future in the analytical context for this approach than for convolution voltammetry.

5.9.5 Subtractive Voltammetry

As in dc polarography, the use of a dual-cell technique to subtract the charging current, solvent impurities, etc., has also been employed in LSV [91,92] at a dme. Close examination of many of the references given in this chapter reveals that dual-cell subtractive or

differential techniques have in fact been used on many occasions,
although this has not been stated explicitly in earlier sections.
For example, optimum performance with derivative LSV methods [63,64]
at a dme required the use of a subtractive mode of operation, and
the NBS work [47,54,93] referred to in Sec. 5.8 frequently employed
this variation of LSV. While it is probably easier to use a dual-
cell method in LSV than in dc polarography, the cautions given in
Chap. 4 with respect to the latter technique still apply. Florence
[94] reports that synchronization of twin electrodes in LSV at a
dme is a tedious affair and requires much patience and considerable
experience to achieve good results. He states that in his laboratories, most analyses are performed using the single electrode mode.
Likewise, the present author would contemplate the use of only dual-
cell LSV if determinations had to be undertaken at a low level that
precluded the possibility of an alternative approach. Figure 5.31
provides an example for which a subtractive experiment is essential.
(The determination of tin in beer was naturally carefully selected
by the author as the example to illustrate the practical importance
of subtractive LSV in industrial laboratories!)

The use of subtractive twin electrode techniques at solid electrodes, while they would seem to be simple in principle, is in practice exceedingly difficult. The problems associated with constructing

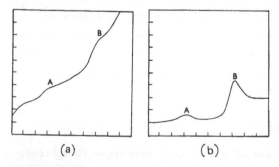

(a)   (b)

FIGURE 5.31   (a) Linear sweep voltammogram at a dme containing
(A) 0.031 µg of tin per milliliter; (B) 0.18 µg of cadmium per
milliliter.   (b) Subtractive linear sweep voltammogram of (a).
[Reproduced from Analyst *88*, 959 (1963).]

two identical solid electrodes are bad enough, but the difficulty in maintaining them in an identical state over long periods of time virtually excludes the possibility of twin solid electrode work.

As stated in Sec. 4.6 on subtractive dc polarography (Chap. 4), the use of a computerized experiment in which the background is recorded and stored in memory and subsequently subtracted from the test solution data, is by far the most successful approach in subtractive voltammetry, because all data are obtained from the same electrode. There is no doubt that if one has access to computerized instrumentation, then this is the way to perform subtractive voltammetry.

5.9.6 Interrupted Voltage Ramps

With mixtures of electroactive species, the problem of distortion or concealment of one peak current by the current peak of another species is shown in Sec. 5.8 to be a serious problem in LSV. However, for reduction, the current at potentials more negative than the peak decays as a function of time, so that if on passing the first peak the ramp is held at a constant potential for a sufficiently long time before being continued at the normal rate (i.e., one uses an interrupted voltage ramp), considerable improvement will occur in definition of more negative neighboring peaks. Figure 5.32 shows the effect of having an interrupted potential sweep rather than the more usual linear one. Several signal generators that offer a multitude of ramp rates and hold times achieved by controlled switching have been reported recently [95-99]. Improved versatility of the potential-time ramp function seems to be a characteristic of modern instruments capable of performing voltammetric methods. This is obviously advantageous, since a linear relationship does not always provide optimum performance (see also Sec. 5.9.1 on staircase voltammetry).

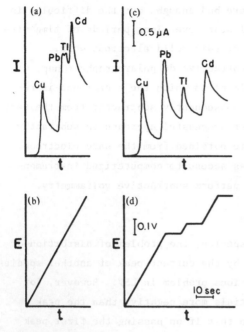

FIGURE 5.32 Normal and interrupted voltammetry of 5×10^{-5} M
Cu(II), Pb(II), Tl(I), and Cd(II) at a hanging mercury drop electrode
in 0.1 M KNO$_3$: (a) normal linear sweep voltammogram; (b) voltage
ramp (40 mV sec^{-1}; (c) interrupted linear sweep voltammogram;
(d) interrupted voltage ramp. [Reproduced from Anal. Chem. *45*, 437
(1973). © American Chemical Society.]

5.10 VOLTAMMETRY WITH FORCED CONVECTION

Voltammetric methods considered so far are applicable to stationary
electrodes in unstirred solutions, where the time scale of the experiment is governed by the scan rate in most instances. If the
electrode is rotated, or alternatively, the solution stirred, then
the mass transfer process occurs by forced convection rather than
solely by diffusion, and current-potential curves recorded under
conditions of convective mass transport are relatively insensitive
to scan rates. Electrodes operating under such conditions include
rotating disk and wire electrodes, streaming mercury electrodes, and
conical and tubular solid electrodes. They are sometimes called
hydrodynamic electrodes, and the measurement of i-E curves is then

referred to as *hydrodynamic voltammetry*. Such methods are of interest in the continuous analysis of flowing streams and in electrolytic synthesis in flowing cells.

Construction of hydrodynamic electrodes is usually more difficult than stationary ones, and obtaining reproducible mass transfer conditions in, say, a flowing stream is not a simple problem. Quantitative solutions applicable to forced convection conditions are difficult to obtain. Intuitively, it can be seen that the flux equation should involve the sum of terms due to diffusion and terms due to convection (migration effects are ignored as before by having an excess of supporting electrolyte). Under conditions where the convective terms are dominant, the limiting current would be expected to be governed by the rate of rotation of the electrode or flow rate of stream and to be relatively insensitive to scan rate. Detailed theoretical treatments of hydrodynamic voltammetry are available in the literature [1,100-116]. Sigmoidal-shaped curves as in dc polarography are usually obtained under conditions of forced convection.

5.10.1 Stationary Electrodes in Flowing Solutions

Examples of results for stationary electrodes in a flowing solution are given below.

Plate Electrodes. According to Levich, the limiting current i_L for a plate electrode in a laminar flow of liquid is given by

$$i_L = 0.68nFCD^{2/3}b1^{1/2}U^{1/2}\nu^{-1/6} \tag{5.35}$$

where

  b = width normal to the flow direction

  1 = length of the plate in the direction of liquid flow

  U = flow velocity

  ν = kinematic viscosity

Other symbols are as used throughout this book.

Conical Electrodes.

$$i_L = 0.77nFACD^{2/3}U^{1/2}v^{-1/6}l^{-1/2} \tag{5.36}$$

where 1 on this occasion is the slant height of the cone and other
symbols are as before.

5.10.2 Rotated Disk Electrodes

For rotated disk electrodes and for a reversible electrode process,
Levich derived the following equation for the limiting current:

$$i_L = 0.62nFACD^{2/3}\omega^{1/2}v^{-1/6} \tag{5.37}$$

where ω is the angular velocity of the disk given by $\omega = 2\pi N$ with
N revolutions per second.

The above examples are given to illustrate the analytically
important result that i_L is a linear function of concentration.
Laminar flow conditions may not be correct under some circumstances,
and turbulent flow or hybrid models may need to be invoked. However,
a linear dependence of i_L on concentration is the prediction of all
equations for a reversible electrode process.

5.10.3 Rotating Ring–Disk Electrode

Reversal techniques equivalent to cyclic voltammetry are obviously
not available at the rotating disk electrode, etc., since the product
of the electrode reaction is continuously swept away from the surface.
The equivalent of cyclic voltammetry is achieved by adding an inde-
pendent ring electrode outside the disk. For the reaction $A \underset{\leftarrow}{\overset{ne}{\rightarrow}} B$,
product B formed at the disk is swept over to the ring where it is
collected and re-oxidized back to A. Therefore, the rotating ring
disk has three zones: (1) the disk, (2) insulating space, and
(3) a ring. This technique has been used sparingly in analytical
voltammetry, and the literature can be consulted for further details
[1,106-115].

Figure 5.33 shows the i-E curves for the reversible one-electron
reduction $U(VI) \underset{\leftarrow}{\overset{e}{\rightarrow}} U(V)$ in KCl at a rotating gold disk electrode with

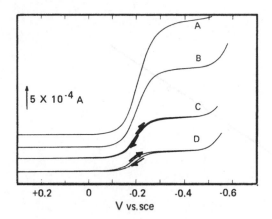

FIGURE 5.33 Voltammetric curves at a rotating gold disk electrode
for 9.9 × 10⁻⁴ M UO_2Cl_2 in 0.02 M HCl/1 M KCl. Rotation rate:
(a) 8100, (b) 3600, (c) 900, (d) 225 rpm. [Reproduced from J.
Electroanal. Chem. *31*, 119 (1971).]

forward and reverse scans almost without hysterisis (cf. cyclic
voltammetry). The dependence of i_L on scan rate is clearly seen.
Pungor et al. [117] have investigated the use of silicone rubber-
based graphite electrodes for continuous flow measurements. Figures 5.34(a) and (b) summarize some of the results of this work, and
good agreement with Levich's theoretical results was obtained.
This work, together with the work of Davenport and Johnson [118]
and references cited therein, demonstrates the considerable scope
for continuous monitoring measurements by hydrodynamic voltammetry.

There seems to be a considerable number of important analytical
problems to which hydrodynamic voltammetry could be applied, particularly the problem of the continuous monitoring of flowing solutions [117-120]. However, actual practical examples are still extremely limited. Whether this is an inherent problem of solid electrodes in which dc polarographic sigmoidal-type i-E curves result
or simply a lack of endeavor or awareness on the part of workers in
the field remains to be seen. The author would like to believe it
is the latter, and that if the same endeavor were undertaken in this
field as with other popular techniques of voltammetry, considerable
success would be achieved.

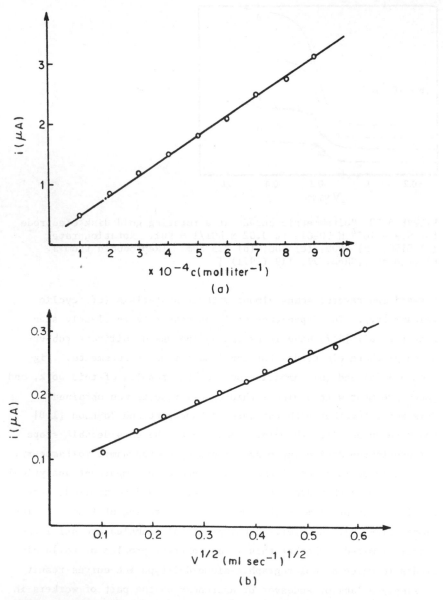

FIGURE 5.34 (a) Voltammetric calibration curve for Propylon in flowing media. Current measured at constant potential in limiting current region at a flow rate of 1.0 ml sec^{-1}. Silicon rubber-based graphite electrode used. (b) Dependence of voltammetric current on the square root of the flow rate. Concentration of Propylon = 10^{-4} M. Same electrode as in (a). [Reproduced from Anal. Chim. Acta *51*, 417 (1970).]

NOTES

1. R. N. Adams, *Electrochemistry at Solid Electrodes*, Dekker, New York, 1969.

2. G. C. Barker and I. L. Jenkins, Analyst 77, 685, (1952).

3. C. Auerbach, H. L. Finston, G. Kissel, and J. Glickstein, Anal. Chem. 33, 1480 (1960).

4. R. C. Propst and M. H. Goosey, Anal. Chem. 36, 2382 (1964).

5. J. W. Hayes, D. E. Leydon, and C. N. Reilley, Anal. Chem. 37, 1444 (1965).

6. B. Nygard, E. Johansson, and J. Olofsson, J. Electroanal. Chem. 12, 564 (1966).

7. G. Papeschi, M. Costa, and S. Bordi, Electrochim. Acta 15, 2015 (1970).

8. G. Willems and R. Neeb, J. Electroanal. Chem. 21, 69 (1969).

9. Ch. Yarnitzky, J. Electroanal. Chem. 51, 207 (1974).

10. P. Courbusier and L. Gierst, Anal. Chim. Acta 15, 254 (1956).

11. J. Riha, Advan. Polarogr. 1, 210 (1960).

12. H. P. Raaen and H. C. Jones, Anal. Chem. 34, 1594 (1962).

13. E. Verdier, R. Graud, and P. Varel, J. Chim. Phys. 66, 376 (1969).

14. B. K. Hahn and C. G. Enke, Anal. Chem. 46, 802 (1974).

15. V. Cermak, Collect. Czech. Chem. Commun. 19, 39 (1954).

16. J. E. B. Randles and W. White, Z. Elektochem. 59, 669 (1955).

17. W. Kemula and Z. Kublik, Anal. Chim Acta 18, 104 (1958).

18. S. Roffia and E. Vianello, J. Electroanal. Chem. 23, App. 9 (1969).

19. G. Piccardi and R. Guidelli, Anal. Lett. 1, 771 (1968).

20. C. A. Streuli and W. D. Cooke, Anal. Chem. 25, 1961 (1953).

21. V. S. Griffiths and W. J. Parker, Anal. Chim. Acta 14, 194 (1956).

22. R. N. Adams, Progr. Polarogr. 2, 503 (1962).

23. J. Heyrovský and J. Kůta, *Principles of Polarography*, Academic Press, New York, 1966, Chap. 2.

24. J. F. Alder and B. Fleet, J. Electroanal. Chem. 30, 427 (1971).

25. R. S. Nicholson and I. Shain, Anal. Chem. 36, 706 (1964).

26. S. W. Feldberg, in *Electroanalytical Chemistry* (A. J. Bard, ed.), Dekker, New York, vol. 3, pp. 199-296.

27. H. Matsuda and Y. Ayabe, Z. Electrochem. *59*, 494 (1955).

28. R. S. Nicholson, Anal. Chem. *37*, 1351 (1965).

29. J. C. Imbeaux and J. M. Saveant, J. Electroanal. Chem. *28*, 325 (1970).

30. W. H. Reinmuth, Anal. Chem. *32*, 1891 (1960).

31. J. M. Saveant and E. Vianello, Advan. Polarogr. *1*, 367 (1960).

32. J. M. Saveant and E. Vianello, Compt. Rend. *256*, 2597 (1963).

33. J. M. Saveant and E. Vianello, Anal. Chim. Acta *8*, 905 (1963).

34. R. S. Nicholson and I. Shain, Anal. Chem. *37*, 178 (1965).

35. L. Nadjo and J. M. Saveant, J. Electroanal. Chem. *44*, 327 (1973).

36. L. Nadjo and J. M. Saveant, J. Electroanal. Chem. *48*, 113, 146 (1973).

37. R. A. Osteryoung, G. Lauer and F. C. Anson, J. Electrochem. Soc. *110*, 926 (1963).

38. C. A. Streuli and W. D. Cooke, Anal. Chem. *26*, 963 (1954).

39. R. H. Wopschall and I. Shain, Anal. Chem. *39*, 1514 (1967).

40. R. H. Wopschall and I. Shain, Anal. Chem. *39*, 1527 (1967).

41. A. M. Miri and P. Favero, Ric. Sci. *28*, 2307 (1958).

42. W. Kemula, Z. Kublick, and A. Axt, Rocz. Chem. *35*, 1009 (1961).

43. J. J. Lingane and L. W. Niedrach, J. Amer. Chem. Soc. *71*, 196 (1949).

44. R. A. Jamieson and S. P. Perone, J. Electroanal. Chem. *23*, 441 (1969).

45. M. Shinagawa, N. Yano, and T. Kurosu, Talanta *19*, 439 (1972).

46. H. Schmidt and M. von Stackelberg, J. Polarogr. Soc. *8*, 49 (1962).

47. E. J. Maienthal and J. K. Taylor, Anal. Chem. *37*, 1516 (1965).

48. G. F. Reynolds and E. A. Terry, J. Polarogr. Soc. 7, 2 (1961).

49. R. Dewolfs and F. Verbeek, Z. Anal. Chem. *269*, 349 (1974).

50. J. W. Latimer, Talanta *15*, 1 (1968).

51. D. C. Graham, J. Amer. Chem. Soc. *63*, 1207 (1941); *68*, 301 (1946); *74*, 4422 (1953); *76*, 4819 (1959); Chem. Rev. *41*, 447 (1947); Chem. Phys. *18*, 903 (1950); Z. Electrochem. *59*, 740 (1950); J. Electrochem. Soc. *99*, 370 (1952); J. Phys. Chem. *61*, 701 (1957); Anal. Chem. *30*, 1736 (1958), and references cited therein.

52. A. N. Frumkin, V. S. Bagotsku, B. N. Kabanov, and Z. A. Ioffa, *Kinetica Electrod nykh*, Protsesov, Izd. Mosk. Univ., Moscow, 1952, and references cited therein.

53. J. W. Loveland and P. J. Elving, J. Phys. Chem. *56*, 250, 255, 935, 941, 945 (1952).

54. E. J. Maienthal, Amer. Lab. 25, June 1973, and references cited therein.

55. G. C. Barker, Advan. Polarogr. *1*, 144 (1960).

56. C. K. Mann, Anal. Chem. *33*, 1484 (1961); *35*, 326 (1965); *36*, 2424 (1966).

57. R. S. Nigmatullin and M. R. Vyaselev, Zh. Anal. Khim. *19*, 545 (1964).

58. J. H. Christie and P. J. Lingane, J. Electroanal. Chem. *10*, 176 (1965).

59. D. R. Ferrier and R. R. Schroeder, J. Electroanal. Chem. *45*, 343 (1973).

60. D. R. Ferrier, D. H. Chidester, and R. R. Schroeder, J. Electroanal. Chem. *45*, 361 (1973).

61. J. J. Zipper and S. P. Perone, Anal. Chem. *45*, 452 (1973).

62. R. C. Rooney, J. Polarogr. Soc. *9*, 45 (1963).

63. S. P. Perone and T. R. Mueller, Anal. Chem. *37*, 2 (1965).

64. L. Ya Shekun, Russ. J. Phys. Chem. *36*, 239 (1962).

65. F. B. Stephens and J. E. Harrar, Chem. Instr. *1*, 169 (1968).

66. S. P. Perone, J. E. Harrar, F. B. Stevens, and R. E. Anderson, Anal. Chem. *40*, 899 (1968).

67. T. R. Mueller, Chem. Instr. *1*, 113 (1968).

68. W. F. Gutnecht and S. P. Perone, Anal. Chem. *42*, 906 (1970).

69. L. B. Sybrandt and S. P. Perone, Anal. Chem. *43*, 383 (1971).

70. C. P. Andrieux, L. Nadjo, and J. M. Saveant, J. Electroanal. Chem. *26*, 147 (1970).

71. J. C. Imbeaux and J. M. Saveant, J. Electroanal. Chem. *44*, 169 (1973).

72. K. B. Oldham, Anal. Chem. *44*, 196 (1972).

73. M. P. Grenness and K. B. Oldham, Anal. Chem. *44*, 1121 (1972).

74. K. B. Oldham, Anal. Chem. *45*, 39 (1973).

75. L. Nadjo, J. M. Saveant, and D. Tessier, J. Electroanal. Chem. *52*, 403 (1974).

76. P. E. Whitson, H. W. Van den Born, and D. H. Evans, Anal. Chem. *45*, 1298 (1973).

77. H. W. Van den Born and D. H. Evans, Anal. Chem. *46*, 643 (1974).

78. M. Goto and K. B. Oldham, Anal. Chem. *45*, 2043 (1973).

79. M. Goto and K. B. Oldham, Anal. Chem. *46*, 1522 (1974).

80. J. M. Saveant and D. Tessier, J. Electroanal. Chem. *65*, 57 (1975).

81. J. M. Saveant and D. Tessier, J. Electroanal. Chem. *61*, 251 (1975).

82. S. C. Lamey, R. D. Grypa and J. T. Maloy, Anal. Chem. *47*, 610 (1975).

83. G. H. Brilmyer, S. C. Lamey and J. T. Maloy, Anal. Chem. *47*, 2304 (1975).

84. K. B. Oldham, J. Electroanal. Chem. *72*, 371 (1976).

85. M. Goto and K. B. Oldham, Anal. Chem. *48*, 1671 (1976).

86. R. S. Nicholson, Anal. Chem. *44*, 478R (1972).

87. K. B. Oldham and J. Spanier, J. Electroanal. Chem. *26*, 331 (1970).

88. M. Goto and K. Ishii, J. Electroanal. Chem. *61*, 361 (1975).

89. H. L. Surprenant, T. H. Ridgway and C. N. Reilley, J. Electroanal. Chem. *75*, 125 (1977).

90. D. E. Smith, Anal. Chem. *48*, 517A (1976).

91. H. M. Davis and J. E. Seaborn, Advan. Polarogr. *1*, 239 (1960).

92. H. M. Davis and H. I. Shalgosky, Advan. Polarogr. *2*, 640 (1960).

93. E. J. Maienthal, Anal. Chem. *45*, 644 (1973).

94. T. M. Florence, Proc. Roy. Australian Chem. Inst. *39*, 211 (1972).

95. J. L. Huntington and D. G. Davis, Chem. Instr. *2*, 83 (1969).

96. R. L. Meyers and I. Shain, Chem. Instr. *2*, 203 (1969).

97. J. S. Springer, Anal. Chem. *42*(8), 22A (1970).

98. R. H. Bull and G. C. Bull, Anal. Chem. *43*, 1342 (1971); C. Li, D. Ferrier and R. R. Schroeder, Chem. Instr. *3*, 333 (1972).

99. G. I. Connor, G. H. Boehme, C. J. Johnson, and K. H. Pool, Anal. Chem. *45*, 437 (1973).

100. J. Newman, in *Advances in Electrochemistry and Electrochemical Engineering* (P. Delahay and C. W. Tobias, eds.), Interscience, New York, vol. 5, 1967, pp. 87-135, and references cited therein.

101. V. G. Levich, *Physicochemical Hydrodynamics*, Prentice-Hall, Englewood Cliffs, N.J., 1962.

102. J. N. Agar, Disc. Faraday Soc. *1*, 26 (1947).

103. B. Bird, W. E. Stewart, and E. N. Lightfoot, *Transport Phenomena*, Wiley, New York, 1960, and references cited therein.

104. H. Matsuda, J. Electroanal. Chem. *15*, 109 (1967); *16*, 153 (1968).

105. H. Matsuda and Y. Yamada, J. Electroanal. Chem. *30*, 261, 271 (1971); *44*, 189 (1973).

106. W. J. Albery and S. Bruckenstein, Trans. Faraday Soc. *62*, 1920 (1966).

107. Yu. B. Ivanov and V. G. Levich, Dokl. Akad. Nauk SSSR, *126*, 1029 (1959).

108. A. N. Frumkin and L. N. Nekrasov, Dok.. Akad. Nauk USSR, *126*, 115 (1959).

109. A. N. Frumkin, L. N. Nekrasov, B. Levich, and Yu. B. Ivanov, J. Electroanal. Chem. *1*, 84 (1959/60).

110. V. A. Vicente and S. Bruckenstein, Anal. Chem. *44*, 297 (1972).

111. S. H. Cadle and S. Bruckenstein, Anal. Chem. *44*, 1993 (1972).

112. B. Miller, M. I. Bellavance, and S. Bruckenstein, Anal. Chem. *44*, 1983 (1972).

113. A. C. Riddiford, in *Advances in Electrochemistry and Electrochemical Engineering* (P. Delahay and C. W. Tobias, eds.), Interscience, New York, 1966, vol. 4, and references cited therein.

114. K. B. Prater and A. J. Bard, J. Electrochem. Soc. *117*, 207 (1970), and references cited therein.

115. W. J. Albery and M. L. Hitchman, *Ring Disc Electrodes*, Clarendon Press, Oxford, 1971, and references cited therein.

116. F. Opekar and P. Beran, J. Electroanal. Chem. *69*, 1 (1976), and references cited therein.

117. E. Pungor, Zs. Fekes, and G. Nagy, Anal. Chim. Acta *51*, 417 (1970), *52*, 47 (1970).

118. R. J. Davenport and D. C. Johnson, Anal. Chem. *45*, 1979 (1973), *46*, 1971 (1974).

119. P. E. Sioda and T. Kambara, J. Electroanal. Chem. *38*, 51 (1972), and references cited therein.

120. J. V. Kenkel and A. J. Bard, J. Electroanal. Chem. *54*, 47 (1974), and references cited therein.

Chapter 6

PULSE POLAROGRAPHY

6.1 NOMENCLATURE

Historically, pulse polarographic methods were developed by Barker
[1,2] as an extension of his work on alternating current (ac)
(square-wave) methods. However, since this technique may be described
using many of the concepts already developed for direct current (dc)
polarography and dc linear sweep voltammetry, it is logically considered prior to the ac methods where several new theoretical concepts need to be developed.

In dc polarography, an (approximately) constant potential is applied to the cell and the resultant current measured. In *pulse polarography*, as the name implies, the potential is applied periodically
during short time intervals. Both the format for application of the
pulse and the current readout may be varied in many ways, and as with
many modern polarographic methods, a nomenclature problem has arisen
with the prefixes "normal," "integral," "derivative," and "differential"
being bandied about somewhat indiscriminantly on occasion.

As stated elsewhere, the advantages of modern polarographic
methods frequently have their origin in an improved faradaic-to-
charging current ratio. Pulse polarography is an excellent example
of this, although it also has several important additional advantages over conventional dc polarography as is seen in subsequent
discussion.

236

In order to understand how pulse polarography is useful in minimizing the measurement of capacitive current, let us consider an electrode maintained at a potential at which no faradaic reaction occurs. The only current flowing at a dropping mercury electrode will be that caused by the increase in the double layer capacitance as the mercury drop grows. However, as shown in Chap. 4 (current-sampled dc polarography), at the end of the drop life when the rate of drop growth is minimal, this residual current will be small and changing only very slowly with time. Thus, prior to the application of the pulse, a small but finite charging current attributable to the dc terms exists in pulse polarography. Such dc effects are frequently ignored in discussions of pulse polarography, but as Christie and Osteryoung [3] have shown, it is often these small dc currents that limit the sensitivity of the pulse polarographic technique.

If a pulse is applied to the electrode so that the potential is suddenly increased to a new value, but one at which there is still no faradaic reaction occurring, then current must flow to charge the double layer to the new potential. At this point in time both dc and pulse charging currents are present. Assuming the model of an ideal capacitor, the pulse charging current will be largest immediately after application of the pulse and then will decay exponentially with time. Now consider what happens when the amplitude of the applied pulse is sufficiently large so that reduction of the electroactive species occurs and a faradaic current can flow. If the potential of the pulse corresponds to a point on the rising part of the dc polarographic wave, the magnitude of the faradaic current will depend on the charge transfer kinetics (that is, the k_s value) or other rate-determining steps of the electrode process. Initially, there will be a large current jump, and then the current will decay as a function of time in a similar fashion to a controlled-potential electrolysis experiment. Since the pulse faradaic current decays at a much slower rate than the charging current, measurement of the current near the end of the pulse duration provides very substantial discrimination against charging current (see also staircase voltammetry in Chap. 5 and Fig. 5.19).

Thus, in pulse polarography only a single pulse is applied to
the system per mercury drop, late in the drop life. Approximately
20 to 40 msec after the application of the pulse, it is assumed
that the charging current has decayed to almost zero. The exact
time for this to occur depends on the RC nature of the electrochemical cell. Measurement of the current remaining after this time
(faradaic) is then made and ignoring the small dc effects mentioned
above, the polarogram is a plot of the faradaic current produced
by the pulse versus the applied potential.

In normal pulse polarography, potential pulses of gradually increasing amplitude are applied to an electrode, starting from an
initial potential where no faradaic current flows. The potential
pulses are of approximately 40 to 60 msec duration, but the potential
between pulses always returns to the initial value. If the area
of the electrode is changing [e.g., at dropping mercury electrode
(dme)], the pulses are always applied at a fixed time in the life
of the drop, so a constant electrode area can be maintained. Current
values (either "instantaneous" or averaged) are measured toward the
end of each pulse application. In the case of measuring currents
over a finite period toward the end of the pulse duration, an electronic integration procedure is used, and hence, the name *integral
pulse polarography* is sometimes employed for what is referred to as
normal pulse polarography in this book.

For operation in the normal pulse mode, two types of instrument
must be distinguished:

1. Those in which the output is simply the current measured
    at a selected time after the pulse duration
2. Those in which the output is the difference between the
    current measured at the selected time interval after the
    pulse duration and the current measured just prior to the
    pulse application when the electrode is at the initial
    potential.

The latter form of instrumentation is the better form of normal
pulse polarography because it subtracts dc effects from the readout [3].

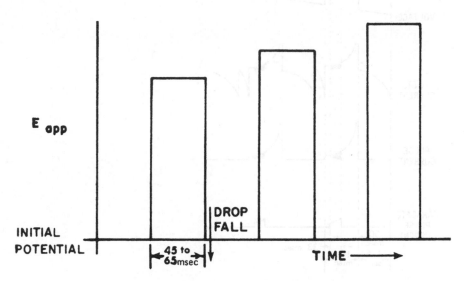

FIGURE 6.1 Applied potential waveform in normal pulse polarography consists of a series of pulses of increasing height.

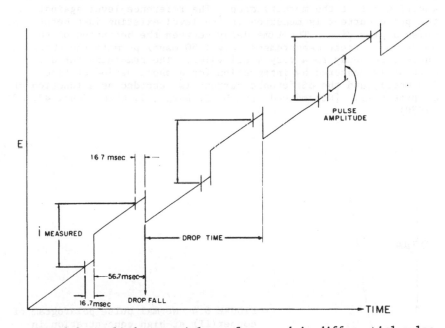

FIGURE 6.2 Applied potential waveform used in differential pulse polarography. [Reproduced from Anal. Chem. *44*, 75A (1972). © American Chemical Society.]

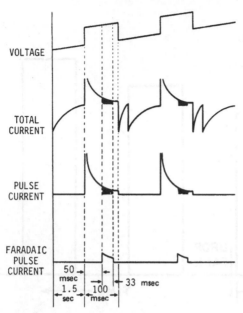

VOLTAGE

TOTAL
CURRENT

PULSE
CURRENT

FARADAIC
PULSE
CURRENT

50 msec

1.5 sec | 100 msec | 33 msec

FIGURE 6.3 Measurement sequence and responses in differential pulse
polarography. Two consecutive mercury drops are considered. The
unmarked small peak in the total current curve corresponds to the
knocking off of the mercury drop. The reference level against which
the pulse current is measured is the level existing just before the
pulse application. The time delay between the beginning of the
pulse and current measurement (e.g., 50 msec) permits the charging
current to decay to a very small value. The remaining faradaic
current is measured by integration for a short period of time
(33 msec), and the difference current is recorded as a function of
dc potential. [Reproduced from D. E. Burge, J. Chem. Educ. *47*, A81
(1970).]

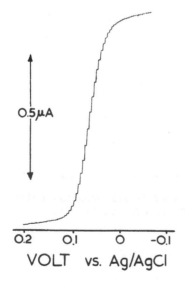

0.5 μA

0.2   0.1   0   -0.1

VOLT vs. Ag/AgCl

FIGURE 6.4 Normal pulse polarogram of
copper(II) at high concentration in
1 M $NaNO_3$. Drop time = 2 sec;
[Cu(II)] = 1×10^{-4} M. Note that a
well-defined sigmoidal-shaped curve is
observed at this concentration. [Repro-
duced from Anal. Chem. *44*, 721 (1972).]
© American Chemical Society.]

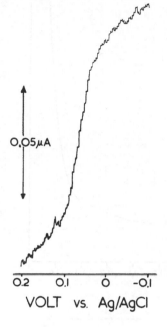

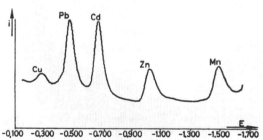

FIGURE 6.5 Normal pulse polarogram of 5×10^{-6} M copper(II) in 1 M $NaNO_3$. Drop time = 2 sec. Note sloping base line resulting from charging current places a limitation on the method, as is the case in dc polarography. [Reproduced from Anal. Chem. *44*, 721 (1972). © American Chemical Society.]

FIGURE 6.6 Determination of trace impurities in high-purity sodium chloride by differential pulse polarography. Pulse amplitude = -25 mV. [Reproduced from Z. Anal. Chem. *264*, 173 (1973).]

In differential pulse polarography, a normal dc voltage ramp is applied to the system. Near the end of the drop life, a small-amplitude pulse of approximately 50 mV is superimposed onto the ramp. As the measured signal is the difference in current measured before and after the application of the pulse, i.e., the change in current produced by the perturbation of the system, a peak-shaped curve is obtained with the peak maximum occurring near $E_{1/2}$ if the

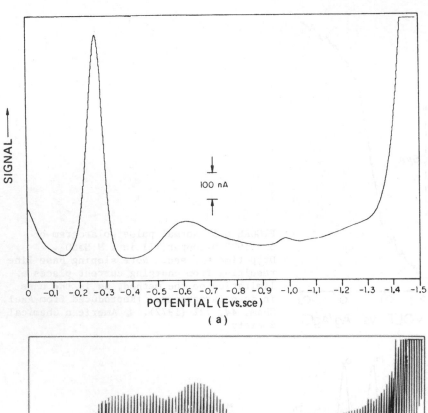

(a)

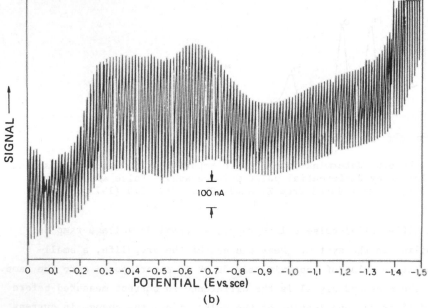

(b)

FIGURE 6.7   (a) Differential pulse and (b) dc polarograms of
1.3×10^{-5} M chloramphenicol in 0.1 M acetate buffer.   Pulse ampli-
tude = –50 mV.   [Reproduced from Anal. Chem. 44, 75A (1972).
© American Chemical Society.]

perturbation (pulse amplitude) is sufficiently small. The term
differential pulse polarography is therefore self-explanatory.

Variations of the above methodology are also available in pulse
polarography, but they have not had widespread use and can await
the end of the chapter for brief discussion. Figures 6.1 to 6.3
summarize diagrammatically the above discussion. Figures 6.4 to
6.7 show the two kinds of pulse polarogram. The considerable advantage over dc polarography is clearly seen in Fig. 6.7. Typical
values of drop time, pulse time, and time of current measurement are
given with the figures. It is important to notice that the pulse
is applied for an appreciable length of time, e.g., 50 msec. This
time scale of the pulse is only the length of the shortest drop time
used in rapid dc polarography. Hence, the pulse method as used analytically is related to dc methods and is not expected to be as
strongly dependent on the electrode kinetics as for ac polarography
and other techniques discussed in subsequent chapters. This implies
that pulse polarography retains a high sensitivity for electrochemically irreversible systems. This important conclusion needs to be
kept in mind when comparing pulse polarography with other methods.

6.2 THEORY AND ANALYTICAL IMPLICATIONS

6.2.1 Normal Pulse Polarography

In the normal pulse mode the i-E relationships are relatively simple
and are akin to the dc polarographic case. However, currents per
unit concentration are of course larger.

If an initial potential is chosen well before the onset of the
wave, that is, where the faradaic current can be assumed to be zero,
the current-potential curve for the reversible $A + ne \overset{\rightarrow}{\leftarrow} B$ process
is given by the approximate expression [4-6]

$$i = nFCA \sqrt{\frac{D}{\pi t_m}} \left(\frac{1}{1 + P} \right) \tag{6.1}$$

where $P = \exp (nF/RT)(E - E^r_{1/2})$ and t_m = the time interval between
the pulse application and current measurement.

As the pulse potential becomes more negative than $E_{1/2}^r$, P approaches zero; so the limiting current i_1 is given by the Cottrell equation

$$i_1 = nFCA \sqrt{\frac{D}{\pi t_m}} \qquad (6.2)$$

The equation to the i-E curve for the reversible case may therefore be written in an analogous fashion to the Heyrovský-Ilkovic equation in dc polarography:

$$E = E_{1/2}^r + 2.303 \frac{RT}{nF} \log \frac{i_1 - i}{i} \qquad (6.3)$$

As the Ilkovic equation applies to reversible and irreversible processes alike, so does the Cottrell relationship [Eq. (6.2)] in pulse polarography. By dividing the Cottrell equation by the Ilkovic equation, the following result is obtained [6]:

$$\frac{i_1 \text{ (normal pulse)}}{i_d \text{ (dc pol)}} = \frac{t^{1/2}}{\sqrt{7/3}\ t_m^{1/2}} \qquad (6.4)$$

where t is the drop time used to obtain the dc polarographic diffusion-controlled current and t_m is the pulse duration as defined above. If usual values for t and t_m are used, the ratio i_1 (normal pulse)/i_d (dc pol) is in the range of 6 to 7. This ratio indicates the sensitivity increase over dc polarography which can be obtained with normal pulse polarography.

In dc polarography it is well recognized that the Ilkovic equation is only approximate. Similarly in normal pulse polarography more exact solutions for the diffusion-current problem are available [7-9] to give equations of the kind

$$i_1 = 4.62 \times 10^4\ nCD^{1/2} m^{2/3} t_m^{1/6} F(\alpha)[1 + 1.354 g(\alpha)\gamma_k] \qquad (6.5)$$

where

$$\alpha = \frac{t_m}{t_p + t_m}$$

t_p = time during which mercury drop has grown prior to pulse application

$$F(\alpha) = \alpha^{-1/3}(1 + \frac{1}{3}\alpha + \frac{7}{54}\alpha^2 + \frac{4}{81}\alpha^3 + \frac{11}{648}\alpha^4 + \frac{77}{17496}\alpha^5)$$

$$g(\alpha) = \frac{\alpha^{1/3}}{F(\alpha)}(1 - \frac{5}{48}\alpha^2 - \frac{25}{432}\alpha^3 - \frac{65}{5184}\alpha^4)$$

$$\gamma_k = 5.04D^{1/2}t_m^{1/6}m^{-1/3}$$

The numerical factors are valid at 25°C. All units are in cgs while i is in amperes. If $t_p = 0$ ($\alpha = 1$), then $t_m = t$ = drop time, and Eq. (6.5) becomes the Matsuda or extended Ilkovic equation valid for dc polarography as is expected. However, the Cottrell equation is most conveniently used for discussion of analytical work, and Eq. (6.5) will not be mentioned in subsequent discussion.

The time domain in normal pulse polarography is governed essentially by the pulse duration, and so when considering quasi-reversible electrode processes, a qualitative picture of phenomena is gained by substituting ideas relative to drop time variation in dc polarography with pulse duration. Thus, in the chapter in dc polarography, the Zn(II) electrode process is used as an example of a quasi-reversible electrode process. The expected result with pulse polarography is shown in Fig. 6.8. The electrode process appears to be more reversible the longer the pulse duration. Thus, with a pulse duration of 0.5 sec, the curve approaches the shape predicted by Eq. (6.3). For the totally irreversible case, completely analogous equations to those given in Chap. 3 for dc polarography can be derived [10] and for reduction,

$$E_{1/2} = E_{1/2}^r + \frac{2.303RT}{\alpha nF} \log 2.31k_s\sqrt{\frac{t_m}{D}} \tag{6.6}$$

Finally, as expected, conclusions and results pertinent to electrode processes exhibiting coupled chemical reactions or adsorption, etc., [11-20] still apply in normal pulse polarography, but of course if

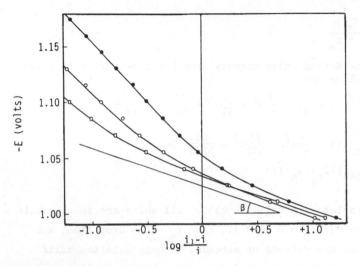

FIGURE 6.8a Normal pulse plots of log $[(i_1 - i)/i]$ vs. E for Zn(II) in 1 M KCl/5 × 10^{-4} M HCl. Pulse times (sec) are (●), 0.036; (o), 0.090; (□), 0.540. β is the theoretical slope angle for a reversible two-electron polarogram. [Reproduced from J. Electroanal. Chem. *14*, 43 (1967).]

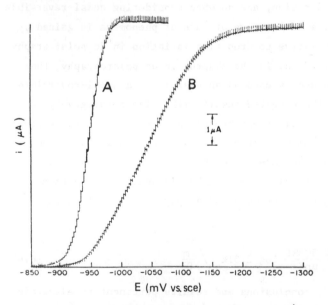

FIGURE 6.8b Normal pulse polarogram of 10^{-4} M Zn(II) in (A) 0.014 M NaNO$_3$ (reversible case) and in (B) 1.05 M NaNO$_3$ (quasi-reversible case). [Reproduced from Electrochim. Acta *11*, 1525 (1966).]

the initial potential cannot be set at a potential where no faradaic current flows as assumed in the above and in all subsequent examples then this has to be considered and taken into account [21]. Normal pulse polarography often gives a severely distorted view of solution composition under such circumstances, because the working electrode is active during the interval between pulses. Extensive discussion of the response of the different electrode processes in normal pulse polarography under the usual situation is therefore not required. All that needs to be recognized in most instances is that the time scale of pulse polarography is somewhat shorter than dc polarography and that due allowance for this should be made. An example of the direct relationship between the two techniques is given below.

In Chap. 4 it is shown that short drop times (rapid polarography) improve the analytical usefulness of waves complicated by adsorption and/or film formation. The papers by Canterford and Osteryoung [22, 23] on anodic processes involving mercury compound formation illustrate that equivalent improvements are also obtained in the normal but not differential pulse polarographic response. The improvement may be attributed to the decreased time scale and equally short electrolysis times of both short drop time or pulse methods. The normal pulse method, having a waveform which only causes electrolysis to occur periodically, will be shown later (in discussion of work of stationary electrodes) to be ideal for minimizing the undesirable influences of certain surface phenomena. Of course, the pulse method has a distinct advantage with respect to charging current, counterbalanced by the time saving inherent in the rapid dc method enabling the use of faster scan rates. The two polarographic methods are therefore by no means equivalent in their analytical utility.

Work by O'Deen and Osteryoung [24] further confirms that many systems which give erratic and ill-defined waves under conditions of conventional dc polarography can give extremely well-defined waves under pulse polarographic conditions. Figure 6.9(a) and (b) shows a comparison of the oxidation of mercury in the presence of halides under normal pulse and dc polarographic conditions. The pulse work

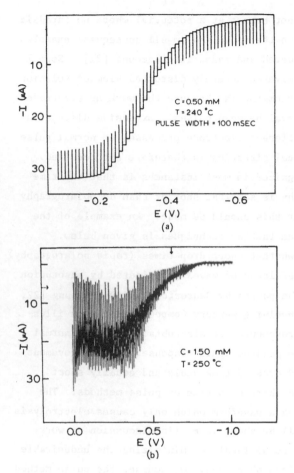

FIGURE 6.9   (a) Normal pulse polarogram of bromide in fused NaNO$_3$
NaNO$_3$-KNO$_3$ at 240°C.   (b) DC polarogram for iodide in a NaNO$_3$-KNO$_3$
melt showing complex behavior.   [Reproduced from Anal. Chem. *43*,
1879, (1971). © American Chemical Society.]

readily enabled the electrode process to be assigned as a reversible
one-electron oxidation step

$$Hg + Br^- \rightleftharpoons HgBr\text{(soluble)} + e$$

whereas studies on the complex wave observed under dc polarographic
conditions led to ambiguous or erroneous conclusions being made con-
cerning the nature of the electrode process.

6.2.2 Differential Pulse Polarography

If the reversible polarographic i-E relationship written in Eq. (6.3) is differentiated and the Cottrell equation substituted for the limiting diffusion-controlled current, the expression

$$\Delta i = \frac{n^2 F^2}{RT} AC \, (-\Delta E) \sqrt{\frac{D}{\pi t_m}} \frac{P}{(1 + P)^2} \tag{6.7}$$

is obtained where Δi = differential pulse current and DE = pulse amplitude. This equation is valid only for the small amplitude case because a differential method is being approximated by the derivative.

A solution [6], valid for all values of ΔE, gives

$$\Delta i = nFAC \sqrt{\frac{D}{\pi t_m}} \frac{P_A \sigma^2 - P_A}{\sigma + P_A \sigma^2 + P_A + P_A^2 \sigma} \tag{6.8}$$

where

$$P_A = \exp \frac{nF}{RT} \left[\frac{E_1 + E_2}{2} - E_{1/2}^r \right]$$

$$\sigma = \exp \frac{nF}{RT} \left[\frac{E_2 - E_1}{2} \right]$$

$E_2 - E_1 = \Delta E$, the pulse amplitude

E_2 = the potential at which current i_2 is measured after the application of the pulse

E_1 = the potential at which the current i_1 is measured in the absence of the pulse

Note that for reduction this means that ΔE should be negative, but in practice the sign is often erroneously omitted in discussion of differential pulse polarography. On other occasions it is omitted for simplicity.

P_A = 1 when Δi is a maximum, so that the expression for the peak or maximum current $(\Delta i)_{max}$ is given by

$$(\Delta i)_{max} = nFAC \sqrt{\frac{D}{\pi t_m}} \frac{\sigma - 1}{\sigma + 1} \tag{6.9}$$

If $-\Delta E/2$ is smaller than RT/nF, this equation simplifies to the small amplitude case $(\Delta i)_{max} = (n^2F^2/4RT)AC (-\Delta E) - \sqrt{D/\pi t_m}$. When $-\Delta E/2$ becomes very large with respect to RT/nF, $(\sigma - 1)/(\sigma + 1)$ approaches unity and $(\Delta i)_{max}$ is simply the Cottrell expression.

From Eq. (6.9) it is apparent that the larger the value of $-\Delta E$, the larger the value of $(\Delta i)_{max}$. In practice, however, it is also obvious that increasing the pulse amplitude increases the width (decreases the resolution), which is undesirable. The *peak half width* is defined as the width of the peak (mV) at the point where the peak current is half its maximum height. The derivative of a dc polarogram or the small-amplitude differential pulse wave has a half width $W_{1/2}$ of $3.52RT/nF$ which gives a value of $90.4/n$ mV at 25°C. For large values of $-\Delta E$, $W_{1/2} \rightarrow -\Delta E$. In differential pulse polarography, the value of $W_{1/2}$ for all values of $-\Delta E$ and with various values of n is summarized in Fig. 6.10. $W_{1/2}$ in this figure was calculated from the expression $RT/nF \ln [(y + \sqrt{y^2 - 4\sigma^2})/(y - \sqrt{4 - 4\sigma^2})]$ with $y = \sigma^2 + 4\sigma + 1$.

In practice, values of ΔE between 10 and 100 mV are used, as a compromise between adequately large values of $(\Delta i)_{max}$ and adequate resolution.

From the above relationships it can be shown that the peak potential E_{peak} is given by

$$E_{peak} = E_{1/2}^r - \frac{\Delta E}{2} \tag{6.10}$$

Thus for a reduction process, the peak potential is shifted in a positive direction as the pulse amplitude increases. (ΔE is negative for reduction.)

Equation (6.9) shows that $(\Delta i)_{max}$ is a linear function of concentration. This is true of many electrode processes other than the reversible case, as is the case in dc polarography. However, catalytic and other perturbations on the electrode process which result in i_d-vs.-C plots being curved in dc polarography have an analogous effect in both normal and differential pulse polarography with respect to i_1 and $(\Delta i)_{max}$, respectively. For the quasi-reversible or totally irreversible electron transfer case, $(\Delta i)_{max}$ is of course a function of k_s and the current per unit concentration is lower than for the

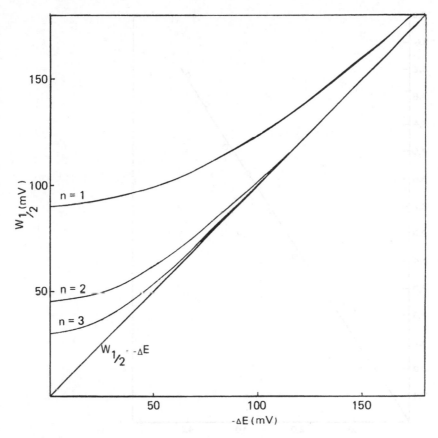

FIGURE 6.10 Plot of peak half width as a function of pulse amplitude. (Provided by courtesy of J. Osteryoung and R. A. Osteryoung.)

reversible case [4]. Presently available theory for nonreversible
processes is not very extensive or rigorous.

Figure 6.11 verifies the dependence of $(\Delta i)_{max}$ on area. Other
aspects of the theory have also been verified in the literature.
Because of the complexity of the instrumentation in differential
pulse polarography, awareness of possible instrumental artifacts [25]
needs to be borne in mind when undertaking theory-vs.-experiment
correlations and these experimentally based errors rather than inadequate theory may account for some reported anomalies. However, it
is worth noting the assumptions made in deriving the above equations
because (as will be shown subsequently) ultimately the breakdowns in
two approximations place limits on the technique.

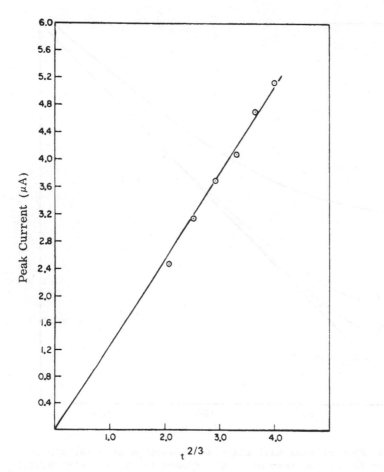

FIGURE 6.11 Plot of peak current versus area of drop ($\propto t^{2/3}$) for
1×10^{-3} M Tl(I) in 0.1 M KNO$_3$. [Reproduced from Anal. Chem. *37*,
1634 (1965). © American Chemical Society.]

It is assumed first that the area growth during the difference
measurement is zero, and second that the ratio t_m/t_p is small.
More rigorous solutions not containing these restrictions are available [3,26-28], which demonstrate the consequences of invoking these
approximations, and these are referred to in Sec. 6.4. Unlike the
normal pulse method, the differential pulse mode has the dc ramp
present, and therefore, it is a dual time domain technique. That is,
it contains characteristics of the dc response and pulse response,
and both can be important when considering rigorous treatments of

electrode processes. Theoretically, the two terms are only strictly
additive for the reversible electrode process. The simplified treatment essentially neglects to include contributions from the dc terms.

6.2.3 Influence of Resistance

The influence of resistance is qualitatively the same as in dc
polarography, but particularly because of the much lower detection
limit available in differential pulse polarography, some new options
requiring renewed assessment of the Ohmic (iR) drop problem are
available. In polarography, the concentration of supporting electrolyte ideally must be at least 25 to 50 times as great as the concentration of reducible species in order to suppress the migration current (Chap. 3). At the 10^{-5} M level, differential pulse polarograms
are still extremely well defined and a supporting electrolyte of
less than 10^{-3} M could still be sufficient for suppression of migration current, *provided a three-electrode system can be used to overcome the ir drop problem.* Figure 6.12 shows curves for 2×10^{-5} M
Cd(II) in 0.01 M and 0.001 M KNO_3. Only a slight difference in peak
height is observed and this may be attributed to differences in diffusion coefficient [6] rather than iR drop. The insensitivity of
pulse polarography with respect to iR drop, relative to some other
modern polarographic techniques, is discussed by Parry and
Osteryoung [6]. However, the analytical chemist should not become
blasé about decreasing the supporting electrolyte concentrations
too much because the value of $(\Delta i)_{max}$ in the presence of uncompensated
resistance is dependent on ohmic losses, as is the decay time of the
charging current. Vigilance toward iR drop, detected as a negative
shift in wave position, decrease in peak height, or presence of significant charging current, etc., should always be undertaken under
conditions of high-resistance, even with pulse polarography. Subsequent discussion (pp. 262-263) with respect to the charging current
response as a function of electrolyte concentration should also be
consulted for further limitations. The advantage of being able to
work with low concentrations of supporting electrolyte is of course
very useful, since the added electrolyte is always a potential source
of impurity.

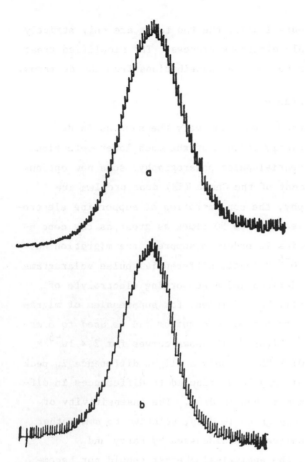

FIGURE 6.12 Effect of supporting electrolyte concentration:
(A) 2×10^{-5} M Cd(II) in 0.001 M KNO_3, (B) 2×10^{-5} M Cd(II) in
0.01 M KNO_3. [Reproduced from Anal. Chem. *37*, 1634 (1965).
© American Chemical Society.]

6.3 CHARACTERIZATION OF ELECTRODE
REVERSIBILITY BY PULSE POLAROGRAPHY

Normal Pulse Polarography. From above equations and discussion,
diagnostic criteria for reversibility under normal pulse polarographic
conditions are shown to be similar to dc polarography. Thus, for
example, plots of E versus $\log[(i_1 - i)/i]$ should be linear with slope
$2.303RT/nF$ with normal pulse polarography. For an irreversible reduction in normal pulse polarography, $E_{1/2}$ will be a function of t_m

as indicated by Eq. (6.6), i.e., drop time governs the time scale in dc polarography, where t_m is the equivalent parameter in normal pulse polarography. The classification of waves into reversible, quasi-reversible, or irreversible therefore applies equally to dc and pulse polarographic waves. However, the status of a given electrode process may be different in the sense that a reversible reaction in dc polarography may be classified as quasi-reversible when examined by pulse polarography and one which is quasi-reversible may be observed as totally irreversible with the shorter time scale of pulse polarography. There are, however, some unique features in normal pulse polarography having no direct analogy in dc polarography which readily permit the characterization of the electrode process. These emanate from a technique [29] called *scan reversal pulse* polarography.

The Cottrell equation applies equally well to reversible or irreversible processes, and for a reduction,

$$(i_1)_{red} = nFAC \sqrt{\frac{D}{\pi t_m}} \tag{6.11}$$

However, $(i_1)_{ox}$, the limiting current obtained by commencing the scan on the reduction process diffusion plateau and proceeding to more positive potentials, depends markedly on the reversibility of the electrode reaction. For a totally irreversible process, no reoxidation of the reduction product is possible. However, since in a commonly used normal pulse mode the current measured represents a difference in current measured after and before pulse application, the positive scan will yield a limiting current corresponding to the dc current flowing prior to pulse application. This current is given by the Ilkovic equation, and the contribution of dc terms in pulse polarography turns out to be important, as is so often the case. Thus,

$$(i_1)_{ox} = -nFAC \sqrt{\frac{7D}{3\pi t}} \tag{6.12}$$

where t is the drop time.

From Eqs. (6.11) and (6.12), the ratio of the limiting currents for the normal and reverse scans for an irreversible reduction has magnitude

$$\frac{(i_1)_{red}}{(i_1)_{ox}} = \sqrt{\frac{3t}{7t_m}} \tag{6.13}$$

If t is greater than t_m by a factor of 100, the preceding ratio will have a magnitude of about 7.

For the totally reversible case it can be shown [29] that

$$\frac{(i_1)_{red}}{(i_1)_{ox}} = \left[1 - \sqrt{\frac{7t_m}{3t + 7t_m}} + \sqrt{\frac{7t_m}{3t}} \right]^{-1}$$

$$\approx 1 \qquad \text{since } t \gg t_m \tag{6.14}$$

The distinction between the 7:1 wave height ratio for irreversible reduction and 1:1 ratio for the reversible process provides a simple and definitive indication of the nature of the electrode process. Figures 6.13 and 6.14 show the situation diagrammatically. For the reversible case, the $E_{1/2}$ values for both the forward and reverse scans are the same, that is,

$$(E_{1/2})_{red} = (E_{1/2})_{ox} = E_{1/2}^r + \frac{RT}{nF} \ln \frac{D_{ox}}{D_{red}} \tag{6.15}$$

For the irreversible case

$$(E_{1/2})_{red} - (E_{1/2})_{ox} \approx \frac{RT}{\alpha nF} \left(0.574 + \frac{1.49t_m}{t} \right) \tag{6.16}$$

and the $E_{1/2}$ value on the reverse scan is several millivolts more negative than for the forward scan.

The use of irreversible processes in process analysis for determining metals in two oxidation states has been described in the literature [30]. In dc polarography, reversible electrode reactions can be used for determining both species involved in the redox couple, because the position of zero current can be used as a reference point (see Chap. 3). In normal pulse polarography, at least for a common instrumental form, differences rather than absolute values of current are measured and no zero-current point is available as a reference. For determining both species in the redox couple by normal pulse polarography, the k_s values must be such that two waves at

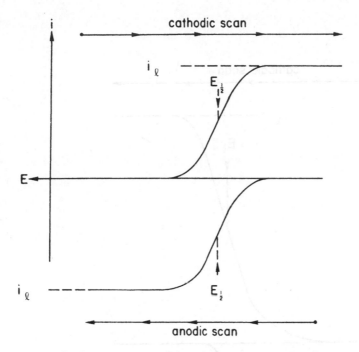

FIGURE 6.13 Scan reversal pulse polarography for a reversible reduction. [Reproduced from Anal. Chem. *42*, 229 (1970). © American Chemical Society.]

considerably different potentials can be observed with one wave corresponding to an irreversible oxidation and the other to an irreversible reduction. The choice of supporting electrolyte and k_s values for this use of pulse polarography are discussed by Parry and Anderson [30], and the ability of the analytical chemist to be able to characterize the nature of the electrode process is again shown to be important in normal pulse polarography.

Differential Pulse Polarography. In differential pulse polarography, measurement of the half width and comparing the value with theory is probably the simplest way to assess the reversibility or otherwise of the electrode process. Other tests based on Eq. (6.8) can also be used with more difficulty. Equations (6.9) and (6.10) are possibilities for simple tests.

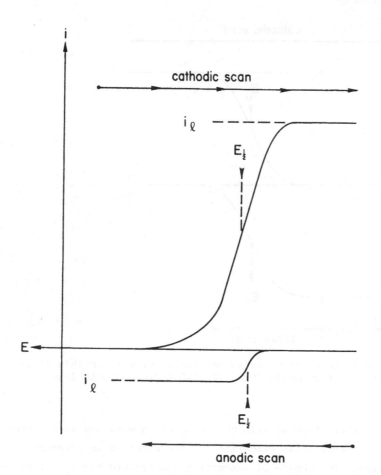

FIGURE 6.14 Scan reversal pulse polarography for a totally irreversible reduction. [Reproduced from Anal. Chem. *42*, 229 (1970).
© American Chemical Society.]

6.4 CHARGING CURRENT AND DIRECT
CURRENT EFFECTS IN PULSE POLAROGRAPHY

The charging current has already been considered qualitatively in
the introduction. The finer aspects of the theory [3] must now be
examined in detail to ascertain the analytical problems likely to be
encountered in pulse polarography. Similarly, it has been hinted
on several occasions that direct current effects can also place
important limitations on the technique. The two phenomena of

charging current and direct current effects are often closely related
and can be considered together conveniently.

6.4.1 Differential Pulse Polarography

At a fixed time t_p in the life of each drop, the current flowing i_1
is measured and a small-amplitude potential step is applied to the
electrode. At a fixed time, t_m, after pulse application, the current
is again sampled. The difference between these two currents
$i_2 - i_1$ is the parameter recorded. Using the above mode of measurement there must remain two current components which are not compensated by measuring the differences in the currents

$$i_1 \ (E, t_p) \qquad \text{and} \qquad i_2 \ (E + \Delta E, \ t_p + t_m)$$

1. In general, the potential E will be in a region in which
faradaic current flows and a dc current $i_{dc}(t_p)$ must be present before
application of the pulse. The current $i_{dc}(t_p + t_m)$ included in i_2
must be different, since the time of measurement in the drop life,
and therefore the electrode area, is different. A faradaic distortion, Δi_{dc}, therefore occurs in differential pulse polarography.
The distortion is concentration independent and depends only on
$n\Delta E$ and t_p / t_m.

2. The measured i_1 also contains a double-layer charging current
$i_c(E, t_p)$ as does i_2, $i_c(E + \Delta E, \ t_p + t_m)$. Since neither the potential
nor the time is the same for these two charging currents, the readout
will contain a further dc distortion, Δi_c due to the difference in
the charging currents. This charging current component represents
the base line on which the pulse current, distorted by the faradaic
dc current, is superimposed, as seen in Fig. 6.5.

Faradaic Effect. During the pulse,

$$i_2(t_p + t_m) = i_{pulse} + i_{dc} \ (E_2, \ t_p + t_m) \tag{6.17}$$

where it is assumed that the current is the sum of the dc current
that would have been flowing without the pulse plus the pulse current.
The faradaic current flowing before the pulse is simply

$i_1(t_p) = i_{dc}(E_1, t_p)$. The measured faradaic current in the differential pulse experiment is, therefore,

$$\Delta i_F = i_2 (t_p + t_m) - i_2 (t_p) = i_{pulse} + \Delta i_{dc}$$

The value of Δi_F is readily calculated, for the reversible case, using previously given equations for both dc and pulse polarography. The result is

$$\Delta i_F \frac{\pi^{1/2}}{nFkm^{2/3}D^{1/2}C} = \frac{(t_p + t_m)^{2/3}}{t_m^{1/2}} \frac{(1 + \sigma^2)P_1}{(1 + P_1)(1 + \sigma^2 P_1)}$$

$$+ \left(\frac{7}{3}\right)^{1/2} \frac{(t_p + t_m)^{1/6} - t_p^{1/6}}{1 + P_1} \qquad (6.18)$$

where k = constant.

Figure 6.15 shows the effect of the dc distortion; the peak potential is shifted slightly in the negative direction and the peak current increased. The distortion is increased by a decrease in the pulse amplitude and/or by an increase in the t_m/t_p ratio. The latter feature of the technique means that the dc distortion can be quite pronounced at short drop times as recently has been demonstrated

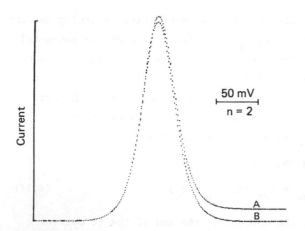

FIGURE 6.15 Calculated faradaic effect on differential pulse polarograms. $t_p + t_m = 0.5$ sec; $n\Delta E = -10$ mV: (A) with distortion, (B) undistorted. [Reproduced from J. Electroanal. Chem. 49, 301 (1974).]

by Bond and Grabaric [32]. Note from Fig. 6.15 that measurement of $(\Delta i)_{max}$ is best made using the base line generated at potentials more positive than the first reduction wave when faradaic distortion terms are operative. If this procedure is not followed, it is conceivable that in some circumstances errors could be introduced into a determination.

Charging Current Effect. Using the dc equations for the charging current [3] gives

$$\Delta i_c = -\frac{2}{3} km^{2/3} \left[\frac{q'(E_2)}{(t_p + t_m)^{1/3}} - \frac{q'(E_1)}{t_p^{1/3}} \right] \tag{6.19}$$

(q' = charge density, and even if $t_m/t_p \to 0$, a term due to the different potentials results.

As the faradaic pulse current decreases with decreasing concentration of electroactive species the charging current ultimately will mask the faradaic current. Figures 6.16 and 6.17 show this effect. The previous claims that low supporting electrolyte

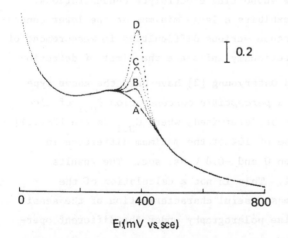

FIGURE 6.16 Simulated differential pulse polarograms for Pb(II) reduction in 0.1 M KCl. $(t_p + t_m) = 0.5$ sec; $\Delta E = -25$ mV. [Pb] is (A) 0, (B) 0.5×10^{-6}, (C) 1.0×10^{-6}, (D) 2.0×10^{-6} M. Current units are normalized values. [Reproduced from J. Electroanal. Chem. *49*, 301 (1974).]

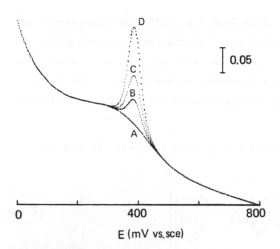

E (mV vs. sce)

FIGURE 6.17 Simulated differential pulse polarograms for Pb(II)
reduction in 0.1 M KCl. $(t_p + t_m)$ = 5.0 sec; ΔE = -25 mV:
[Pb(II)] is (A) 0, (B) 2×10^{-8}, (C) 5×10^{-8}, (D) 1×10^{-7} M.
Current units are normalized values. [Reproduced from J. Electroanal.
Chem. *49*, 301 (1974).]

concentrations are tolerable by differential pulse polarography may
now be further examined. Figure 6.18 shows simulated differential
pulse polarograms with low supporting electrolyte concentrations.
Note that the background exhibits a large minimum at the lower concentrations, and this would cause serious difficulties in measurement of
peak height and also deleteriously influence the limit of detection.

Sensitivity. Christie and Osteryoung [3] have used the above type
of calculations to define a *perceptible concentration* $C_{0.1}$ of electroactive species that can be determined, where $C_{0.1}$ is the [Pb(II)]
which would give a response of 10% of the maximum difference in
background response between 0 and -0.8 V vs. sce. The results
are presented in Table 6.1. This is not a calculation of the
detection limit but is a most useful characterization of the sensitivity of differential pulse polarography under the different operating conditions.

Equation (6.18) shows that a differential pulse polarogram in
the absence of faradaic reaction is essentially a differential
capacity curve [31]. Therefore, the background current will depend

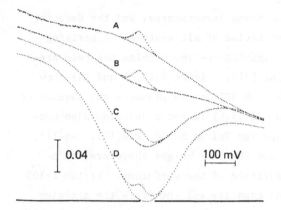

FIGURE 6.18 Simulated differential pulse polarograms for 10^{-8} M Pb(II) in (A) 0.916 M, (B) 0.100 M, (C) 0.01M, and (D) 0.001 M NaF solutions as well as the associated backgrounds in the absence of Pb(II). Prop time = 5 sec; ΔE = -25 mV. Current units are normalized values. [Reproduced from J. Electroanal. Chem. *49*, 301 (1974).]

TABLE 6.1 Calculated Perceptible Concentrations of Pb(II) by Differential Pulse Polarography

Supporting Electrolyte	$t_p + t_m$ (sec)	Perceptible concentration[a] ($\times 10^9$ M) ΔE (mV)				
		5	10	25	50	100
0.916 M NaF	0.5	620	344	185	150	178
	1.0	166	99.4	61.8	56.6	74.3
	2.0	48.8	32.2	23.5	23.9	33.7
	5.0	11.4	8.8	7.7	8.5	12.7
0.1 M KC1	0.5	956	590	374	332	395
	1.0	290	200	147	142	179
	2.0	100	77.4	63.8	65.0	84.8
	5.0	29.5	25.6	23.2	24.6	32.9
0.1 M NaF	0.5	548	302	161	130	152
	1.0	146	86.6	53.2	48.4	63.2
	2.0	42.5	27.8	20.0	20.2	28.5
	5.0	9.8	7.5	6.5	7.2	10.7
0.001 M NaF	0.5	304	199	142	143	196
	1.0	97.1	71.8	59.7	64.6	92.9
	2.0	35.4	29.4	27.1	30.6	45.2
	5.0	11.1	10.2	10.2	11.9	17.8

[a]Data taken from Note 3.

on the nature of the solution being investigated, and the data for
Pb(II) is by no means representative of all systems. A striking
example of the differential capacitance-type problem has been presented by Myers and Osteryoung [31]. Figure 6.19(a) and (b) show
differential pulse curves for 1 M HCl in the presence and absence of
20 µg liter^{-1} As(III). Figure 6.19(b) is for a solution also containing the surface-active species Triton X-100 (0.001%). As(III)
reduction is not reversible and the peak height therefore can be
strongly affected by the adsorption of the surfactant. Triton X-100
also changes the differential capacity and therefore the charging
current. If the background changes in an unknown and unpredictable
fashion, the use of a calibration curve for the determination of a
species would of course be dubious. The sensitivity of the background current to surface-active materials is further illustrated
in Fig. 6.20. Peptone solutions are prepared from an enzyme digest
of proteins and represent a reasonably simplified model for analytical samples with a complex matrix of surface-active compounds.
The change in charging current would seriously interfere with many
determinations.

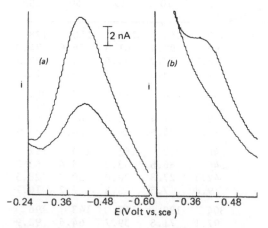

FIGURE 6.19   Effect of the surfacant Triton X-100 on differential
pulse polarographic backgrounds. (a) 20 µg As(III) in 1 M HCl and
the background in 1 M HCl. (b) Same as (a) but with 0.001% Triton
X-100. Scan rate = 2 mV sec^{-1}; drop time = 2 sec; ΔE = -100 mV.
[Reproduced from Anal. Chem. *46*, 356 (1974). © American Chemical
Society.]

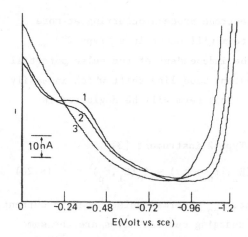

E(Volt vs. sce)

FIGURE 6.20 Effect of Peptide surfacants on charging current in differential pulse polarography. Scan rate = 10 mV sec^{-1}; drop time
= 0.5 sec; ΔE = -100 mV. (1) 0.1 M KCl + 0.1 M HCl; (2) as for (1)
+ 0.001% peptone; (3) as for (1) + 0.005% peptone. [Reproduced
from Anal. Chem. *46*, 356 (1974). © American Chemical Society.]

If the effect of surfacants is so severe as to make direct
determinations impossible, Myers and Osteryoung [31] have shown that
amperometric titrations employing differential pulse polarography
for the endpoint detection are an attractive alternative. Their
work should be consulted for further details. Amperometric titrations provide a powerful adjunct to polarographic methodology, and
so this analytical approach should not be neglected, since it substantially increases the scope of all polarographic techniques, not
just differential pulse polarography.

6.4.2 Normal Pulse Polarography

For operation in the normal pulse mode, as mentioned earlier, the two
types of instrument must be distinguished: (1) those in which the
readout is simply the current measured at time t_m after pulse application and (2) those which measure the difference in current at
time $(t_p + t_m)$ and t_p.

Faradaic Effect. If the initial potential E_1 is chosen so that
no faradaic current flows, there is no dc faradaic component in the
pulse current measured [3]. If the initial potential is on the

limiting current region of an electrode process occurring at more positive potentials, a faradaic term will occur in a "Type 2" instrument. However, this will be independent of the pulse potential E_2 and will only appear as a constant base line shift which is easily offset by a recorder adjustment. This term will be neglected in subsequent discussion.

Charging Current Effect. For a Type 2 instrument [3]

$$i_T = i_{np}(E_2,\ t_p + t_m) + i_c(E_2,\ t_p + t_m) - i_c(E_1, t_p) \qquad (6.20)$$

where i_T = total measured current, i_{np} = normal pulse Faradaic current, and i_c = charging current. The charging current terms are the same as for differential pulse polarography, and the normal pulse current is given by

$$i_{np} = \frac{knF\ D^{1/2} Cm^{2/3}(t_m + t_p)^{2/3}}{\pi^{1/2} t_m^{1/2}(1 + P)} \qquad (k = constant) \qquad (6.21)$$

For a "Type 1" instrument, the last term in Eq. (6.20) is absent. Figure 6.21 shows simulated normal pulse polarograms, and Table 6.2 gives minimum perceptible concentrations. The values of $C_{0.1}$ are much larger than for differential pulse polarography because of the appreciably larger absolute background in the normal pulse mode.

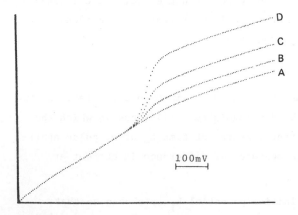

FIGURE 6.21 Simulated normal pulse polarograms for Pb(II) reduction in 0.1 M KCl. $(t_p + t_m)$ = 5.00 sec. [Pb(II)] is (A) 0, (B) 0.2×10^{-6} (C) 0.5×10^{-6}, (D) 1.0×10^{-6} M. [Reproduced from J. Electroanal. Chem. *49*, 301 (1974).]

TABLE 6.2 Calculated Perceptible
Concentrations of Pb(II) in 0.1 M
KCl with Normal Pulse Polarography[a]

$(t_p + t_m)$ sec	$C_{0.1} \times 10^6$ M
0.5	2.57
1.0	1.27
2.0	0.63
5.0	0.25

[a]Data taken from Note 3.

6.5 PULSE VOLTAMMETRY AT STATIONARY ELECTRODES

6.5.1 Normal Pulse Voltammetry

Normal pulse voltammetry can be used to advantage with solid electrodes [33]. Since the time scale of the pulse experiment is predominantly governed by the pulse duration at both stationary and nonstationary electrodes, essentially the same theory as given above is still applicable. Note the distinction here between the dc techniques where in changing from a dme to a stationary electrode, the time domain governing the experiment changed from drop time to scan rate, and different theoretical concepts had to be invoked for the two cases.

When the normal pulse method is employed to study the electrode reaction A + ne $\rightleftharpoons$ B, advantages at stationary electrodes originate from the potential format. At the commencement of a scan, the initial potential chosen is inadequate to cause significant reduction of A. However, at more negative potentials reduction occurs and B is produced at the electrode surface. Between pulses, the electrode potential is returned to its initial value where the reaction does not proceed from left to right. In fact, if the system is reversible, or k_s sufficiently fast, oxidation of B to A will occur at the rest potential, and the electrode will be "cleaned" periodically. The renewing of the surface is invaluable when B is a solid deposited on the electrode, or an adsorbed species, and the nature of the pulse potential format provides the advantages associated with the dme.

Of course, if the electrode process is totally irreversible, the
oxidation of B will not occur to any significant extent and the above
advantage is lost.

In the Ag(I)/Ag(0) electrode process at a platinum electrode
[34], the silver is reduced for only the very brief period of the
pulse, but the electrode is kept for a much longer period at a potential where the silver metal is reoxidized and no net deposit of metal
accumulates on the electrode surface.   In this situation, excellent
i-E pulse curves are obtained under normal pulse conditions.   However, for a sufficiently irreversible system such as the Au(III)/Au(0)
process at a pyrolytic graphite electrode [33], a rest potential cannot be chosen at which oxidation of deposited metallic gold takes
place.   Hence, a net depletion will occur during the very short pulse,
and the depletion will not be eliminated during the much longer
waiting period between pulses, despite the fact that the interval
between pulses exceeds the pulse duration 50 to 100 times.   This
depletion effect is quite significant. Since under many operating
conditions the diffusion layer is considerably thinner during pulses
than the shear layer in a stirred solution, stirring has essentially
no effect on the normal pulse voltammogram obtained for the reversible reduction of silver. However, stirring can eliminate the depletion effect for reduction of gold, and it has therefore been
recommended by Oldham and Parry [33], that pulse voltammetric
analyses of irreversible systems should be carried out in the presence of stirring.   Figure 6.22 shows the depletion effect and its
elimination by stirring for the reduction of gold.   It has also
been the author's experience that normal pulse voltammetry is a most
useful technique at stationary electrodes [35], and the use of this
potential format in solid electrode work could probably be exploited
a great deal more. Even endeavoring to assign the reversibility or
otherwise of an electrode process by pulse voltammetry is relatively
simple at a stationary electrode, because analogous theory to that
for the dme is frequently applicable. Thus, a linear plot of
E vs. $\log[(i_1 - i)/i]$ with slope $2.303RT/nF$ would be indicative of
a reversible process from a pulse voltammogram.

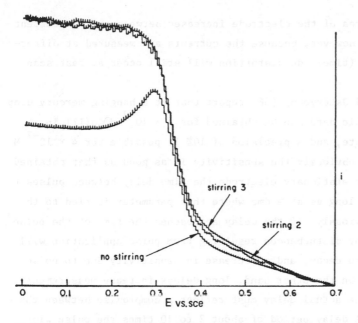

FIGURE 6.22 Normal pulse voltammogram for the reduction of Au(III) at a pyrolytic graphite electrode without stirring, with mild stirring ("stirring 2") and with faster stirring ("stirring 3"). [Reproduced from Anal. Chem. *38*, 867 (1966). © American Chemical Society.]

6.5.2 Differential Pulse Voltammetry

For a reversible process, the theory for differential pulse voltammetry at stationary electrodes, as expected, is essentially the same as given previously for the dme. Keller and Osteryoung have presented a limiting case of the theory [36], and recently Rifkin and Evans [37] have provided a more general description of the reversible case. The potential format, because it still includes the dc ramp, does not have the unique feature of the normal pulse method which enables elimination of certain analytically undesirable phenomena. The advantages that can accrue with this method at a stationary electrode may only arise from increased electrode area and the concomitant ease of measuring larger currents, increased speed of analysis, and constant area. The latter may be an advantage because, as seen previously, faradaic and charging current distortion terms occur at a

dme when the area of the electrode increases between the two current
measurements. However, because the currents are measured at different potentials (time), dc distortion will still occur at fast scan
rates [38].

Keller and Osteryoung [36] report that at a hanging mercury drop
electrode, usable data can be obtained for 4×10^{-8} M Cd(II) in
potassium nitrate, and a precision of 10% is possible for 4×10^{-7} M
Cd(II). Quite obviously the sensitivity is as good as that obtained
at a dme. At a stationary electrode the time delay between pulses
need not be as long as at a dme where this parameter is tied to the
drop time. Obviously, if the delay approaches the time of the pulse
width, transient disturbances resulting from pulse application will
not have time to decay, and a decrease in sensitivity due to noise
could result. On the other hand, long delays increase measurement
time, and so the actual delay must reflect a compromise between these
two aspects. A delay period of about 2 to 10 times the pulse width
seems to be most satisfactory [36-40]. A detailed analytical evaluation of differential pulse voltammetry has been reported by Rifkin
and Evans [39] using a platinum electrode and computer-based instrumentation. Less than 10^{-6} M of reversibly oxidized species could be
determined in acetonitrile and about 10^{-6} M concentrations of irreversibly oxidized species in the same solvent. Differential pulse
voltammetry is obviously an excellent method to use at stationary
electrodes.

6.5.3 Pulse Voltammetry at a
Dropping Mercury Electrode

As in dc linear sweep voltammetry, it is possible to synchronize the
pulse technique to a dme and obtain pulse voltammograms on a single
mercury drop [40]. This approach is particularly successful in the
differential pulse mode. Depletion effects limit this variation of
the normal pulse mode. The fastest scan rate that can be used is
governed by the necessity of having a delay of about 10 msec between
pulses so that this period is at least equal to or, preferably,
longer than the pulse width. Scan rates of around 100 mV/sec

recorded under these conditions still permit sufficient data points to be obtained to give accurate i-E curves. This is a most attractive voltammetric method, having the advantages of the dme, extremely high sensitivity and rapid measurement time. The fact that the highly convenient dme is still an integral part of the experiment means that the high reproducibility, not always associated with stationary electrodes, is retained. Looked at from the overall viewpoint, and even allowing for the author's biased position, the technique of differential pulse voltammetry at a dme can be postulated as having most of the desirable features ideally associated with the analytical use of a polarographic method [40].

6.5.4 Pulse Voltammetry at Rotated Electrodes

The relative insensitivity of the pulse methods to convective mass transport was noted previously with respect to stirring. Under conditions where the Nernst layer thickness is small relative to the convective sheer layer, pulse methods also would be expected to be independent of the rotation rate of the electrode. This assumption obviously cannot be true in the limit of very high rotation rates, and the theoretical problem with respect to both normal and differential pulse voltammetry at rotated electrodes has been examined by Myers et al. [41].

At short pulse widths and low rotation rates, the current will obey the Cottrell equation, while at long pulse widths and rapid rotation rates the current should be given by the Levich equation (see Chap. 5). In the case where the convective contribution is relatively small, the instantaneous limiting current in the normal pulse mode is given by

$$i_1 = nFACD \left[\frac{1}{\sqrt{\pi D t_m}} + \frac{1.02\omega^{3/2} t_m}{\gamma^{1/2}} \right] \tag{6.22}$$

where

t_m = time of measurement after application of the pulse

ω = angular velocity

γ = kinematic viscosity

The first term in this expression is the Cottrell expression and the perturbation due to rotation of the electrode can be assessed by the inequality

$$\frac{f}{\sqrt{\pi D t_m}} > \frac{1.02\omega^{3/2} t_m}{\gamma^{1/2}} \tag{6.23}$$

where f is the fraction by which the current is increased by rotation.

It can be seen that extremely short pulse widths are needed to obtain currents independent of rotation rate over the entire range of rotation rates available to the analytical chemist. For currents to be within 10% of the Cottrell value, the Nt_m factor (rps × sec) must be less than 1.4. Analytical work using pulse voltammetry under turbulent conditions or at rotating electrodes does not appear to be prevalent, and while theoretically it appears likely to be most successful, the absence of practical evaluation makes detailed assessment difficult. However, it is clear from the data in Note 41, and from theoretical expectations, that if the turbulance is too great, the independence of the limiting current on stirring, rotation rate, etc., no longer holds, and in this situation, allowance for this needs to be made as is the case for dc work. The advantages of being able to work under the conditions where the current is independent of the turbulance would of course be invaluable in analytical applications.

The differential pulse current at a rotated electrode [41] is given by the normal pulse current multiplied by a constant, $(\sigma - 1)/(\sigma + 1)$, which depends only on pulse height. It should depend therefore on rotation rate in exactly the same manner as does the normal pulse current for a given pulse width. Experimental data support this contention [41].

6.6 SOME OTHER VARIATIONS IN PULSE POLAROGRAPHY

6.6.1 Pseudoderivative Pulse Polarography

In Chap. 4 the technique of pseudoderivative dc polarography is described. The same approach has been used in the pulse method [42,43].
Voltage pulses of increasing amplitude are used, just as in the normal
mode, but a difference current measurement is used. The current
measured for a given pulse is subtracted from that for the following
pulse and the difference plotted as a function of potential.

In dc polarography, considerable improvement was noted in
the use of derivative techniques. This should be the case also in
pulse polarography but the closely related differential technique
has been more widely used. However, there are some advantages in
the derivative of pseudo-derivative approach. If one wishes to use
very short drop times in pulse polarography so as to permit fast scan
rates, the faradaic distortion effects described earlier in this
chapter restrict the use of the differential pulse method. The normal
pulse or pseudo-derivative pulse methods to not suffer from such distortions, and at very short drop times the pseudoderivative method is
marginally superior to the differential method [43]. The elimination
of undesirable phenomena attributable to adsorption can be achieved
with pseudoderivative pulse polarography [43] and it is in this area
that the major use of the technique should be expected. The minimization of adsorption can be exploited very successfully at stationary
electrodes, as described in the following section, in the closely
related technique of differential double-pulse voltammetry.

6.6.2 Differential Double-Pulse Voltammetry

The technique of differential double-pulse voltammetry is closely
related to the pseudoderivative technique and has been applied to
the in vivo determination of catecholamines [44]. Problems with film
formation on platinum electrodes precludes the use of linear sweep
or differential pulse voltammetry for determining catecholamines in

physiological media. Methods based on the normal pulse wave form,
such as pseudoderivative pulse voltammetry, or the one presently
under discussion, are ideally suited for minimizing effects of film
formation, because the electrode is periodically returned to an
initial value where "electrode cleaning" occurs, as noted elsewhere.
Figure 6.23 shows the wave form used in differential double pulse
voltammetry. Two simultaneously varying unequal square-wave potential
pulses are applied to the electrode, and the current difference
between them is measured as a function of potential.

The potential applicability of linear sweep voltammetric techniques to analytical and mechanistic charactization of biogenic
catecholamines such as dopamine and norepinephrine in vivo had been
clearly demonstrated by Adams and co-workers [45,46]. However, the

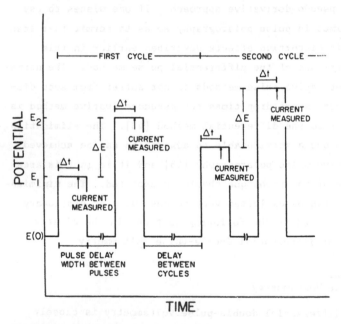

FIGURE 6.23 Potential-time waveform used in differential double-
pulse voltammetry. Typical values for the parameters would be
Δt = 15 msec; pulse width = 20 msec; delay time between pulses
= 200 msec; time for one complete cycle = 1 sec; ΔE = 0 to 200 mV.
[Reproduced from Anal. Chem. 48, 1287 (1976). © American Chemical
Society.]

difficulties of lack of sensitivity, susceptibility to electrode
poisoning, and interference from competing reactions meant the method
had considerable restrictions. Attempts to employ differential pulse
voltammetry [44,47] failed for two reasons: first, the products of
catecholamine oxidation undergo subsequent chemical reactions to
form solid deposits on the electrode, resulting in a continuous
decline in the peak height with use (Fig. 6.24); and second, chemi-
sorption of numerous species on conventional platinum electrodes
caused marked and variable inhibition of the electrode process.
Figure 6.25 shows the first scan and 10th scan for oxidation of
dopamine when using the differential double pulse voltammetric
waveform as an alternative to the usual differential pulse one
(Fig. 6.24). The value of the method is clearly revealed by comparing
the two figures. Chemical modification of the platinum electrode
[44] coupled with the differential double pulse method enables even
better performance as the curve for 8×10^{-7} M dopamine shown in
Fig. 6.26 demonstrates.

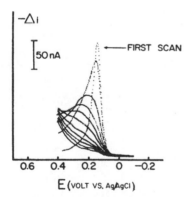

FIGURE 6.24 Differential pulse curves for oxidation of 1×10^{-4} M
dopamine at an initially clean platinum electrode. $\Delta E = 25$ mV;
$\Delta t = 50$ msec; A = 1.3×10^{-3} cm^2. Successive scans are shown from
top to bottom. [Reproduced from Anal. Chem. *48*, 1287 (1976).
© American Chemical Society.]

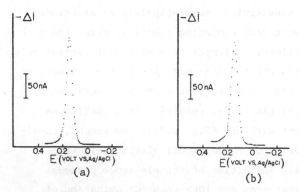

FIGURE 6.25 Differential double-pulse voltammetry curves for oxidation of 1×10^{-4} M dopamine at a platinum electrode. $\Delta E = 10$ mV; $\Delta t = 15$ msec; $A = 1.3 \times 10^{-3}$ cm^2. (a) First scan; (b) 10th scan. [Reproduced from Anal. Chem. 48, 1287 (1976). © American Chemical Society.]

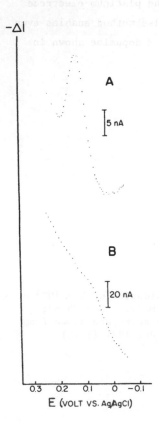

FIGURE 6.26 Differential double-pulse curves for oxidation of 8×10^{-7} dopamine: (A) iodide-treated platinum electrode; (B) untreated platinum electrode. $\Delta E = 50$ mV; $\Delta t = 15$ msec. [Reproduced from Anal. Chem. 48, 1287 (1976). © American Chemical Society.]

6.6.3 Alternate Drop
Differential Pulse Polarography

Alternate or dual-drop differential pulse polarography [48-50] is a
variant of differential pulse polarography, differing in that the
pulse, once applied, does not return to the base line until the end
of the next drop (Fig. 6.27). The important difference is that the
current samplings are taken at exactly the same time in the drop
life and at the same potential. The same procedure can, of course,
be applied also to the normal pulse mode [49].

As can be seen from Fig. 6.27 the samplings differ in time as
well as in the voltage applied compared with the conventional method.
It was shown previously that faradaic and charging current distortion in differential pulse polarography results from taking the two
measurements at different times (areas) and potentials on the same
drop. Both factors account for uncompensated residuals if the conventional difference $S_1 - S_2$ is taken, but are totally corrected
for if the difference $S_2 - S_3$ is taken instead. The sensitivity of
the instrumentation in two-drop operation is about 10^{-8} M for a

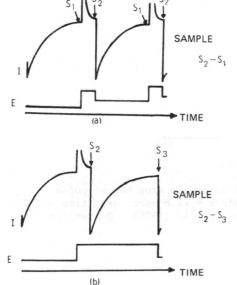

FIGURE 6.27 Comparison of sampling procedures for (a) differential pulse polarography and
(b) the alternate drop mode.
[Reproduced from Chem. Instrum.
5, 257 (1973-74).]

two-electron reduction step, and the range can be extended into the 10^{-9} M region if the output is heavily filtered. Additional filter-ing invariably introduces large distortions in the output, but for strictly analytical work with reference to a calibration curve, this can be tolerated. Future use of the alternate drop or equivalent approaches could be expected to become widespread for solutions below the 10^{-6} M level. Figure 6.28 shows the background current with ordinary and alternate drop normal pulse polarography in 1 M Hcl and the reduction in charging current is obvious. Figures 6.29 and 6.30 show some differential pulse curves for Pb(II) in 1 M KCl with both ordinary and alternate drop methods. Clearly, the decrease in charging current is accompanied by a small diminution in the faradaic current value [49], but this is not a serious problem, being less than a factor of 0.5 with usual values of t_m and t_p.

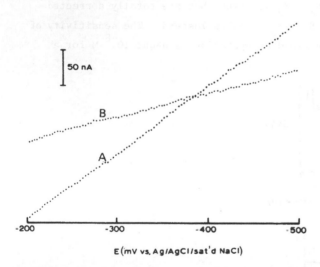

FIGURE 6.28   (A) Ordinary and (B) alternate drop normal pulse polarograms for 1 M HCl.   Pulse width = 12.6 msec; drop time = 1.0 sec.   [Reproduced from Anal. Chem. *48*, 242 (1976). © American Chemical Society.]

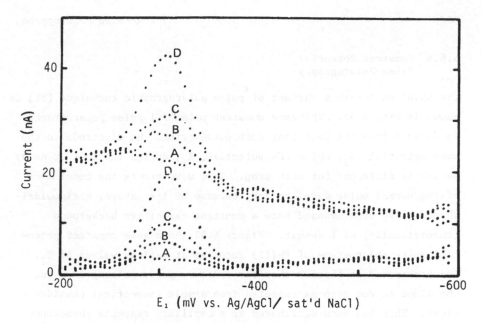

FIGURE 6.29 Ordinary (upper curves) and alternate drop (lower curves) differential pulse polarograms for Pb(II) in 0.1 M KCl. Pulse width = 19.4 msec; drop time = 1.0 sec; ΔE = −50 mV; [Pb(II)] = (A) 0, (B) 40, (C) 80, (D) 160 ng/ml. [Reproduced from Anal. Chem. 48, 242 (1976). © American Chemical Society.]

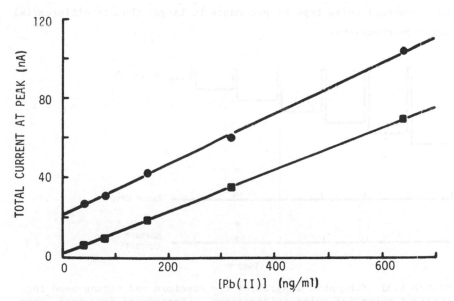

FIGURE 6.30 Calibration curves for Pb(II) in 0.1 M KCl (●) ordinary and (■) alternate drop differential pulse polarography. Pulse width = 19.4 msec; drop time = 1.0 sec; ΔE = −50 mV. [Reproduced from Anal. Chem. 48, 242 (1976). © American Chemical Society.]

6.6.4 Constant Potential
Pulse Polarography

The waveform for this variant of pulse polarographic technique [51] is
shown in Fig. 6.31. The name *constant potential pulse polarography*
is derived from the fact that each pulse brings the electrode to the
same potential, E_2, while the potential, E_1, during the rest or delay
period is different for each drop. This waveform is the complement
of the normal pulse waveform and, because of its nature, a sigmoidal-
shaped curve superimposed onto a constant capacitive background
(theoretically) will result. Figure 6.32 shows some constant potential pulse polarograms of Pb(II) recorded at various values of E_2.
The background is flatter than for ordinary pulse polarograms, but
the slope is not zero as expected from simple theoretical considerations. This has been attributed to a capillary response phenomenon
[51] first described by Barker and Gardner [4]. This variant of
pulse polarography does not appear to be as sensitive as either
ordinary or alternate drop differential pulse polarography because
the capillary response is larger, simply because the pulse amplitude
in any normal pulse type of procedure is larger than in differential
pulse polarography.

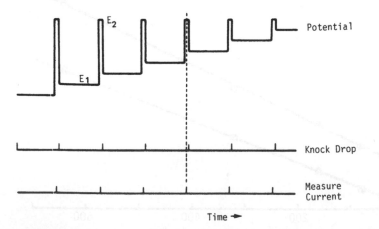

FIGURE 6.31 Schematic diagram of the waveform and timing used for
constant potential pulse polarography. [Reproduced from Anal. Chem.
48, 561 (1976). © American Chemical Society.]

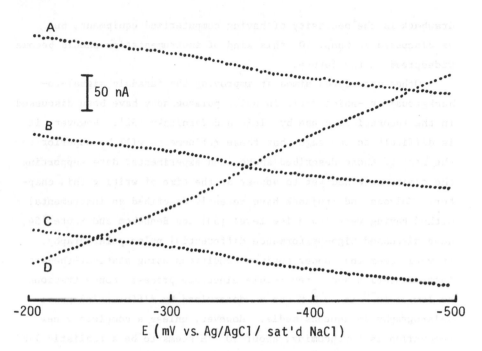

FIGURE 6.32 Constant potential pulse polarograms for 100 ng/ml of
Pb(II) in 1 M HCl. Pulse width = 10 msec; drop time = 0.5 sec.
The values of E_2 are (A) -0.50, (B) -0.35, (C) -0.20 V vs. Ag/AgCl
(sat'd NaCl). Curve D is a normal pulse polarogram (initial potential = -0.20 V) recorded under the same conditions. [Reproduced
from Anal. Chem. *48*, 561 (1976). © American Chemical Society.]

6.6.5 Miscellaneous

As with all modes of polarography, the possibility exists of using
subtractive techniques to eliminate the charging current. In
Chap. 4 on advances in dc polarography, the twin-electrode, dual-cell
method, where one cell contains the solvent/electrolyte and the other
the test solution, was considered to be too difficult to operate on
a routine basis because of the problem of maintaining two capillaries
in an identical state; the same reasoning holds in pulse polarography.
However, computerized instrumentation where the blank is stored in
memory and subtracted from the test curves using the same cell and
capillary for both measurements is readily implemented [32] and
represents practical means of obtaining high-quality data. A

drawback is the necessity of having computerized equipment, but
as discussed in Chap. 10, this kind of instrumentation should become
widespread in the future.

Other procedures aimed at improving the faradaic signal-to-
background or -noise ratio in pulse polarography have been discussed
in the theoretical sense by Klein and Yarnitzky [52]. However, it
is difficult to envisage that these refinements will be superior to
the best of those described above, and experimental data supporting
the claims made had yet to appear at the time of writing this chapter. Kalvoda and Trojanek have recently described an instrumental
method having very low noise level [53] and Bennekom and Shute [54]
have discussed high-performance differential pulse polarography.
It would seem that under optimum conditions using state-of-the-art
instrumentation and a reversible electrode process, concentrations
as low as 10^{-8} to 10^{-9} M can be determined by differential pulse
polarography in aqueous media. However, unless a completely new
innovation is forthcoming, about 10^{-8} M seems to be a realistic limit
to aim for in routine analytical work with differential pulse polarography. With normal pulse methods, around 10^{-7} M would be the kind
of detection limit that can be expected in the analytical laboratory
for an electrode process whose limiting current is diffusion-
controlled in aqueous media.

6.7 ASSESSMENT OF ANALYTICAL USEFULNESS

The relatively recent advent of high-quality three-electrode pulse
polarographic instrumentation has led to a remarkable upsurge of
analytical work, particularly in the differential pulse mode. The
normal pulse method in the future probably will have restricted use
in specialized applications of the kind mentioned in passing in this
chapter (e.g., process analysis [30], adsorption at stationary electrodes, etc.). There is no doubt in the author's mind that the differential pulse method must ultimately become one of the most widespread polarographic methods. The general applicability and high
sensitivity have already been used in analytical problems covering

a wide area of chemical endeavor, as perusal of Notes 55 to 75 demonstrates.

Figure 6.7(a) is the differential pulse polarogram of a
1.3×10^{-5} M solution of the antibiotic chloramphenicol. Figure
6.7(b) is a dc polarogram of the identical solution. This concentration represents the approximate lower limit for the dc method.
The wave is clearly discernible, but accurate evaluation of the wave
height and $E_{1/2}$ would be almost impossible. In the differential pulse
case, however, the clearly defined sharp peak allows precise measurement of the peak height and exact location of peak potential.

Figure 6.33 shows the i-E curves which can be obtained with a
variety of polarographic methods, from the nitrosated insecticide
carbaryl ("Sevin"). The polarographic determination of nitrosated
carbaryl is an accepted technique for determining trace residues
of this material. Curve A shows the dc polarogram of a 6.4 ppm solution. No detectable wave is observed. Curve B shows the dc linear

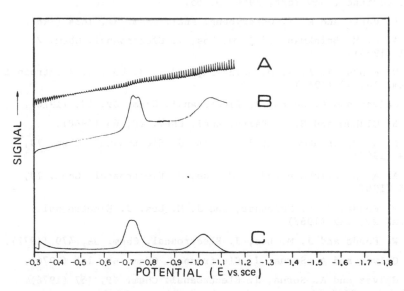

FIGURE 6.33 Comparison of dc, linear sweep, and differential pulse
polarograms for 6.4 ppm nitrosated carbaryl ("Seven"): (A) dc polarogram, drop time = 1 sec; (B) linear sweep voltammogram, scan rate
= 500 mV sec^{-1}; (C) differential pulse polarogram, ΔE = -50 mV,
drop time = 1 sec. [Reproduced from Anal. Chem. *44*, 75A (1972).
©American Chemical Society.]

sweep voltammogram at a dme using a scan rate of 500 mV sec^{-1}. The
wave is now clearly discernible and the determination readily carried
out. Curve C shows the differential pulse polarogram. In the
example being considered, the differential pulse method proved
to be five times more sensitive, but linear sweep voltammetry was
more rapid. The value of modern polarographic analytical methods is
most effectively demonstrated in Figs. 6.7 and 6.33.

NOTES

1. G. C. Barker, *Proc. Congr. Mod. Anal. Chem. Ind.*, St. Andrews,
    1957, p. 199.

2. G. C. Barker and A. W. Gardner, Z. Anal. Chem. *173*, 79 (1960).

3. J. H. Christie and R. A. Osteryoung, J. Electroanal. Chem. *49*,
    301 (1974).

4. G. C. Barker and A. W. Gardner, At. Energ. Res. Estab. (Brit.)
    AERE Harwell, C/R, 2297 (1958).

5. P. Delahay, *New Instrumental Methods in Electrochemistry*,
    Interscience, New York, 1954, p. 55.

6. E. P. Parry and R. A. Osteryoung, Anal. Chem. *37*, 1634 (1965).

7. A. A. A. M. Brinkman and J. M. Los, J. Electroanal. Chem. *7*,
    171 (1964).

8. A. W. Fonds, A. A. A. M. Brinkman, and J. M. Los, J. Electroanal.
    Chem. *14*, 43 (1967).

9. J. Galvez and A. Serna, J. Electroanal. Chem. *69*, 133 (1976).

10. K. B. Oldham and E. P. Parry, Anal. Chem. *40*, 65 (1968).

11. A. A. A. M. Brinkman and J. M. Los, J. Electroanal. Chem. *14*,
     269 (1967).

12. A. A. A. M. Brinkman and J. M. Los, J. Electroanal. Chem. *14*,
     285 (1967).

13. A. W. Fonds, J. L. Molenaar, and J. M. Los, J. Electroanal.
     Chem. *22*, 229 (1967).

14. A. W. Fonds and J. M. Los, J. Electroanal. Chem. *36*, 479 (1972).

15. J. Galvez and A. Serna, J. Electroanal. Chem. *69*, 145 (1976).

16. J. Galvez and A. Serna, J. Electroanal. Chem. *69*, 157 (1976).

17. L. F. Roeleveld, B. S. C. Wetsemar, and J. M. Los, J.
     Electroanal. Chem. *75*, 839 (1977), and references cited therein.

18. G. C. Barker and J. A. Bolzan, Z. Anal. Chem. *216*, 215 (1966).

19. J. Flemming, J. Electroanal. Chem. *75*, 421 (1977).

20. J. H. Christie, E. P. Parry, and R. A. Osteryoung, Electrochim. Acta *11*, 1525 (1966).

21. J. L. Morris, Jr., and L. R. Faulkener, Anal. Chem. *49*, 489 (1977).

22. D. R. Canterford, J. Electroanal. Chem. *52*, 144 (1974).

23. J. A. Turner, R. H. Abel, and R. A. Osteryoung, Anal. Chem. *47*, 1343 (1975).

24. W. O'Deen and R. A. Osteryoung, Anal. Chem. *43*, 1871 (1971).

25. J. H. Christie, J. Osteryoung, and R. A. Osteryoung, Anal. Chem. *45*, 210 (1973).

26. G. J. M. Heijne and W. E. Van Der Linden, Anal. Chim. Acta *82*, 231 (1976).

27. J. W. Dillard and K. W. Hanck, Anal. Chem. *48*, 218 (1976).

28. I. Ruzić, J. Electroanal. Chem. *75*, 25 (1977).

29. K. B. Oldham and E. P. Parry, Anal. Chem. *42*, 229 (1970).

30. E. P. Parry and D. P. Anderson, Anal. Chem. *45*, 458 (1973).

31. D. J. Myers and J. Osteryoung, Anal. Chem. *46*, 356 (1974).

32. A. M. Bond and B. S. Grabaric, Anal. Chim. Acta *88*, 227 (1977).

33. K. B. Oldham and E. P. Parry, Anal. Chem. *38*, 867 (1966).

34. E. P. Parry and R. A. Osteryoung, Anal. Chem. *36*, 1366 (1964).

35. A. M. Bond, T. A. O'Donnell, and R. J. Taylor, Anal. Chem. *46*, 1063 (1974).

36. H. E. Keller and R. A. Osteryoung, Anal. Chem. *43*, 342 (1971).

37. S. C. Rifkin and D. H. Evans, Anal. Chem. *48*, 1616 (1976).

38. K. F. Drake, R. P. Van Duyne, and A. M. Bond, J. Electroanal. Chem. *89*, 231 (1978).

39. S. C. Rifkin and D. H. Evans, Anal. Chem. *48*, 2174 (1976).

40. H. Blutstein and A. M. Bond, Anal. Chem. *48*, 248 (1976).

41. D. J. Myers, R. A. Osteryoung, and J. Osteryoung, Anal. Chem. *46*, 2089 (1974).

42. D. E. Burge, J. Chem. Educ. *47*, A81 (1970).

43. A. M. Bond and R. J. O'Halloran, J. Electroanal. Chem. *68*, 257 (1976).

44. R. F. Lane and A. T. Hubbard, Anal. Chem. *48*, 1287 (1976).

45. P. T. Kissinger, J. B. Hart, and R. N. Adams, Brain Res. *55*, 209 (1973).

46. R. L. McCreery, R. Dreiling, and R. N. Adams, Brain Res. *73*, 15 (1974).

47. R. F. Lane, A. T. Hubbard, K. Fukunago, and R. J. Blanchard, Brain Res. *114*, 346 (1976).

48. B. H. Vassos and R. A. Osteryoung, Chem. Instrum. *5*, 257 (1973-74).

49. J. H. Christie, L. L. Jackson, and R. A. Osteryoung, Anal. Chem. *48*, 242 (1976).

50. J. G. Osteryoung, J. H. Christie, and R. A. Osteryoung, Bull. Soc. Chim. Belg. *84*, 647 (1975).

51. J. H. Christie, L. L. Jackson, and R. A. Osteryoung, Anal. Chem. *48*, 561 (1976).

52. N. Klein and Ch. Yarnitzky, J. Electroanal. Chem. *61*, 1 (1975).

53. R. Kalvoda and A. Trojanek, J. Electroanal. Chem. *75*, 151 (1977).

54. W. P. Van Bennekom and J. B. Schute, Anal. Chim. Acta *89*, 71 (1977).

55. E. Temmerman and F. Verbeek, J. Electroanal. Chem. *12*, 158 (1966); *19*, 423 (1968); Anal. Chim. Acta *43*, 263 (1968); *50*, 505 (1970).

56. A. Lagrou and F. Verbeek, J. Electroanal. Chem. *19*, 125, 413 (1968).

57. E. P. Parry, T. H. Johnston, and H. J. Goldner, J. Electroanal. Chem. *13*, 177 (1967).

58. P. L. Cox, J. P. Heotis, D. Polin, and G. M. Rose, J. Pharm. Sci. *58*, 987 (1969).

59. D. J. Myers and J. Osteryoung, Anal. Chem. *45*, 267 (1973).

60. M. A. Brooks, J. A. F. deSilva, and L. M. D'Arcoute, Anal. Chem. *45*, 263 (1973).

61. J. A. F. de Silva and M. R. Hackman, Anal. Chem. *44*, 1145 (1972).

62. R. Impens, Z. M'vunzu, and P. Nangniot, Anal. Lett. *6*, 253 (1973).

63. E. Palecek and J. Doskosil, Anal. Biochem. *60*, 518 (1974).

64. L. De Galan, C. Erkelem, C. Jongerius, W. Maertens, and C. I. Morring, Z. Anal. Chem. *264*, 173 (1973).

65. R. W. Garber and C. E. Wilson, Anal. Chem. *44*, 1357 (1972).

66. E. P. Parry and K. B. Oldham, Anal. Chem. *40*, 1031 (1968).

67. G. Wolff and H. W. Nurnberg, Z. Anal. Chem. *216*, 169 (1966).

68. D. G. Prue, C. R. Warner, and B. T. Kho, J. Pharm. Sci. *61*, 249 (1972).

69. M. I. Abdulah and L. G. Royle, Anal. Chim. Acta *58*, 283 (1972).

70. R. Michielli and G. Bowning, Jr., J. Agr. Food Chem. *22*, 449 (1974).

71. K. M. Kadish and V. R. Spiehler, Anal. Chem. *47*, 1714 (1975).

72. K. Hasebe and J. Osteryoung, Anal. Chem. *47*, 2412 (1975).

73. J. F. Coetzee, G. H. Kazi, and J. C. Spurgeon, Anal. Chem. *48*, 2170 (1976).

74. E. Palecek, V. Brabec, and F. Jelen, J. Electroanal. Chem. *75*, 471 (1977).

75. W. M. Chey, R. N. Adams, and M. S. Yllo, J. Electroanal. Chem. *75*, 731 (1977).

Chapter 7

SINUSOIDAL ALTERNATING CURRENT POLAROGRAPHY

7.1 NOMENCLATURE

Any periodic voltage waveform can be used to generate a technique which can be called *alternating current* (ac) *polarography*. Figure 7.1 shows well-known examples of periodic waveforms available with most function generators. In view of the essentially limitless number of possible waveforms, one of the major difficulties many authors have found in describing the method of ac polarography is to define exactly what techniques are covered by the term. For example, Fleet and Jee [1] introduce their section on ac polarography in the following way.

> The field of A.C. polarography is becoming increasingly complex owing to the large variety of techniques which it now encompasses. The A.C. signals may be applied to the cell as either a voltage or a current and may be sinusoidal, square-wave, triangular wave, or amplitude-modulated form. The measured signal can be at the same frequency as the applied signal or can be a harmonic or intermodulation harmonic.

Figure 7.2 illustrates some of the waveforms used in various polarographic methods that broadly can be called alternating current. Note that the method of differential pulse polarography, already considered in Chap. 6, is contained in this figure. In keeping with the aim of this book, greatest detail and treatment is given to the most commonly used methods of polarography in the area

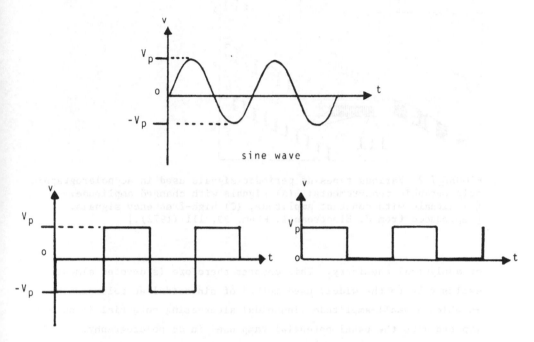

sine wave

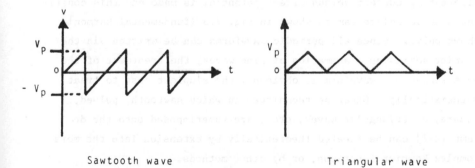

FIGURE 7.1 Some periodic waveforms available for use in alternating current polarography. [Reproduced from Anal. Chem. *47*, 2321 (1975). ©American Chemical Society.]

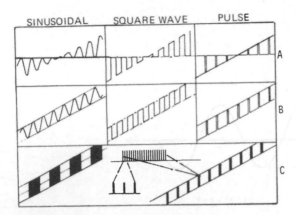

FIGURE 7.2 Various types of periodic signals used in ac polarographic
polarographic measurements: (A) signals with changed amplitude,
(B) signals with constant amplitude, (C) high-frequency signals.
[Reproduced from J. Electroanal. Chem. *39*, 111 (1972).]

of analytical chemistry. This chapter therefore is devoted almost
exclusively to the widely used method of sinusoidal ac polarography
in which a small-amplitude sinusoidal alternating potential is superimposed onto the usual potential ramp used in dc polarography.
After filtering out the dc component of the experiment, a plot of
alternating current versus direct potential is made and this constitutes an ac polarogram as shown in Fig. 7.3 (fundamental harmonic
shown only). Since all periodic waveforms can be written via the
Fourier series as a summation of sine waves, the treatment of the
pure sinusoidal waveform is obviously the simplest case to treat
mathematically. Other ac techniques in which sawtooth, pulsed,
square, or triangular waves, etc., are superimposed onto the dc
ramp [2-7] can be treated theoretically by extension into the more
complex Fourier functions, or by other methods.

Instead of measuring the alternating current as a function of
dc potential with completely automated instrumentation, one could
measure the impedance as a function of dc potential with an impedance
bridge. Measurement of impedance via a bridge circuit normally
requires a tedious, manual, point-by-point operation and the technique

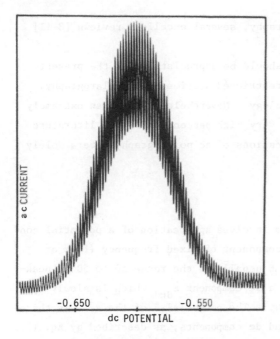

FIGURE 7.3 Fundamental harmonic ac polarogram of 3×10^{-3} M Cd(II) in 1 M Na_2SO_4. Applied potential is a 10 mV peak-to-peak sine wave at 320 Hz. [Reproduced from Anal. Chem. 35, 1811 (1963). ©American Chemical Society.]

referred to as *faradaic impedance measurement*, despite the fact that it is essentially equivalent to ac polarography, does not have the required degree of automation essential for its use in a modern analytical laboratory [8]. On these grounds, this alternative mode of measurement in ac polarography is excluded from subsequent discussion in this book. However, to get a balanced perspective, it should be noted that for studies in the area of electrode kinetics, as distinct from analytical applications, impedance measurements have proved extremely valuable [8-11]. The impedance characteristics as a function of direct potential can of course be transformed into an ac polarogram using the generalized form of Ohm's law, and vice versa, so that the coverage of ac polarography also really encompasses the impedance method. For additional discussion on terminology and nomenclature as well as further aspects of the broad topic of alternating current polarography, several excellent reviews [8-11] are available.

With this in mind, it should be appreciated that the present chapter encompasses a very restricted section of ac polarography, in terms of possible methodology. Nevertheless, it is an extremely important chapter because a very high percentage of the literature devoted to analytical applications of ac polarography refers solely to the sinusoidal version.

7.2 BASIC PRINCIPLES

A conventional ac polarogram involves application of a potential containing a small sinusoidal component of fixed frequency (f Hz or ω rad sec^{-1}) and amplitude ΔE usually in the range 10 to 50 mV peak-to-peak (p-p) together with a dc component E_{dc}, which is slowly scanned as in dc polarography. The potential E of the cell is therefore given by a sum of ac and dc components, as described by Eq. (7.1).

$$E = E_{dc} - \Delta E \sin \omega t \qquad\qquad (7.1)$$

Note that ΔE must have the magnitude of half the peak-to-peak value in this equation and in all subsequent theoretical work, although this fact has not always been appreciated by some workers who have mistakenly used peak-to-peak or root-mean-square (rms) values, or else have omitted units in discussions of the theory.

The solution conditions are normally the same as those used in conventional dc polarography, and a supporting electrolyte is employed whose concentration substantially exceeds that of the electroactive species of interest (Chap. 2). Such conditions enable the theory to be solved in terms of the usual rate-determining steps which may be diffusion control, heterogeneous electron transfer, homogeneous chemical reactions coupled to the electron transfer step, or charging of the electrode double layer, etc., as is the case with previously considered polarographic methods.

The alternating current whose frequency is the same as that of the applied alternating potential is measured as a function of dc potential. The resulting plot of the fundamental harmonic alternating current [$I(\omega t)$] versus dc potential constitutes the conventional ac polarogram. As with all polarographic methods, $I(\omega t)$ contains both faradaic and charging current components. The faradaic current provides a peak-shaped curve which coincides with the rising part of the dc polarographic wave (Fig. 7.4) for a reversible process. The peak current I_p of the ac polarogram is generally a linear function of

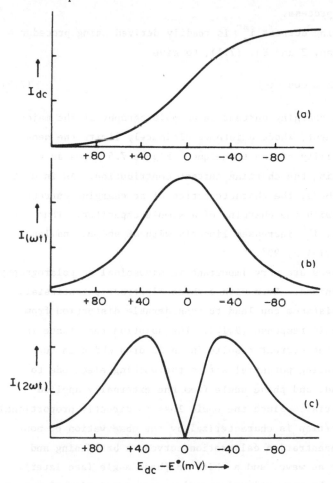

FIGURE 7.4 (a) DC, (b) fundamental, and (c) second harmonic ac polarograms for a reversible (diffusion-controlled) electrode process. AC polarograms are total current amplitude polarograms. [Reproduced from Computers in Chemistry and Instrumentation *2*, 369 (1972).]

concentration and is normally used as the basic parameter in quantitative analytical applications. The peak potential E_p corresponding
to the value of I_p is closely related to $E_{1/2}$, and is characteristic
of the electroactive species and the medium, as $E_{1/2}$ was shown to be
in dc polarography. Kinetic applications of ac polarography are
based on the fact that the magnitude, shape, and position of the
faradaic wave is governed by the kinetic and thermodynamic properties
of the electrode process.

The ac charging current i_c^{ac} is readily derived using procedures
established in Chap. 2 and Eq. (7.1), to give

$$i_c^{ac} = AC_{dl} \, \Delta E \, \omega \, \cos \omega t \tag{7.2}$$

In most cases the charging current in ac polarography is the major
nonfaradaic component, whose existence ultimately limits the sensitivity and versatility of the technique. Figure 7.5 shows an ac
polarogram including the charging current contribution. As in most
polarographic methods, the characteristics of ac charging current are
those associated with the charging of a simple capacitor. Thus, as
seen in Eq. (7.2), i_c^{ac} increases linearly with ΔE and ω, and leads
the applied potential by 90°.

Ohmic iR losses are very important in sinusoidal ac polarography
and must be recognized. Even with a three-electrode potentiostat,
uncompensated resistance can lead to considerable distortion from
the desired faradaic response [9,12]. The solution resistance in
combination with the current results in an iR drop which in turn
causes the alternating potential across the working electrode to
differ in amplitude and phase angle from the externally applied
alternating potential. Since the ohmic loss is directly proportional
to current, distortion is characterized by the observation of nonlinear I_p-vs.-concentration calibration curves, a broadening and
flattening of the ac wave, and a shift in phase angle (see later).

So far only the conventional ac polarogram consisting of a plot
of total fundamental harmonic current as a function of applied dc
potential has been considered. In any ac technique, components at

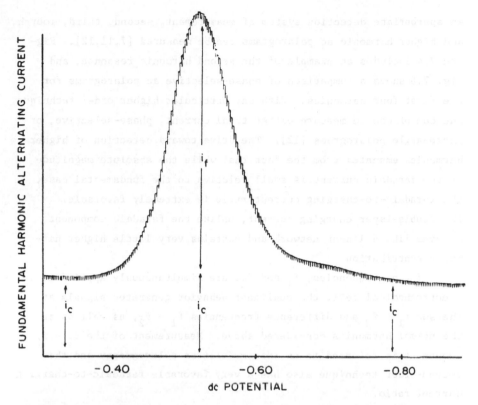

FIGURE 7.5 Fundamental harmonic ac polarogram including charging current. i_c = double-layer charging current; i_f = faradaic current. [Reproduced from Computers in Chemistry and Instrumentation 2, 369 (1972).

various phase angles with respect to the applied alternating potential rather than total current may also be measured. This leads to development of phase-sensitive ac polarography and phase-angle measurements which will be considered later in this chapter.

In addition to the fundamental harmonic current, the ac response includes components which occur because the electrochemical cell is a nonlinear network. In the sinusoidal experiment being considered in this chapter, higher order current components are found at frequencies which are integral multiples of the fundamental frequency, as well as at zero frequency (dc) of course. Thus, by developing

an appropriate detection system of measurement, second, third, fourth,
and higher harmonic ac polarograms can be measured [7,11,12]. Figure 7.4 includes an example of the second harmonic response, and
Fig. 7.6 shows a comparison of phase-selective ac polarograms for
the first four harmonics. With any particular higher order technique
one can of course measure either total current, phase-selective, or
phase-angle polarograms [12]. The drive toward detection of higher
harmonics emanates from the fact that while the absolute magnitude
of the faradaic current is small relative to the fundamental case,
the faradaic-to-charging current ratio is extremely favorable.
The double-layer charging current, unlike the faradaic component,
behaves like a linear network and contains very little higher harmonic contribution.

If two frequencies, f_1 and f_2, are simultaneously applied to an
electrochemical cell, the nonlinear behavior generates signals at
the sum $f_1 + f_2$ and difference frequencies $f_1 - f_2$, as well as at
the normal harmonics considered above. Measurement of the $f_1 - f_2$
component is referred to as *intermodulation* polarography and this
second-order technique also has a very favorable faradaic-to-charging
current ratio.

Measurement of the zero-order term, which is the direct current
induced by the applied sinusoidal ac potential, is referred to as
faradaic rectification polarography. Table 7.1 summarizes some of
the polarographic techniques which can be derived by using a superimposed sinusoidal wave. It now should be obvious that the cell
current in ac polarography contains a great deal more information
than that conveyed by recording a conventional ac polarogram, and
some of this can be useful to the analytical chemist. The following
discussion is aimed at showing how the various ac techniques can
be used systematically.

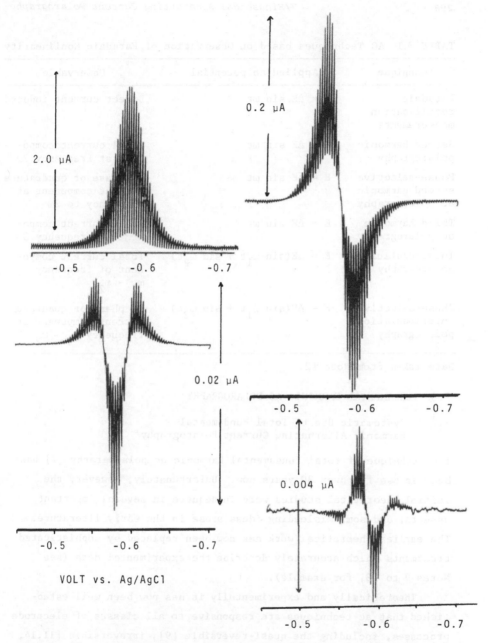

FIGURE 7.6 Phase-selective ac polarograms for the fundamental, second, third, and fourth harmonics. Solution is 5×10^{-4} M cadmium(II) in 1 M HCl. Applied potential is a 10 mV peak-to-peak sine wave at $f = 156$ Hz. Measured response is top left (1f), top right (2f), bottom left (3f), bottom right (4f). [Reproduced from Anal. Chem. *47*, 2321 (1975). © American Chemical Society.]

TABLE 7.1 AC Techniques based on Observation of Faradaic Nonlinearity

Technique	Applied ac potential	Observable
Faradaic rectification measurements	$E = \Delta E \sin \omega t$	Direct current induced
Second harmonic polarography	$E = \Delta E \sin \omega t$	Total current component at frequency 2ω
Phase-selective second harmonic ac polarography	$E = \Delta E \sin \omega t$	In phase or quadrature current component at frequency to 2ω
Third harmonic ac polarography	$E = \Delta E \sin \omega t$	Total current component at frequency 3ω
Intermodulation polarography	$E = \Delta E(\sin \omega_1 t + \sin \omega_2 t)$	Total current component of frequency $\omega_1 - \omega_2$
Phase-selective intermodulation polarography	$E = \Delta E(\sin \omega_1 t + \sin \omega_2 t)$	In phase or quadrature current component at frequency $\omega_1 - \omega_2$

Data taken from Note 12.

7.3 FUNDAMENTAL HARMONIC AC POLAROGRAPHY

7.3.1 Systematic Use of Total Fundamental Harmonic Alternating Current Polarography*

The technique of total fundamental harmonic ac polarography [7] has been in use for over 30 years now. Unfortunately, however, the initial theoretical studies were inadequate in several important aspects, and some misleading ideas arose in the early literature. The earlier theoretical work has now been replaced by sophisticated treatments which accurately describe the experimental data (see Notes 9 to 13, for example).

Theoretically and experimentally it has now been well established that ac techniques are responsive to all classes of electrode processes, including the quasi-reversible [9], irreversible [11,14, 15], and those involving complex mechanistic schemes [11]. Early

*Parts of the discussion in this section are reproduced by permission from a review by the author [13].

generalizations that ac polarography can be applied only to "reversible" electrode processes should be absent from current literature. However, it has been emphasized elsewhere in this book that modern polarographic methods generally provide optimum analytical performance for reversible processes, and this is especially true in ac polarography.

To establish the reversibility of a dc electrode process, normally one would establish rigorously that certain criteria hold (see Chap. 2). Fundamentally, the reversibility or otherwise of an ac electrode process should be defined even more rigorously than for the dc case. The time scale for ac polarography is governed predominantly, although not completely, by the frequency of the experiment, and it is possible that a reversible dc electrode process is nonreversible on the ac time scale. Furthermore, with the wide frequency range available, ac electrode processes can be reversible at, say, low frequency (for example, 20 Hz) and irreversible at higher frequencies (for example, 1000 Hz).

For convenience, the faradaic ac electrode processes can be considered in the usual four classes:

1. Reversible (diffusion-controlled, Nernstian) ac electrode process
2. Quasi-reversible ac electrode process
   a. Reversible dc charge transfer
   b. Quasi-reversible dc charge transfer
3. Irreversible dc electrode process
4. AC electrode processes with coupled chemical reactions or adsorption

The subdivision of the quasi-reversible category reflects the dual time domain of an ac polarogram. That is, since both dc and ac potentials are applied, the time scale is governed by a dc term (drop time) and an ac term (frequency). The need to consider the dc aspect of an ac experiment is frequently overlooked but, as will be shown subsequently, some of the more recent developments in ac techniques utilize advantages gained by varying the dc component of the experiment.

7.3.2 Reversible AC Electrode Processes

Reversible ac waves, being defined as those controlled solely by diffusion at all potentials, are relatively rare. The value of k_s has to be extremely large (for example 0.5 to 1 cm sec^{-1}) and the frequency low (for example, 100 Hz). For the purpose of the present discussion, the theory for the reversible ac electrode process is taken from Note 9. Thus, linear diffusion to a planar electrode is assumed.

The same boundary value problem is involved as for linear sweep voltammetry (LSV) and other techniques, but the potential E is given by Eq. (7.1) ($E = E_{dc} - \Delta E \sin \omega t$). The dc potential term is considered constant. This requires that the scan rate should be slow relative to the rate of change of alternating potential and that the dc potential does not change significantly over the life of a single mercury drop. The latter restriction also applies to dc polarography.

Substitution of Eq. (7.1) into the Nernst equation leads to exponential terms which can be developed into a power series [9]. Equating coefficients gives a system of integral equations whose solutions correspond to the direct current, dc faradaic rectification, fundamental harmonic ac, second harmonic ac, etc., reversible waves. Assuming the ac signal is of small amplitude ($\leq 8/n$ mV), the current produced by a fundamental harmonic reversible wave $I(\omega t)$, is given by the expression

$$I(\omega t) = \frac{n^2 F^2 A C_A (\omega D_A)^{1/2} \Delta E}{4RT \cosh^2(j/2)} \sin(\omega t + \frac{\pi}{4}) \qquad (7.3)$$

where

$\qquad$ A = electrode area

$\qquad$ C_A = concentration of electroactive species

$\qquad$ ω = angular frequency

$\qquad$ D_A = diffusion coefficient of electroactive species

$\qquad$ ΔE = amplitude of applied alternating potential

$\qquad$ t = time

$$j = \frac{nF}{RT} (E_{dc} - E^r_{1/2})$$

E_{dc} = dc component of potential

$E_{1/2}$ = reversible half-wave potential

The shape of the ac reversible wave can be described by an expression

$$E_{dc} = E^r_{1/2} + \frac{2RT}{nF} \ln \left[\left[\frac{I_p}{I}\right]^{1/2} \pm \left[\frac{I_p - I}{I}\right]^{1/2} \right] \qquad (7.4)$$

and it can be shown easily that the peak potential of the wave, E_p, is equal to $E^r_{1/2}$. In Eq. (7.4), I_p is the peak current corresponding to the value $\cosh^2(j/2) = 1$, i.e., the maximum peak faradaic alternating current $[I(\omega t)]_{max}$ at E_p and I is $I(\omega t)$. Equation (7.4) in fact has the same shape as the derivative of the reversible dc wave. This result also extends to higher harmonics because it has been shown that for a reversible process (but no other class) the nth harmonic has the shape of the nth derivative of the dc wave [9].

Examination of Eq. (7.4) shows that E_{dc} at half the wave height, where $I = I_p/2$, is given by

$$E_{dc} = E^r_{1/2} + \frac{2RT}{nF} \ln (\sqrt{2} - 1) \qquad (7.5a)$$

or

$$E_{dc} = E^r_{1/2} + \frac{2RT}{nF} \ln (\sqrt{2} + 1) \qquad (7.5b)$$

where the two solutions correspond to the two equivalent parts of the symmetrical ac wave. Subtraction of Eq. (7.5a) from (7.5b) gives the width of the ac wave at half its height.

$$\text{Half width} = \frac{2RT}{nF} \ln \frac{(\sqrt{2} + 1)}{(\sqrt{2} - 1)} = \frac{4RT}{nF} \ln (\sqrt{2} + 1)$$

$$= 1.52 \left(2.303 \frac{RT}{nF}\right) \qquad (7.6)$$

At 25°C, Eq. (7.6) has a value close to 90/n mV, and experimental measurement of the half width provides a convenient check on the

reversibility or otherwise of the fundamental harmonic ac electrode
process.

The half width is an extremely useful guide to the assignment
of a reversible ac electrode process, but since it is only a two-
point analysis of the ac wave, it is not entirely satisfactory. The
complete wave can be analyzed using Eq. (7.4). A graphic plot of

$$E_{dc} \text{ versus } \log \left[\left(\frac{I_p}{I}\right)^{1/2} \pm \left(\frac{I_p - I}{I}\right)^{1/2} \right] \qquad (7.7)$$

should be a straight line of slope 2 (2.303 $\frac{RT}{nF}$) for a reversible
process, where $\Delta E \leq 8/n$ MV. Figure 7.7 shows such a plot for the
reduction process Tl(I) + e $\rightleftharpoons$ Tl(amalgam) in 0.5 M NaClO$_4$ at
25°C. A straight line of slope (118 ± 2) mV is observed, in excellent agreement with the theoretical value of 118.2/n mV.

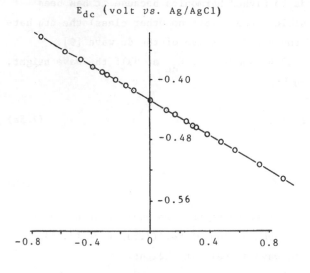

FIGURE 7.7 A test for reversibility of the ac electrode process
Tl(I) + e $\rightleftharpoons$ Tl (amalgam) in 0.5 M NaClO$_4$. [Reproduced from Anal.
Chem. 44, 315 (1972). © American Chemical Society.]

The reversible fundamental harmonic ac electrode process of the type $A + ne \rightleftharpoons B$ is also characterized by several other features. The peak potential should be independent of concentration and drop time as should the shape of the wave. Absence of dependence of E_p and wave shape on these variables is not in itself an unambiguous test for reversibility, but as will be shown later, the shapes of some quasi-reversible processes and those coupled with adsorption or chemical reactions are often remarkably sensitive to changes in drop time. Examination of the drop time dependence therefore can be very useful sometimes in confirming the nature of the electrode process.

Although it has been established that one requisite of a reversible process is that $E_{1/2}$ and E_p should be equal, close or exact agreement between the dc and ac parameter can also be obtained for quasi-reversible or other classes of electrode process. The mere equivalence of these two parameters does not on its own define an electrode process as being reversible (as has sometimes been assumed in the literature). For this observation needs to be coupled with considerably more evidence before any firm conclusions about reversibility can be reached.

Equation (7.3) obviously also has important implications in analytical work. The parameter of interest in establishing a calibration curve is I_p, where

$$I_p = \frac{n^2 F^2 A C_A (\omega D_A)^{1/2} \Delta E}{4RT} \sin\left(\omega t + \frac{\pi}{4}\right) \tag{7.8}$$

A linear dependence of I_p on concentration, area, and ΔE, and a squared dependence on n are therefore predicted. Since $A \propto m^{2/3} t^{2/3}$, then I_p should be independent of mercury column height for a gravity-controlled drop time experiment. This is different from diffusion-controlled waves where i_d is proportional to the square root of mercury column height. For reversible redox processes in which both the oxidant and reductant are soluble in solution, this prediction is observed experimentally. For reversible amalgam-forming systems, spherical diffusion terms lead to a small column height dependence, particularly at long drop times. Smith and co-workers [16] in a series of papers have discussed this influence of spherical diffusion. For non-amalgam-forming systems, a column height dependence is indicative of kinetic complications.

Figure 7.8 shows a plot of I_p vs. concentration for the reduction of In (n = 3), Cd (n = 2) and Tl (n = 1) in 1 M NaCl. Apart from the linearity, a second feature is the dependence of I_p per unit concentration on n. In dc polarography a linear dependence of i_d on n is predicted, but the n^2 factor in ac polarography [Eq. (7.8)] accentuates differences in sensitivity for the three species being considered. Reversible three-electron reductions such as In(III), Sb(III), or Bi(III) in hydrochloric acid therefore are among the most sensitive in ac polarography. Under optimum conditions, the limits of detection for the species in Fig. 7.8 are approximately 5×10^{-6} M for Tl(I), 2×10^{-6} M for Cd(II), and 1×10^{-6} M for Sb(III), using total alternating current measurements.

It should also be noted that I_p is independent of k_s for reversible processes. This is particularly important because any variation of k_s resulting from a slight change in matrix will not alter I_p. By comparison, as will be seen later, I_p is dependent on k_s for quasi-reversible electrode processes, and the analytical use of ac waves derived from nonreversible electrode processes are far more

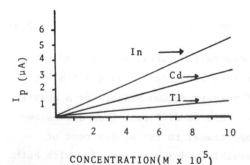

FIGURE 7.8 Plots of peak height vs. concentration for In(III), Cd(II), and Tl(I) in 1 M NaCl. ΔE = 10 mV rms; f = 50 Hz. [Reproduced from Anal. Chem. *44*, 315 (1972). ©American Chemical Society.]

susceptible to interference. This and other reasons will be seen
to lead to the conclusion that ac polarography is a technique generally best suited to reversible electrode processes, and that the
characterization of the nature of the ac electrode process is therefore essential.

7.3.3 Quasi-Reversible AC Electrode Processes

Reversible DC Charge Transfer. Quasi-reversibility of ac electrode
processes is more common than complete reversibility. Quasi-reversible
electrode processes are described in part, but not completely, by
Eqs. (7.3) and (7.4), which apply to the reversible ac electrode
process. Such electrode processes are not, however, completely diffusion controlled in the ac sense. The first type of quasi-reversible
behavior in which reversible dc charge transfer is observed, or very
nearly so, is sometimes difficult to distinguish simply from the
fully reversible electrode process. In fact, in analytical and
electroanalytical applications with low-frequency ac polarography,
any distinction is quite slight.

Theoretical considerations of the quasi-reversible electrode
reaction [9] show that E_p varies with frequency but approaches $E_{1/2}^r$
at low frequency and for $k_s \geq 10^{-2}$ cm sec^{-1}. With a k_s value of this
order, the dc electrode process is certainly reversible or close to
reversible, and so the equivalence

$$E_{1/2} = E_{1/2}^r = E_p$$

can apply for such systems, as is also the case with reversible ac
electrode processes. However, the ac quasi-reversible wave will be,
in general, slightly broader than that of the reversible electrode
process, and even slight departures from the theoretical Nernstian
slope (2.303RT/nF) of the dc plot of

$$E_{dc} \text{ vs. } \log \frac{i_d - i}{i}$$

will be exhibited in the ac wave as broadening.

For the purposes of this chapter, reversible dc charge transfer
is considered to be applicable if the dc plot of E_{dc} vs. log $[(i_d - i)/i]$
is linear with slope close to 2.303RT/nF. If a curved plot or a linear
plot of slope greater than 2.303RT/nF is observed, then dc charge
transfer will be considered to be quasi-reversible and the corresponding ac quasi-reversible electrode process will be considered in the
following section.

The important difference in distinguishing a quasi-reversible
electrode process from a reversible one is the magnitude of $I(\omega t)$.
The value of $I(\omega t)$ for the quasi-reversible electrode process is
complex [9]. However, it contains terms for k_s and α, and therein
lies a major difference from the reversible case. It is also appreciably less in magnitude than for the reversible electrode process.

A useful illustration of a quasi-reversible ac electrode process
with reversible dc charge transfer may be seen in the reduction of
bismuth(III), [Bi(III) + 3e $\rightleftharpoons$ Bi(amalgam)] in perchloric acid in
the presence of various concentrations of chloride. Table 7.2 gives
some data obtained by Bauer and Elving [17]. In the absence of
chloride, the electrode process for Bi(III) is irreversible by the
usual dc criteria. However, with addition of as little as 2.2×10^{-3} M
chloride, or greater, the dc plot of E_{dc} vs. log$[(i_d - i)/i]$ is

TABLE 7.2 Influence of Chloride on the Reduction of Bismuth(III)[a]

Chloride concn (M)	Rate constant (cm sec^{-1})	Slope of dc plot of E_{dc} vs. log$[(i_d - i)/i]$ (mV)
0.0000	0.023	40
0.0022	0.030	23
0.0062	0.079	23
0.018	0.21	23
0.060	0.20	21
0.180	0.20	20

[a]Bismuth concentration = 2.53×10^{-4} M in 0.5 M perchloric acid.
Data from Note 17.

consistent with what would be considered a reversible dc electrode process. The rate constant, however, increases with increasing chloride concentration.

Data given by Bond and Waugh [18] on the dependence of the I_p value as a function of chloride concentration, are given in Fig. 7.9. The I_p value increases markedly with chloride ion concentrations up to about 0.1 M and then becomes virtually independent of chloride concentration. At chloride concentrations below 0.1 M, the electrode process has reversible dc charge transfer, but the ac electrode process is quasi-reversible and I_p increases as k_s increases. Above 0.1 M chloride, the value of k_s is sufficiently high for the electrode process to be reversible even in the ac sense and I_p becomes independent of k_s. In concentrated chloride media, the Bi(III) electrode process attains k_s values ≥ 1 cm sec^{-1} [19,20]. Values of k_s of this order are necessary for an ac electrode process to be diffusion controlled and, hence, completely reversible.

Quasi-Reversible DC Charge Transfer. If k_s lies between 2×10^{-2} and 5×10^{-5} cm sec^{-1}, the dc electrode processes may be defined as quasi-reversible (Chap. 2).

Reduction and oxidation of tin(II) in fluoride media [21] provide excellent examples of such a system. A dc polarogram of Sn(II) in 0.8 M NaF is given in Fig. 7.10. Two quasi-reversible dc electrode processes are observed. The oxidation (more positive) wave corresponds to the electrode process Sn(II) $\rightleftharpoons$ Sn(IV) + 2e and the reduction wave to Sn(II) + 2e $\rightleftharpoons$ Sn(amalgam). The heights of dc waves are independent of kinetics of the electrode process and

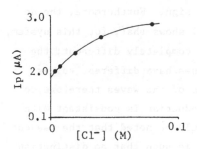

FIGURE 7.9 Dependence of peak height on the concentration of chloride for the electrode process Bi(III) + 3e $\rightleftharpoons$ Bi(0). ΔE = 10 mV rms; f = 50 Hz. [Reproduced from *Anal. Chem.* 44, 315 (1972). © American Chemical Society.]

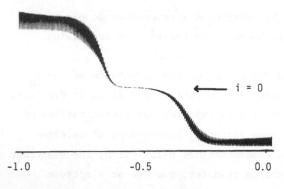

FIGURE 7.10 Quasi-reversible dc electrode processes for Sn(II)
in 0.8 M NaF. More positive wave is the oxidation electrode process
Sn(II) $\rightleftharpoons$ Sn(IV) + 2e. More negative electrode process is the
reduction Sn(II) + 2e $\rightleftharpoons$ Sn(0). [Reproduced from Anal. Chem. 44,
315 (1972). © American Chemical Society.]

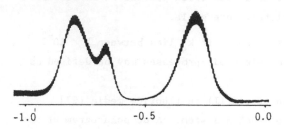

FIGURE 7.11 The quasi-reversible ac electrode processes for Sn(II)
in 0.8 M NaF. Same solution as Fig. 7.10. ΔE = 10 mV rms; f = 50 Hz.
[Reproduced from Anal. Chem. 44, 315 (1972). © American Chemical
Society.]

i_d for both waves is the same except for sign. Furthermore, the dc
shapes are almost identical. Figure 7.11 shows that, for this system,
the two ac electrode waves are obviously completely different; the
two quasi-reversible ac electrode processes have different values
of k_s and of α and the shapes and heights of the waves therefore do
not coincide. The double peak for the reduction is consistent with
an α value far removed from 0.5. It should be noted that the readout
of ac polarography with most instruments is such that no distinction

is possible between oxidation and reduction waves, and dc polarography
is necessary to make the distinction.

Theoretically, quasi-reversible electrode processes have E_p values
quite close to $E_{1/2}^r$ [9]. Whether the values are more positive or
negative than $E_{1/2}^r$ depends upon the values of k_s and α. Although
intuitively it may seem strange to have E_p more positive than $E_{1/2}^{..}$
and therefore, of course, also more positive than $E_{1/2}$, this theoretical observation is in agreement with experiment.

Results are given at various concentrations of sodium fluoride
for zinc(II) reduction in Table 7.3; $E_{1/2}^r$ was calculated from the
dc polarograms from extrapolation of the limiting slope of the curved
plot of

$$E_{dc} \text{ vs. } \log \frac{i_d - i}{i}$$

For this set of data, the shapes of the ac waves remain essentially the same for all concentrations of fluoride, although I_p
decreases with increase in fluoride, and k_s presumably decreases
slightly with increasing fluoride concentration.

With examples such as for tin(II) and zinc(II), the departure
from reversibility is small and E_p is also close to $E_{1/2}^r$. Examples
where the k_s value is considerably lower can be given. In 1 M perchloric acid the electrode process

$$Bi(III) + 3e \rightleftharpoons Bi(amalgam)$$

has a k_s value of 3.8×10^{-4} cm sec^{-1} [20]. The dc wave is reasonably
well defined, but considerably drawn out, the value of $E_{1/4} - E_{3/4}$
being 64 mV. At 50 Hz the ac wave is slightly asymmetrical with a
half width of 96 mV, and obviously the electrode process is not reversible. E_p is -0.004 V vs. Ag AgCl compared with the dc value for
$E_{1/2}$ of 0.000 V vs. Ag AgCl [13].

In 1 M HNO_3 the bismuth electrode process has a k_s value of
3.7×10^{-3} [20]. An extremely well-defined ac wave is observed as
shown in Fig. 7.12 with a half width of 64 mV. As in 1 M perchloric
acid, reduction is not reversible, although an extremely symmetrical

TABLE 7.3 AC and DC Polarographic Parameters for Reduction of Zinc in Fluoride Media[a]

[NaF] (M)	$-E_{1/2}$ (V vs. Ag/AgCl)	i_d (μA)	$-E^r_{1/2}$ (V vs. Ag/AgCl)	$-E_p$ (V vs. Ag/AgCl)	I_p (μA)
0.00	0.9547	1.535	0.9505	0.9443	0.975
0.16	0.9636	1.495	0.9611	0.9508	0.785
0.24	0.9651	1.481	0.9625	0.9533	0.690
0.40	0.9731	1.421	0.9705	0.9610	0.585
0.56	0.9764	1.403	0.9726	0.9655	0.520
0.80	0.9821	1.367	0.9780	0.9716	0.440

[a][Zn(II)] = 2 × 10^{-4} M. Ionic strength of 1.0 monitored by sodium perchlorate. ΔE = 10 mV rms. F = 50 Hz.

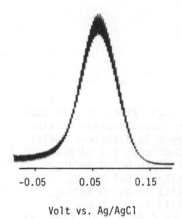

-0.05   0.05   0.15

Volt vs. Ag/AgCl

FIGURE 7.12 An ac polarogram of bismuth(III) in 1 M HNO_3. ΔE = 10 mV rms; f = 50 Hz. [Reproduced from Anal. Chem. *44*, 315 (1972). © American Chemical Society.]

well-defined wave is observed which without mathematical analysis, might be assumed to be reversible. Furthermore, using a frequency of 50 Hz, E_p is 0.048 V vs. Ag AgCl, which is very close to the value for $E_{1/2}$ of 0.050 V vs. Ag/AgCl [13]. Obviously, agreement between E_p and $E_{1/2}$ cannot be used to define a reversible electrode process.

The shapes of the polarographic waves for quasi-reversible electrode processes have been shown to be extremely variable in the examples given and ac quasi-reversible electrode processes are typified by this feature. Any variation in k_s or α can cause marked changes in I_p and shape. Consequently, drop time variations are often extremely significant, as are changes in electrolyte, in determining the shape and height of the ac wave observed. Figure 7.13 shows analytical calibration curves of I_p vs. concentration of Cu(II) and Zn(II) in 0.5 M sodium perchlorate and 0.5 M sodium fluoride. Copper (II) and zinc(II) both undergo reduction of the type

$$M(II) + 2e \rightleftharpoons M(amalgam)$$

in these media. The copper(II) electrode process, however, approaches reversibility, but that for zinc is quasi-reversible. Two important features arise from Fig 7.13. First, the much greater sensitivity of the more reversible electrode process is obvious. Second, the value of I_p for copper(II) is virtually the same in both perchlorate

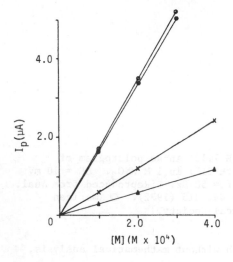

FIGURE 7.13 Plots of peak current vs. concentration of zinc and copper in different media. ΔE = 10 mV rms; f = 50 Hz. (o) Cu(II) in 0.5 M NaClO$_4$. (•) Cu(II) in 0.5 M NaF, (x) Zn(II) in 0.5 M NaClO$_4$, (▲) Zn(II) in 0.5 M NaF. [Reproduced from Anal. Chem. *44*, 315 (1972). © American Chemical Society.]

and fluoride media. However, for zinc, considerable differences in I_p occur between perchlorate and fluoride media. This latter behavior typifies a quasi-reversible electrode process, and even slight changes in electrolyte can often markedly alter I_p, as the value depends on k_s. On the other hand, reversible ac electrode processes are independent of k_s, and I_p does not usually change significantly even with considerable change in the nature of the electrolyte, unless the electrode process is altered substantially by complexation, for example. The only alteration in I_p which can arise with a reversible ac electrode process, according to Eq. (7.8), is a change in $D_A^{1/2}$, and this should be fairly small in most cases.

The analytical use of quasi-reversible ac electrode processes therefore can be extremely unreliable. Measurement of I_p on an unknown solution and comparison with standards should normally be undertaken only if it has been previously established that the composition of the unknown is such that it will not alter k_s, compared with the value obtained in the standards. The usual analytical procedure of ensuring that the wave shape and position of unknown and standards are the same should of course be strictly adhered to as this provides a means for validating data. Any complexing reagent, adsorbable species, or other electroactive species present in the unknown, but not included in the standard, can potentially alter k_s and lead to incorrect results. This feature, combined with the lower sensitivity of quasi-reversible electrode processes, makes it difficult to recommend use of anything but reversible ac electrode processes for analytical purposes, except where one is very careful with the data analysis and validation procedures, or else when one is willing to employ the method of standard additions to each sample, assuming it is applicable.

7.3.4 Irreversible AC Electrode Processes

The early inexact theory of ac polarography indicated that the magnitude of the ac polarographic wave is vanishingly small for irreversible electrode processes. However, contrary to earlier beliefs, Timmer et al. [14] and Smith and McCord [15] have now established by theory and experiment that a finite, measurable ac polarographic wave is obtained with irreversible processes. Despite these results, the idea that ac measurements can be made on irreversible electrode processes has been slow to filter through to electrochemists, and exceedingly few examples of applications have appeared in the literature.

Although quantitative differences are apparent between the theoretical treatments for the irreversible ac electrode processes, the theories are in agreement regarding the very important quantitative prediction that a measurable ac wave of magnitude and shape, which is independent of k_s, will be observed with irreversible systems. The sole influence of k_s is to determine the position of the wave on the dc potential axis [15]. The current magnitude $I(\omega t)$ is a rather complex function. However, unlike in the reversible case, it is proportional to α, the charge transfer coefficient. The position of an irreversible wave is given by Eqs. (7.9) to (7.11).

$$E_p = E_{1/2}^r + \frac{RT}{\alpha nF} \ln \left(\frac{1.349 k_s t^{1/2}}{D^{1/2}} \right) - \frac{RT}{2\alpha nF} \ln Q \tag{7.9}$$

where $Q = 1.907(\omega t)^{1/2} \lambda$, $D = D_A^{\beta} D_B^{\alpha}$, and $\beta = 1 - \alpha$. Since

$$E_{1/2} = E_{1/2}^r + \frac{RT}{\alpha nF} \ln \left(\frac{1.349 k_s t^{1/2}}{D^{1/2}} \right) \tag{7.10}$$

it follows that

$$E_p - E_{1/2} = - \frac{RT}{2\alpha nF} \ln Q \tag{7.11}$$

This result indicates that the ac E_p value with irreversible systems is displaced in the negative direction by a substantial amount from the dc half-wave potential. This criterion characterizes an irreversible ac reduction wave.

Other features also characterize irreversible ac waves. The waves are extremely broad and are of very low sensitivity (i.e., low current per unit concentration) compared with those for reversible ac electrode processes. For totally irreversible electrode processes, dc polarography or, even better, pulse polarography, would be preferred, in general, to ac polarography in analytical applications. Conversely, of course, the low current per unit concentration provides a considerable advantage if a species exhibiting a reversible process is to be determined in the presence of a species giving rise to an irreversible process. The high specificity obtained with ac methods will be discussed in detail at the end of this chapter.

7.3.5 Electrode Processes with Coupled Chemical Reactions or Adsorption

In Secs. 7.3.3 and 7.3.4 the departure from reversibility of electrode processes has been considered in terms of slow charge transfer. However, kinetic effects introduced by coupled chemical reactions or adsorption can be rather prevalent with certain types of systems and frequently these effects can explain the departure from reversibility rather than slow charge transfer. Note 11 provides an excellent example of some of the effects of coupled chemical reactions on electrode processes. Adsorption effects are considered in detail in the same reference.

Examination of the theory for electrode processes with coupled chemical reactions [9,11] shows that the effect on current amplitude, and the shape of the wave are sufficiently similar to those of quasi-reversible charge transfer that one effect could easily be mistaken for the other and unless one is prepared to proceed with measurement of frequency response, preferably the phase angle (see later) and other time consuming measurements and calculations, such as those undertaken by Huebert and Smith in their work on the polarographic behavior of cyclooctatetraene [22], then the reason for departure from reversibility may often be open to question.

Adsorption and related surface phenomena, when coupled to the charge transfer, can lead to highly distorted waves. Such waves are usually not analytically usable, although in some cases extremely high sensitivity can be obtained. The special classification of nonfaradaic processes called *tensammetric waves*, which arise from adsorption-desorption processes, will be considered under a separate section.

With most existing commercial instrumentation, the analytical chemist would probably find it far too time consuming, if not impossible, to undertake studies to determine the mechanism of electrode processes involving coupled chemical reactions and/or adsorption, and in any case there is probably no real advantage in doing this. The peak current being proportional to k_s and α, and probably other parameters [11], means that as for slow electron transfer, analytical use of the electrode process is somewhat "chancey" unless the method of "standard additions" is used (assuming it is applicable) or strict precautions are made to ensure that the standards and unknown solutions are prepared and maintained in exactly the same medium. As is also the case where slow charge transfer gives rise to the quasi-reversible electrode process, the peak current is likely to be extremely sensitive to slight variations in the supporting electrolyte. For instance, if one were to attempt to determine an organic species undergoing an electrode process with coupled chemical reaction or adsorption in a nonaqueous solvent, then one could have to ensure that the most likely impurity, that of water, were completely absent

for all standards and unknowns or remained at the same level in all
solutions, as the electrode process is likely to be dependent on the
water concentration.

Although the more reversible the electrode process, the better
the analytical method is likely to be, it is often necessary to use
electrode processes showing departures from reversibility. It has
been the author's experience that if a nonreversible electrode process
is to be used, then the best check that correct results are being
obtained is to ensure that the peak potentials and shapes of ac
waves from unknown solutions are identical to the standards used for
the determination. Any departure from this agreement is an excellent
indication that interference is being encountered with the analytical
method.

7.4 PHASE-SENSITIVE FUNDAMENTAL
     HARMONIC AC POLAROGRAPHY

In Sec. 7.3, certain theoretical and experimental correlations that
may be made with conventional ac polarographic instrumentation are
given. The extension to the application of phase-sensitive, three-
electrode, variable frequency, and variable amplitude ac polarography
is now considered, both in terms of experimental and theoretical
correlations developed previously and in terms of additional correlations possible with this form of instrumentation, particularly
those directly relevant to the analytical applications.

In Sec. 7.3, it was established and strongly recommended that
analytical use of ac polarography is best confined to fast electrode
processes. That section adequately covers the features of nonreversible electrode processes, from the analytical viewpoint, and it is
considered that any further discussion on totally irreversible electrode processes, and electrode processes with coupled chemical reactions, while being of considerable interest from mechanistic and
other viewpoints, need not be entertained further. Hence, this seg-
ment will be confined mainly to a discussion of features affecting
the study and analytical use of fast electrode processes.

7.4.1 Background or Charging Current

In order to understand the advantages and the use of phase-sensitive instrumentation in ac polarography, it is necessary to introduce a discussion of the charging current, which has virtually been neglected in the first section. Equation (7.2) gives the mathematical formulation of charging current in ac polarography. Implications of this equation are examined in the subsequent discussion.

7.4.2 Dependence of Charging Current and Faradaic Current on Frequency

Charging Current. Figure 7.14 shows the variation of the charging current with frequency in 0.5 M $NaClO_4$. The considerable increase of charging or background current with frequency can be seen in this electrolyte. In agreement with Eq. (7.2), the charging current actually increases directly with frequency, provided no ohmic iR drop or instrumental artifacts are present.

Faradaic Current

Reversible AC Electrode Process. Figure 7.15 shows the variation of the faradaic current with frequency for the electrode process

$$Cd(II) + 2e \rightleftharpoons Cd(amalgam)$$

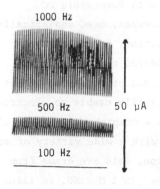

1000 Hz

500 Hz   50 μA

100 Hz

-0.4   -0.5   -0.6

Volt vs. Ag/AgCl

FIGURE 7.14 Variation of charging current with frequency. ΔE = 10 mV p-p; medium = 0.5 M $NaClO_4$. [Reproduced from *Anal. Chem. 44*, 315 (1972). © American Chemical Society.]

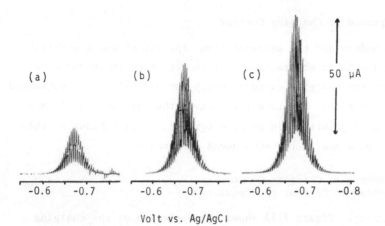

Volt vs. Ag/AgCl

FIGURE 7.15 Variation of faradaic current with frequency for the reversible ac electrode process Cd(II) + 2e $\rightleftharpoons$ Cd(0) in 5 M HCl. ΔE = 10 mV p-p. Phase-sensitive readout: (a) f = 20 Hz; (b) f = 100 Hz; (c) f = 200 Hz. [Cd(II)] = 1 × 10^{-3} M. [Reproduced from Anal. Chem. 44, 315 (1972). © American Chemical Society.]

in 5 M HCl. According to Eq. (7.8), I_p varies with the square root of frequency for a reversible electrode process and this relationship is closely obeyed for cadmium. That is, a plot of I_p vs. $\omega^{1/2}$ is a straight line with intercept at the origin and an additional criterion of ac polarographic reversibility is now introduced.

Quasi-Reversible AC Electrode Process with Reversible DC Charge Transfer. Very few ac electrode processes, used analytically, are likely to be as highly reversible as cadmium, and this type of electrode process therefore cannot be considered typical.

Assuming that the ac copper (II) electrode process was in fact likely to be a typical example of an analytically usable ac electrode process, Bond and Canterford [23] undertook a comprehensive study of the ac polarography of copper in 1 M NaNO$_3$ with a wide variety of ac techniques. In the remainder of this section, data are drawn from this work on the copper(II) electrode process in 1 M NaNO$_3$ to illustrate the various features of ac polarography with phase-sensitive three-electrode instrumentation, in addition to considering the highly reversible cadmium electrode process.

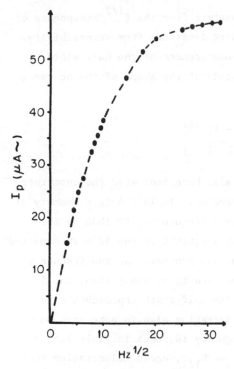

FIGURE 7.16 Variation of I_p with frequency for the Cu(II) + 2e ⇄ Cu(0) quasi-reversible ac electrode process in 1 M NaNO$_3$. ΔE = 10 mV p-p; [Cu(II)] = 1 × 10^{-3} M. [Reproduced from Anal. Chem. *44*, 315 (1972). ©American Chemical Society.]

Figure 7.16 shows a plot of I_p vs. $f^{1/2}$ for copper in 1 M NaNO$_3$. At low frequencies this plot approaches a straight line passing through the origin, and the faradaic current is proportional to $f^{1/2}$. At higher frequencies, the proportionality to frequency becomes considerably less than $f^{1/2}$. A simple manner in which to explain this behavior is to consider the influence of frequency, as altering the time scale of the ac polarographic experiment. At low frequency, a fast electrode process can exhibit behavior approximating that of a reversible ac electrode process. Any increase in frequency of the polarographic experiment decreases the time scale, and eventually a frequency will be reached which does not allow the electrode process to even closely approach diffusion-controlled conditions. Thus, at high frequencies, considerable departure from the $f^{1/2}$ dependence of I_p occurs, coincident with increasing departure from reversibility.

It is also worth noting that measurement of the half width or the more rigorous treatment of a study of the shape of the ac wave, based on a plot of

$$E_{dc} \quad vs. \quad \log \left[\left(\frac{I_p}{I} \right)^{1/2} \pm \left(\frac{I_p - I}{I} \right)^{1/2} \right]$$

as suggested in Sec. 7.3.2, would also have indicated the departure from reversibility at higher frequencies. Table 7.4 is a summary of data obtained from copper at variable frequency. At 1000 Hz the half width of 71 mV is considerably greater than the 90/n mV expected for a reversible two-electron reduction process. As the frequency is decreased, the half width can be seen to decrease also, until at frequencies between 10 and 100 Hz the half width approaches closely to the reversible value. It is interesting also to note that only a very small dependence of E_p on frequency is found in Table 7.4, and that this value is almost the same as $E_{1/2}$, again illustrating that comparison of $E_{1/2}$ and E_p does not provide a very sensitive or satisfactory criterion for ac reversibility.

TABLE 7.4 Variation of E_p and Half Width with Frequency, for the Copper(II)/Copper(0) Quasi-reversible AC Electrode Process in 1 M NaNO₃[a]

Frequency (Hz)	E_p (V vs. Ag/AgCl)	Half width (mV)
10	0.075	48
100	0.076	53
200	0.078	56
500	0.080	63
600	0.081	67
800	0.082	68
1000	0.083	71

[a]ΔE = 10 mV, p-p, [Cu(II)] = 1.00 × 10⁻³ M. Values from Note 23.

Relative Dependence of Faradaic and Charging Currents on Frequency.
From the above, it has been established that the charging current increases directly with ω, and the faradaic current increases at a rate
$\leq \omega^{1/2}$. That is, the charging or background current increases faster
with increasing frequency than does the faradaic current. Thus,
although high-frequency ac polarography would be expected to increase
the sensitivity, by giving higher faradaic currents per unit concentration, the increasingly unfavorable ratio of the charging and faradaic currents outweighs this advantage, and high-frequency ac polarography, in itself, does not achieve improved sensitivity. It is
because of this phenomenon that conventional total alternating
current polarography as considered in Sec. 7.3 is usually carried
out at low frequencies of about 50 Hz rather than at high frequencies.

The use of extremely low frequencies, although theoretically
having the most favorable ratio of faradaic to charging current, is
not favored in analytical applications of ac polarography because
of the relatively high instrumental noise level encountered.

The use of low-frequency ac polarography, in say the 20 to 100
Hz range, as well as being advantageous on the grounds of having a
more favorable faradaic-to-charging current ratio compared to high-
frequency ac polarography, has the added advantage that ac electrode
processes exhibiting quasi-reversible behavior, for example, Cu(II)
in 1 M $NaNO_3$, will be measured at closer to diffusion-controlled or
reversible conditions. In analytical applications of ac polarography,
this feature is desirable, as discussed in Sec. 7.3.

7.4.3 Dependence of Charging and Faradaic
Currents on AC Amplitude

Charging Current. Figure 7.17 shows the ac charging current in
5 M HCl at variable applied ac voltages ΔE over the range of 0.1 to
10 mV. An approximately linear relationship is evident. A considerable decrease in noise with increasing values of ΔE is also evident.

Faradaic Current. Figure 7.18 shows the variation of faradaic current
with ΔE for the reduction of cadmium (II) in 5 M HCl. In accordance
with Eq. (7.8), which applies to a reversible ac electrode process,

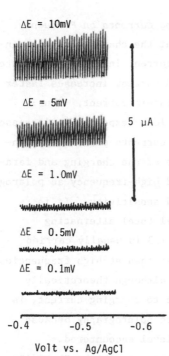

ΔE = 10mV

ΔE = 5mV

5 μA

ΔE = 1.0mV

ΔE = 0.5mV

ΔE = 0.1mV

-0.4   -0.5   -0.6

Volt vs. Ag/AgCl

FIGURE 7.17 Variation of charging current with applied ac voltage. f = 200 Hz; medium = 5 M HCl. [Reproduced from Anal. Chem. **44**, 315 (1972). © American Chemical Society.]

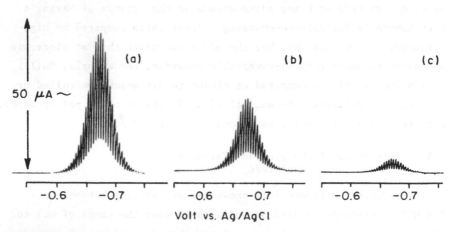

(a)   (b)   (c)

50 μA ~

-0.6   -0.7   -0.6   -0.7   -0.6   -0.7

Volt vs. Ag/AgCl

FIGURE 7.18 Variation of faradaic current with ΔE for the reversible ac electrode process Cd(II) + 2e $\rightleftharpoons$ Cd(0) in 5 M HCl: f = 100 Hz; (a) ΔE = 10 mV; (b) ΔE = 5 mV; (c) ΔE = 1 mV. [Cd(II)] = 1 × 10^{-3} M. [Reproduced from Anal. Chem. **44**, 315 (1972). © American Chemical Society.]

a linear relationship is evident. Figure 7.19 shows plots of I_p vs. ΔE for the quasi-reversible ac copper(II) electrode process in 1 M $NaNO_3$ at variable frequency. Again, a linear relationship is evident.

The half width and peak potential are essentially independent of ΔE for both reversible and quasi-reversible ac electrode processes. From the foregoing it can be seen that changes in ΔE (provided ΔE is not too large, e.g., $\leq \sim 20$ mV p-p) essentially have no influence on the ac electrode process other than to increase the faradaic current in a linear fashion.

Relative Dependence of Charging and Faradaic Currents on AC Amplitude. From the preceding, it follows that the charging and faradaic currents show approximately the same dependence on ΔE. Hence, in principle there should be no particular analytical advantage in choosing any value of ΔE. However, the use of lower ΔE values results in lower faradaic currents per unit concentration, and a consequent increase in instrument noise in detecting the same concentration. Therefore, the higher ΔE values are to be preferred in analytical applications of ac polarography, with all classes of fast electrode processes.

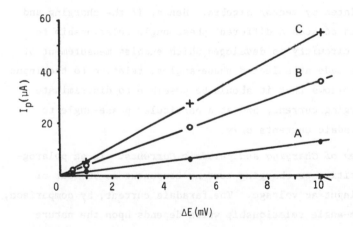

FIGURE 7.19 Variations of I_p with ΔE for the quasi-reversible Cu(II) + 2e $\rightleftharpoons$ Cu(0) ac electrode process in 1 M $NaNO_3$: (A) f = 10 Hz; (B) f = 100 Hz; (C) f = 500 Hz. [Cu(II)] = 1 × 10^{-3} M. [Reproduced from Anal. Chem. *44*, 315 (1972). © American Chemical Society.]

The reason for the use of ΔE values of around 10 to 50 mV p-p in
conventional ac polarographic instrumentation, as used in Sec. 7.3
can now be understood.

7.4.4 Dependence of Charging and
Faradaic Currents on Phase Angle

In Secs. 7.4.1 to 7.4.3, both the charging and faradaic currents
have been considered as a function of ΔE and ω, and it has been shown
why low-frequency ac polarography at ΔE values of around 10 mV are
advantageous. This work, although important as background material
and for other reasons, is applicable to all ac polarographic instrumentation and has not yet specifically introduced the real use and
purpose of phase-sensitive, three-electrode ac polarography. The
advantages of this form of polarographic instrumentation will now
become evident when considering the various phase-angle relationships
of the faradaic and charging currents with respect to the applied
ac voltage.

Theoretical Aspects of Phase-Sensitive Detection. In any ac network,
alternating currents possess a particular phase relationship to the
input voltage, and the magnitude of the currents at different angles
to this, are related by vector algebra. Hence, if the charging and
faradaic currents possess a different phase angle relationship to
each other, and circuitry is developed which enables measurement of
ac current to be made at selected phase-angles, relative to the input
ac voltage, it follows that it should be possible to discriminate
against the charging current, and at a particular phase-angle to
measure pure faradaic currents only.

Phase Angles of Charging and Faradaic Currents. In ac polarography, the capacitive or charging current component is 90° out of
phase with the input ac voltage. The faradaic current, by comparison,
exhibits a phase-angle relationship which depends upon the nature
of the electrode process [9]. In particular, for a reversible ac
electrode process, the faradaic current is 45° out of phase with
the applied ac voltage. For quasi-reversible ac electrode processes,

the phase angle will be close to 45°, with low-frequency ac polarography, where the system approaches closest to diffusion-controlled reversible ac conditions, but significant departures from this occur at high frequencies to give phase angles less than 45°. These phase-angle relationships for the faradaic current again assume that no ohmic iR drop effects are present. Figure 7.20 graphically shows the phase relationships of the charging and faradaic currents. From the above discussion, it follows that measurement of phase angle provides an additional criterion for assigning the reversibility of an ac electrode process. However, the measurement of phase angle is usually too time consuming and a tedious process for the analytical chemist to undertake with most existing commercial instrumentation, and the use of other criteria, as outlined previously, is recommended on the basis of being experimentally simpler.

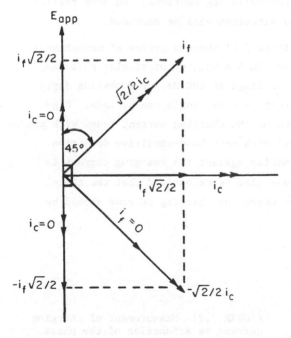

FIGURE 7.20 Phase relationship of charging and faradaic currents, relative to the applied ac voltage. i_f = faradaic current; i_c = charging current. [Reproduced from Anal. Chem. *44*, 315 (1972). © American Chemical Society.]

Discrimination against the Charging Current. From the previous discussion of phase angles, it follows that if ac measurements can be made at 0° or 180°, relative to the applied ac voltage, then the charging current component will be zero and the measured signal will consist solely of $\pm \sqrt{2}/2$ of the faradaic current, for a reversible electrode process. Discrimination against the charging current is therefore achieved, using a phase-sensitive detector capable of detecting only that component of the alternating current which is of a specific phase. The theoretical advantage of using phase-sensitive ac polarography, in preference to the conventional ac polarography, can therefore be appreciated.

Experimental Aspects of Phase-Sensitive Detection. The discussion above concerns the completely idealized, or theoretical situation for phase-sensitive ac polarography, and assumes that no ohmic effects are present. In the following sections, the more realistic, experimentally encountered situation will be examined.

Charging Current. Figure 7.21 shows a series of measurements made of the charging current in 5 M HCl, at particular phase angles relative to the applied ac voltage at 200 Hz. The obvious dependence of the charging current on phase angle can be seen. Figure 7.22 shows a comparison of the charging current found with phase-sensitive readout, compared with non-phase-sensitive detection. The considerable discrimination against the charging current is obvious. However, it should also be recognized that the theoretical or ideal situation where the charging current should be

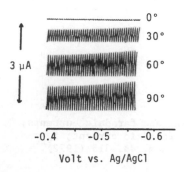

FIGURE 7.21 Measurement of charging current as a function of the phase angle, relative to the applied ac voltage. Medium = 5 M HCl; f = 200 Hz. [Reproduced from Anal. Chem. *44*, 315 (1972). © American Chemical Society.]

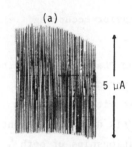

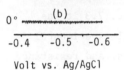

FIGURE 7.22 Comparison of measured charging current with (a) non-phase-sensitive and (b) phase-sensitive readout. Medium = 0.5 M $NaClO_4$; ΔE = 10 mV; f = 100 Hz. [Reproduced from Anal. Chem. *44*, 315 (1972). © American Chemical Society.]

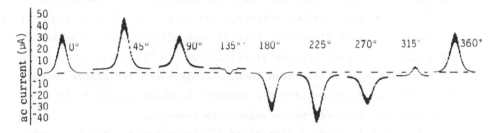

FIGURE 7.23 Measurement of the faradaic current at various phase angles, relative to the applied ac voltage, for the Cd(II) + 2e $\rightleftharpoons$ Cd(0) electrode process in 5 M HCl. [Cd(II)] = 1 × 10^{-3} M. ΔE = 10 mV p-p. f = 100 Hz. [Reproduced from Anal. Chem. *44*, 315 (1972). © American Chemical Society.]

zero, with a phase-angle measurement, ϕ, of 0°, is not obtained. This phenomenon basically arises from resistance effects, as will be discussed later.

Faradaic Current. Figure 7.23 shows the measurement of the cadmium(II) electrode process in 5 M HCl at various values of ϕ, between 0° and 360°. The expected behavior of I_p with ϕ is approximately observed. However, it can be seen in Fig. 7.23 that when ϕ is 135° or 315°, the faradaic current is not zero, as would be expected from vector algebra for a reversible electrode process which is 45° out of phase with the applied voltage. This apparent

anomaly is again mainly a result of nonideal behavior occurring
because of the influence of resistance.

  Influence of Resistance (iR Drop) Effects on Phase Relationships.
The theoretical argument, that the charging and faradaic currents
should be 90° and 45°, respectively, out of phase with the applied
ac voltage, is made assuming resistance effects (iR drop) are absent.
In fact, resistance effects alter the phase relationships of both
the charging and faradaic currents greatly. In Sec. 7.3 it was
also shown that the attainment of other theoretical-experimental
correlations is also hindered by resistance effects, and resistance
can be seen to be an extremely important factor in ac polarography.
It can be seen therefore, that if phase-sensitive ac polarography
is to be used with maximum advantage and in a scientific or logical,
rather than empirical manner, then iR drop effects need to be kept
to a minimum or even preferably eliminated. A three-electrode system
is therefore essential. Even then, uncompensated resistance terms
(Chap. 2) cause nonideality and positive feedback circuitry is
essential for accurate phase-angle measurements.

  Figure 7.24 shows a comparison of the charging current measured
at 0°, relative to the applied ac voltage and at a frequency of
1000 Hz, both with and without positive feedback circuitry. The

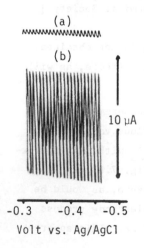

(a)

(b)

10 µA

-0.3   -0.4   -0.5

Volt vs. Ag/AgCl

FIGURE 7.24 Comparison of charging current
as measured (a) with and (b) without positive feedback circuitry to eliminate most
of the uncompensated resistance. ΔE = 10 mV
p-p; f = 1000 Hz; medium = 0.5 M NaClO$_4$.
[Reproduced from Anal. Chem. 44, 315 (1972).
© American Chemical Society.]

considerable decrease in charging current as the iR drop is increasingly eliminated, can be noted, and theoretical expectations are
approached. Figure 7.25 shows that, with this feedback circuit,
the expected zero value of the faradaic current at a phase angle of
135° is now virtually obtained, even with concentrated solutions
of depolarizer. Although such an operation leads to closer attainment of ideal operating conditions, where it should be possible to
measure virtually purely faradaic currents at 0° relative to the
applied ac voltage, it is doubtful whether the analytical chemist,
involved in routine work, would wish to proceed beyond the stage of
using a standard phase-sensitive three-electrode polarograph, unless
an extremely high resistance solvent is being used and/or high-
frequency ac polarography is necessary. Discussion can therefore
return, at this stage, to a consideration of the standard form of
instrumentation.

Optimum Conditions for Use of Phase-Sensitive, Three-Electrode
AC Polarography. Since it has been established that considerable,
but not complete, discrimination against background or charging
current is possible, and readily achieved with phase-sensitive ac
polarography, provided three-electrode instrumentation is used, it
becomes evident from previous discussion that this instrumentation
should provide a considerable improvement in performance over the
type of instrumentation considered in Sec. 7.3.1.

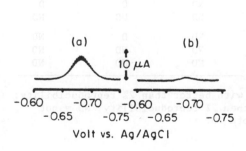

FIGURE 7.25 Comparison of
faradaic current for the
Cd(II) + 2e $\rightleftharpoons$ Cd(0) electrode
process as measured at the
phase angle of 135° (a) without
and (b) with positive feedback
circuitry to eliminate most of
the uncompensated resistance.
$\Delta E = 10$ mV p-p; $f = 100$ Hz;
$[Cd(II)] = 1 \times 10^{-3}$ M;
medium = 5 M HCl. [Reproduced
from Anal. Chem. *44*, 315 (1972).
© American Chemical Society.]

TABLE 7.5 Analytical Use of the Copper(II)/Copper(0) Quasi-reversible
AC Electrode Process in 1 M $NaNO_3$ with Phase-sensitive, Three-
electrode Instrumentation

		Frequency (Hz)		
[Cu] (M)	ΔE (mV, p-p)	10	100	1000
6×10^{-4}	10	QD	QD	QD
	1.0	D	QD	QD
	0.1	ND	QD	QD
4×10^{-4}	10	QD	QD	QD
	1.0	D	QD	QD
	0.1	ND	QD	QD
1×10^{-4}	10	QD	QD	QD
	1.0	D	QD	QD
	0.1	ND	D	QD
6×10^{-5}	10	QD	QD	QD
	1.0	ND	QD	QD
	0.1	ND	D	D
4×10^{-5}	10	QD	QD	QD
	1.0	ND	QD	QD
	0.1	ND	ND	D
1×10^{-5}	10	D	QD	QD
	1.0	ND	QD	QD
	0.1	ND	ND	ND
6×10^{-6}	10	D	QD	QD
	1.0	ND	D	D
	0.1	ND	ND	ND
4×10^{-6}	10	ND	QD	D
	1.0	ND	D	D
	0.1	ND	ND	ND
1×10^{-6}	10	ND	D	ND
	1.0	ND	ND	ND
	0.1	ND	ND	ND

[a]QD = quantitatively detectable with better than 2% reproducibility;
D = detectable, but with less than 2% reproducibility; ND = not
detectable. Data taken from Note 23.

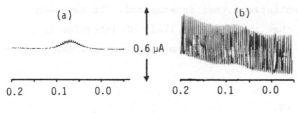

Volt vs. Ag/AgCl

FIGURE 7.26 Comparison of (a) phase-sensitive and (b) non-phase-sensitive ac polarograms for 6×10^{-6} M solution of copper in 1 M NaNO3. $\Delta E = 10$ mV p-p; f = 100 Hz. [Reproduced from Anal. Chem. *44*, 315 (1972). © American Chemical Society.]

It is emphasized strongly that arguments on the optimum frequency and applied ac voltages given previously still apply, because the charging current is not completely discriminated against. Hence, because the charging current increases at a greater rate with frequency than does the faradaic current, the use of low-frequency ac polarography is still favored over the high frequencies with phase-sensitive instrumentation. The fact that ac electrode processes approach more closely to diffusion-controlled conditions at low frequency still applies, of course. Furthermore, the use of the higher ΔE values, for example 10 mV, which give rise to higher faradaic currents per unit concentration and less likelihood of noise problems than low values of ΔE, for example, 0.1 to 1 mV, are still recommended with phase-sensitive instrumentation.

Table 7.5 gives a summary of a survey of the determination of copper in 1 M $NaNO_3$ by phase-sensitive ac polarography at various values of f and ΔE. This table clearly illustrates that optimum conditions are realized with f around 100 Hz and ΔE of about 10 mV.

Figure 7.26 shows comparative polarograms for copper at the 6×10^{-6} M level in 1 M $NaNO_3$. It can be seen that at this concentration, unless phase-sensitive ac polarography is used, the charging current virtually masks the faradaic current. It can also be noted from Fig. 7.26, that it is not only the absolute limit of detection which is altered substantially by using the phase sensitive readout, but rather it is the precision and accuracy with which measurements

can be made at low concentrations that is enhanced. It has been
the author's experience that the absolute limit of detection is
improved by about 15 to 20 times that given with conventional instrumentation in Sec. 7.3.1.

7.5 OTHER DEVELOPMENTS IN
FUNDAMENTAL HARMONIC AC POLAROGRAPHY

Most of the developments in ac polarography considered below are
based on utilizing "tricks" described in earlier sections on dc
polarography. However, the degree of success achieved by undertaking
such ventures in the ac rather than dc mode is frequently quite different, as will become evident from subsequent discussion.

7.5.1 Short Controlled Drop Times

In dc polarography, the use of short drop times in the method of
rapid polarography (Chap. 4) enables fast scan rates to be used.
In ac polarography short controlled drop times coupled with fast
scan rates have also been used widely, particularly in this author's
laboratories [11,24-29]. A number of advantages, apart from time
saving, also accumulate via the use of short drop times in dc polarography as was demonstrated in Chap. 4 (e.g., minimization of problems
arising from film formation), and these are automatically transferred
to the ac method, because this technique includes the dc polarographic
response in its formulation. This general conclusion is frequently
overlooked in considering ac polarography probably because the dc
current terms are filtered out and do not appear in the final readout
form.

    Under rapid ac polarographic conditions, the readout mode can
of course be phase-selective [27] or non-phase-selective [24], and
the theory is essentially predictable in the expected manner [24,25],
except that now it must of course be acknowledged that the area term
in the ac equations can be varied by having either an (almost) fixed
flow rate or mercury and varying the drop time, or vice versa. Thus,
the idea that the peak height of a reversible ac wave should be independent of mercury column height needs to be modified accordingly.

Because of the differing i-t behavior occurring with respect to both faradaic and charging current terms, certain ideas in rapid ac polarography cannot be formulated by direct extrapolation of a knowledge of the dc technique. With rapid dc polarography it was demonstrated both theoretically and experimentally that the ratio of faradaic to charging current becomes less favorable the shorter the drop time. With the ac technique, both the faradaic and charging current show a $t^{2/3}$ dependence [Eqs. (7.2) and (7.3)] for a given flow rate of mercury, and therefore, the faradaic-to-charging current ratio is approximately independent of drop time [29]. Consequently in ac polarography the use of short controlled drop times does not result in a loss of sensitivity, and the use of faster scan rates (time saving) is gained without concomitant losses in another direction.

Finally, several significant instrumental advantages accrue with the use of short drop times [29], particularly with respect to iR drop.

The net uncompensated ohmic loss in any electrochemical experiment is of course governed by Ohm's law, and therefore, the variation in the product of the current and resistance terms as the drop time decreases will determine the magnitude of the iR loss in rapid ac polarography.

The magnitude of the alternating current $I(\omega t)$, depends on several variables, including the nature of the electrode process, the applied alternating potential ΔE, the frequency ω, the concentration C, and the area of the electrode A, as described earlier.

A general result, however, is that

$$I(\omega t) \propto AC\Delta E f(\omega) f(\theta) \tag{7.12}$$

where $f(\omega)$ is a function of frequency and $f(\theta)$ is a function of the phase angle.

Assuming the mercury drop to be spherical, which will be a good approximation with capillaries of internal radius between 0.017 and 0.2 mm [30], allows the calculation of the area of the drop at any instant during its growth. The expression to be used is as already noted in Chap. 3, with m being the flow rate of mercury and t the (drop) time. From Eqs. (7.12) and (7.13) it can be seen that the ratio of the faradaic current for natural drop time $I(\omega t)_{nat}$ to that for rapid drop time $I(\omega t)_{rap}$, under a particular set of conditions and with the same flow rate of mercury for both experiments, is given by

$$\frac{I(\omega t)_{nat}}{I(\omega t)_{rap}} = \frac{t_{nat}^{2/3}}{t_{rap}^{2/3}}$$

To complete the discussion on iR drop, the uncompensated resistance of the electrochemical cell needs to be considered. The dependence of this term on drop time will depend on the experimental arrangement [31–33]. If a simple two-electrode system is used, the total resistance, including that of the reference electrode, solution, and dme, etc., will be uncompensated. Under these conditions, contribution of resistance terms varying with drop time will depend on the exact nature of the experiment. However, with a three-electrode potentiostat system with or without positive feedback circuitry, the uncompensated resistance terms are derived mainly from the solution resistance between the dme and the reference electrode, and the resistance of the dme itself. The uncompensated resistance is then expected to be dependent on drop time. In the limiting case, assuming the appropriate equations to be valid for very short drop times, it can be shown [34] that the uncompensated resistance R varies inversely with the radius of the dme, r; and since

$$R \propto \frac{1}{r} \tag{7.15}$$

and

$$r \propto t^{1/3} \tag{7.16}$$

then

$$R \propto t^{-1/3} \tag{7.17}$$

in this limiting case.

Since the alternating current shows dependence on $t^{2/3}$, the ohmic iR drop should be dependent on drop time raised to a power between the limits 1/3 and 2/3, and it follows that even though the uncompensated resistance increases with decreasing drop time, the iR drop decreases. That is, when uncompensated resistance is significant, the rapid ac method should be affected less than conventional ac polarography by this parameter.

The absolute magnitude of $I(\omega t)$ is also important in a discussion of ohmic losses. At high frequencies, $I(\omega t)$ obviously increases and iR losses must then be more important. Indeed it is well known [9] that the upper limit of usable ac frequency is governed by the magnitude of the iR drop. Thus short drop times should extend the accessible frequency range of ac polarography. One important consequence of iR drop as demonstrated previously is that it invalidates the use of simple theoretical phase angle relationships between the faradaic and charging current components, both to each other and to the applied alternating potential. Ideally, the charging current should be 90° out of phase from the input alternating potential, and measurement of the in-phase component should provide faradaic currents unencumbered by charging current contributions. Ohmic losses however, cause departure from this relationship.

Figure 7.27 shows a series of in-phase ac polarograms of Cd(II) in 1 M hydrochloric acid at 1000 Hz, recorded at natural drop time and at controlled drop times of 5, 2, 1, and 0.5 sec, under the same conditions.

The natural drop time polarogram has a considerable charging current contribution in the in-phase component. The controlled drop time polarograms clearly show the decreasing ratio of charging to faradaic current as the drop time is shortened.

Figure 7.28 shows that the positive feedback circuit can be used to eliminate most of the charging current and achieve similar ideality to that in the short drop time experiment, and this verifies

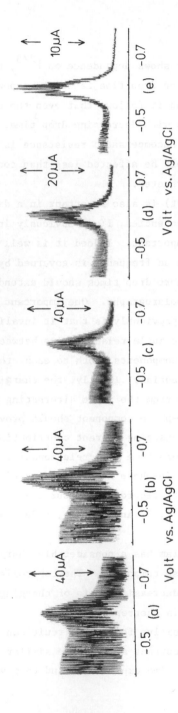

FIGURE 7.27 Minimization of resistance effects in ac polarography by decreasing the drop time. [Cd(II)] = 3×10^{-4} M in 1 M hydrochloric acid. Three-electrode potentiostat system used without positive-feedback circuitry. ΔE = 10 mV p-p; frequency = 1 kHz.

	(a)	(b)	(c)	(d)	(e)
Drop time (sec)	Natural	5	2	1	0.5
Scan rate (mV sec^{-1})	1	1	2	5	10

[Reproduced from Talanta 21, 591 (1974).]

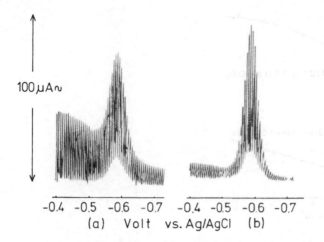

100 μA~

-0.4 -0.5 -0.6 -0.7   -0.4 -0.5 -0.6 -0.7
(a)   Volt   vs. Ag/AgCl   (b)

FIGURE 7.28 Application of positive feedback circuitry to eliminate resistance effects with natural drop time ac polarography. [Cd(II)] = 3 × 10^{-4} M in 1 M hydrochloric acid. ΔE = 10 mV p-p; scan rate = 2 mV sec^{-1}; frequency = 1 kHz: (a) three-electrode and (b) three-electrode with feedback circuitry. [Reproduced from Talanta *21*, 591 (1972).]

that the charging current in the experiments shown in Fig. 7.27 arises from uncompensated resistance effects present even with a three-electrode ac polarograph.

From these observations it is apparent that the operative capacity of any phase-sensitive, three-electrode ac polarograph can be improved considerably by using short controlled drop times, particularly with respect to the upper frequency limit. Indeed any form of ac polarography operated without 100% iR compensation can be improved substantially by using shorter drop times. For the analyst this is an additional bonus to that of being able to use considerably faster scan rates, and, for example, the considerably improved performance of the short drop-time method in high-resistance nonaqueous solvents is therefore not unexpected [29,35].

Considerable curvature in calibration curves resulting from iR drop commonly occurs in two-electrode ac polarography. Figure 7.29 shows a comparison of an analytical calibration curve under natural and controlled drop time conditions; the difference is a direct

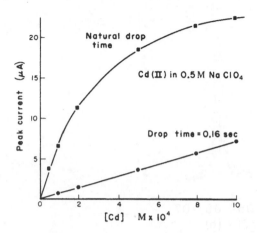

FIGURE 7.29 Improved linearity of calibration curves obtained in ac
polarography with a two-electrode system, using short drop times.
Frequency = 50 Hz. Applied alternating potential = 50 mV rms.
[Reproduced from Talanta *21*, 591 (1974).]

consequence of differences in ohmic losses. The same considerations
apply to three-electrode systems, particularly in nonaqueous solvents,
and linear calibration curves are to be expected over wider concentration ranges with short drop time ac experiments.

     One other possible problem with high-frequency ac polarography
is that it can induce difficulties associated with potentiostat
instability. Examination of the theory [36] predicts that the
shorter drop time, with its associated lower double-layer charge,
should in fact lead to an increase in stability. This has also been
observed experimentally.

     Similarly, examination of current-time curves shows that under
many operating conditions the noise level is minimal early on in
the drop life and maximal late in the drop life. Thus, rapid polarographic measurements are frequently made at a point in the drop life
where noise is at a minimum. In general, it would appear that rapid
polarographic measurements are made under the most favorable conditions of both potentiostat and dme stability as well as with minimum
ohmic losses.

As can be seen from the above, the rapid method in ac polarography is extremely useful and there seems no reason why the controlled
drop time rather than the gravity-controlled approach should not be
used on the majority of occasions.

7.5.2 Current-Sampled AC Polarograms

In dc polarography, the use of the current sampling method in
which only the current at the end of the drop life is plotted,
provided both a means for discrimination against the charging current and an improved readout form (Chap. 4). In ac polarography,
the i-t behavior of both the faradaic and charging currents are the
same and the former advantage cannot be gained by using this method
of current measurement. An improved or simplified readout form,
however, does result from this approach, as shown in Fig. 7.30.
This mode of measurement is now available on several commercially
available ac polarographs and is progressively becoming the standard
format.

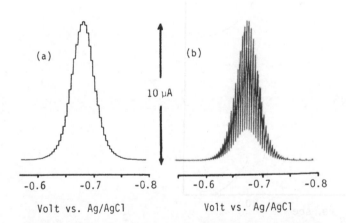

FIGURE 7.30 Comparison of controlled drop time ac polarography
and current-sampled ac polarography: (a) current-sampled ac
polarography with controlled drop time of 5 sec; (b) controlled
drop-time ac polarography with a drop time of 5 sec.
ΔE = 10 mV p-p; F = 100 Hz. In-phase components measured. Solution is 10^{-3} M cadmium in 5 M HCl. [Reproduced from Anal. Chem. 44,
315 (1972). © American Chemical Society.]

7.5.3 AC Polarography
in the Subtractive Mode

As with all polarographic measurements, it is possible to utilize a
dual-cell arrangement or computerized instrumentation and subtract
the background current from the readout (see Chaps 4 to 6). The
former approach has been used in ac polarography [36,37]. Figure 7.31 demonstrates how effectively subtraction of the charging
current can be undertaken with total alternating current measurements. As was noted with the other polarographic methods, the

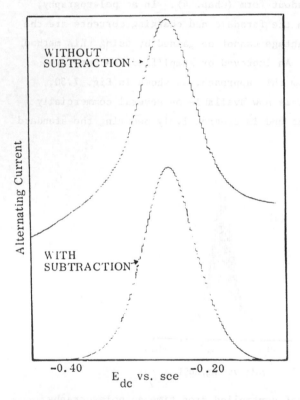

FIGURE 7.31 Fundamental harmonic ac polarograms of the
iron(III)/iron(II) system with and without subtraction of charging
current. System: 0.80×10^{-3} M Fe(III) in 0.50 M $K_2C_2O_4$.
Applied: 160 Hz, 10 mV peak-to-peak sine wave. Measured: 160 Hz,
current component at the end of drop life with compensation for iR
drop. [Reproduced from Anal. Chem. *38*, 1119 (1966). © American
Chemical Society.]

subtractive mode is not readily used in routine analytical work,
unless a computerized system is available, and phase-sensitive detection would be a far more favorable approach to minimize the
charging current in the general situation. However, as Barker and
Faircloth point out [37], while a subtractive experiment may give
no great increase in sensitivity when phase-sensitive detection is
used, there still may be a general improvement in performance resulting from the elimination of responses produced by impurities in
the base solution and the small, uncompensated charging current.

7.5.4 Fast Sweep AC Voltammetry at a Dropping Mercury Electrode

In Chap. 5, the technique of dc linear sweep (fast sweep) voltammetry
at stationary electrodes was described. AC voltammetry at stationary electrodes can be undertaken also at much faster scan rates than
those normally employed in ac polarography, and this approach should
be an important variation in the methodology. Assuming that
$\Delta E\omega \gg v$, where v is the scan rate of dc potential, it turns out
that the theory for the reversible process at a stationary electrode
is the same as under polarographic conditions. That is, Eq. (7.1)
is valid and independent of dc (scan rate) terms. However, if
chemical or other steps are rate-determining, such equivalences
between stationary and dropping electrode theories are not expected,
because the dc terms will exert important and substantially different
influences.

All of the possibilities discussed in Chap. 5 for dc linear
sweep voltammetry can be extended to the ac experiment and, instead
of using genuinely stationary electrodes [38,39], the sweep can be
synchronized to a dme [40-44] to give an excellent electroanalytical
method. Figures 7.32 to 7.34 diagrammatically demonstrate results
using this latter approach and provide comparisons with linear
sweep dc voltammetry at a dme under the same conditions. Second
harmonic curves are also included in these figures for subsequent
discussion under the second harmonic section, toward the end of
this chapter.

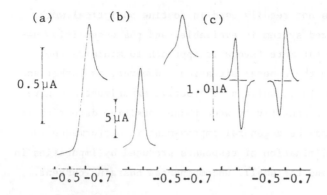

FIGURE 7.32 AC and dc linear sweep voltammograms at a dme of
7.5×10^{-5} M cadmium(II) in 1 M hydrochloric acid. (a) direct
current mode. Presweep delay = 3 sec. v = 100 mV sec^{-1}.
(b) In-phase and quadrature components of the fundamental harmonic
mode. Presweep delay = 3 sec. v = 50 mV sec^{-1}. f = 500 Hz.
ΔE = 10 mV p-p. (c) In-phase and quadrature components of the
second harmonic mode. Presweep delay = 3 sec. v = 50 mV sec^{-1}.
F = 500 Hz. ΔE = 10 mV p-p. [Reproduced from Anal. Chem. *46*, 1934
(1972). © American Chemical Society.]

A brief and approximate treatment of the theory for this method
is presented below.

Faradaic Current. For a reversible reduction, A + ne $\rightleftharpoons$ B, with
A and B both soluble, the equation for the fundamental harmonic
alternating current at a stationary electrode can be derived by
considering the following equations.

As for ac polarography, $E = E_{dc} - \Delta E \sin \omega t$. However, unlike
for polarographic conditions, the dc potential is a function of
time and $E_{dc} = E_{initial} - vt$, where v is the dc scan rate and t is
the experiment time (see Chap. 5). Solution of these two equations
gives

$$I(\omega t) = \frac{n^2 F^2 A C_A (\omega D)^{1/2} \Delta E F'(vt)}{4RT \cosh^2(j/2)} \sin(\omega t + \frac{\pi}{4}) \qquad (7.18)$$

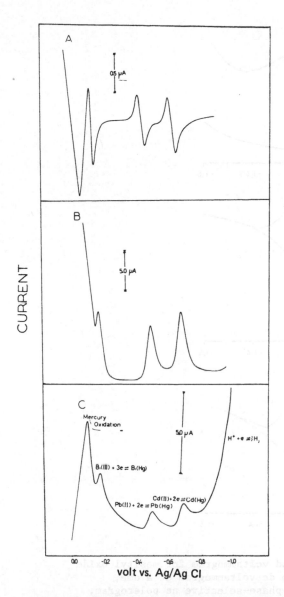

FIGURE 7.33 Multicomponent system containing 1×10^{-4} M Bi(III), Pb(II), and Cd(II) in 5 M HCl. (a) Second harmonic ac voltammogram. Presweep delay = 1 sec. v = 200 mV sec^{-1}. ΔE = 10 mV p-p. f = 400 Hz. In-phase component. (b) Fundamental harmonic ac voltammogram. Presweep delay = 1 sec. v = 200 mV sec^{-1}. ΔE = 10 mV p-p. f = 400 Hz. In-phase component. (c) DC voltammogram. Presweep delay = 1 sec. v = 200 mV sec^{-1}. [Reproduced from Anal. Chem. 46, 1934 (1974). © American Chemical Society.]

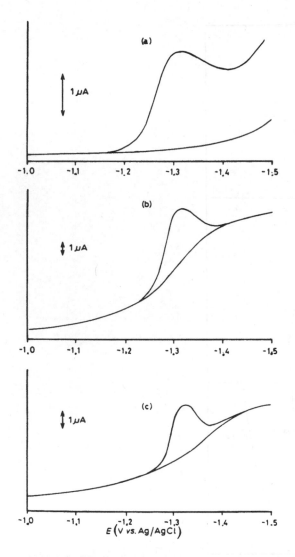

FIGURE 7.34 Polarograms and voltammogram for mesityl oxide
$(2 \times 10^{-4}$ M) (a) Fast sweep dc voltammogram at a dme,
$v = 500$ mV sec^{-1}. (b) Non-phase-selective ac polarogram,
$\Delta E = 50$ mV at 20 Hz. (c) Fast sweep non-phase-selective
ac voltammogram. $v = 500$ mV sec^{-1}; $\Delta E = 50$ mV at 20 Hz.
[Reproduced from J. Electroanal. Chem. 43, 349 (1973).]

Equation (7.18) is the same as Eq. (7.3) except for the time-
dependent function F'(vt), which is essentially a dc scan rate term
[44]. For the condition $\Delta E\omega \gg v$ in which the ac time scale is very
much shorter than the dc one, it turns out that $F(vt) \rightarrow 1$.

At a dme, the electrode area is time-dependent and A in
Eq. (7.18) must be replaced by A_t, where

$$A_t = 0.85m^{2/3} \left[t_0 + \left(\frac{E_t - E_{initial}}{v} \right) \right]^{2/3} \qquad (7.19)$$

In Eq. (7.19), t_0 is the presweep delay and $E_{initial}$ the potential
at which the scan commences. The importance of using A_t instead of
A in equations for work at a dme will become obvious in discussion
of the charging current.

Equation (7.18) can be tested as in polarography by plotting
$[I(\omega t)]_{max} = I_p$ versus the various parameters, and generally excellent theoretical-experiment correlations have been found [38,44].
For example, the in-phase and quadrature components for reduction
of cadmium are shown in Fig. 7.32(b) and the expected phase-angle
relationships are obtained. The quadrature component in this figure
contains considerable charging current contribution, as is the case
in polarography. Consequently, phase-selective detection also provides considerable improvement in ac voltammetry.

The influence of slow electron transfer and other mechanistic
nuances is contained in the work of Bond et al. [45]. In summary,
provided that $\Delta E\omega \gg v$, a direct analogy with the polarographic case
exists when ideas related to drop time in polarography are replaced
by scan rate.

Charging Current. The charging current in ac linear sweep voltammetry
under conditions where $\Delta E\omega \gg v$, essentially gives the same equations
as those found in polarography (E independent of v for theoretical
purposes). The A_t term as in Eq. (7.19), however, provides an important feature of the charging current equations at a dme.

If we assume that the electroactive species does not affect the double layer and the dc and ac components can be considered independently, the expression for the capacitance current can be derived as follows using the procedures of Chaps. 2 and 5.

$$i_c = - \frac{d}{dt} (a_t q') \tag{7.20}$$

$$i_c = - q' \frac{dA_t}{dt} - A_t \frac{dq'}{dt} \tag{7.21}$$

where

q' = charge density on the electrode surface

A_t = surface area of the dme at time t

From the differential capacitance of the double layer per unit area, which is defined as

$$C_{dl} = \frac{dq'}{dE} = \frac{dq'}{dt} \frac{dt}{dE} \tag{7.22}$$

where

E = applied potential given by Eq. (7.1)

Equation (7.21) can now be written as

$$i_c = - q' \frac{dA_t}{dt} + Q_t C_{dl} v + A_t C_{dl} \Delta E \omega \cos \omega t \tag{7.23}$$

Hence, the dc and ac components of charging current can be expressed as

$$(i_c)_{dc} = - q' \frac{dA_t}{dt} + A_t C_{dl} v \tag{7.24a}$$

$$(i_c)_{ac} = A_t C_{dl} \Delta E \omega \cos \omega t \tag{7.24b}$$

The area terms in Eq. (7.24a) and (7.24b) mean that the charging current will show a time dependence giving rise to the undesirable sloping base line found in both the fast sweep dc and fundamental harmonic ac voltammetric methods at a dme.

Equation (7.24a) was considered in Chap. 5 and has been shown to be approximately correct under a range of conditions. Equation (7.24b) can be tested easily in the fundamental harmonic mode by using phase-selective detection, since the capacitance current of the double layer should have a phase angle of $\pi/2$, i.e., quadrature components should consist of pure charging current and the in-phase component should contain no charging current. This equation has been examined in 1 M HCl [44] and close to the expected phase angle found. However, higher frequencies than normally used in polarography generally need to be employed to comply with the condition $\Delta E\omega \gg v$. The magnitude of the charging current is therefore often large and some nonideality is observed in that the in-phase component does contain a small but finite charging current contribution. Because of this nonideality, even when the in-phase component is measured, a sloping base line will be seen, although obviously this will only become important at much lower concentrations than when total alternating currents are being measured. As would be expected if the A_t term is the main contributor to this problem, the slower the scan rate, the more the base-line problem, and this is shown in Fig. 7.35.

Analytical Aspects. The fast sweep ac voltammetric technique, particularly at a dme, provides a powerful analytical tool because, while providing a means for decreasing the analysis time, it still retains all of the advantages of ac techniques. For reduction of cadmium in HCl, linear calibration curves are obtained over at least three orders of magnitude (10^{-3} to 10^{-6} M) [44]. The reproducibility of these voltammograms is better than 1%. The reproducibility can be compared with values generally quoted for polarography at natural drop times of between 1 and 2%, or for voltammetry at other stationary electrodes of between 2 and 5%.

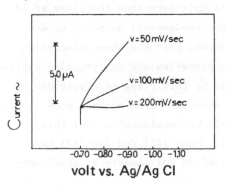

FIGURE 7.35 Dependence of charging current on scan rate in 1 M HCl. Fundamental harmonic mode. Quadrature component. ΔE = 10 mV p-p; f = 600 Hz. Presweep delay = 2 sec. [Reproduced from Anal. Chem. *46*, 1934 (1974). © American Chemical Society.]

Figure 7.33 shows a comparison between dc, phase-selective fundamental and second harmonic ac voltammograms at a dme for a solution containing 1×10^{-4} M Bi(III), Pb(II) and Cd(II) in 5 M hydrochloric acid.

In the dc technique, the waves for the more negatively reduced lead and cadmium species are not easily evaluated because the currents from the preceding electrode processes must be subtracted, for analytically usable peak heights to be obtained. Similarly, the mercury oxidation causes difficulties in evaluating the bismuth peak height.

With the fundamental harmonic technique, the readout form is more conducive to accurate interpretation because the current rapidly decays to zero on both the positive and negative sides of the peak. This facilitates the measurement of the peak height and no difficulties are encountered even in the presence of preceding waves.

Because of the double-layer capacitance current, a slightly sloping base line is observed in the phase-selective fundamental harmonic mode, as can be seen in Fig. 7.33. Consequently, fundamental harmonic ac voltammetry at a dme when nonphase detection is used, suffers greatly from the presence of charging current and is not even as sensitive as the dc method. The use of phase-selective detection with the ac technique is therefore strongly recommended.

The second harmonic ac voltammogram (see later discussion also) obtained at a dme is shown in Fig. 7.33(c). The system exhibits a virtually flat base line compared with the dc or fundamental harmonic ac techniques because little or no charging current is present. The resolution is also excellent and this technique therefore would appear to be more attractive than the fundamental harmonic or dc methods. It should be noted, of course, that the above discussion has applied to reversible electrode processes and analytically applications will generally be restricted to this class of electrode process. However, this means that discrimination against non-reversible electrode processes should be considerable, and this technique should be highly specific. Bond [46] and Jee [47] have recently summarized further aspects of ac voltammetry at stationary electrodes.

7.5.5 AC Cyclic Voltammetry

DC cyclic voltammetry consists of applying a triangular dc voltage
to the electrochemical cell and reading out the resultant dc current
as a function of the applied voltage. This technique enables both
the forward and reverse steps of the redox reaction to be studied.
This method has been extended to the ac format at stationary electrodes by superimposing an ac voltage onto the dc triangular ramp,
and either the fundamental or second harmonic ac response is recorded
as a function of the applied dc potential [45,48-51]. As would be
expected, this gives considerable advantage in readout format because
of both improved wave shape and discrimination against charging
current.

   Figure 7.36 shows dc, fundamental and second harmonic ac cyclic
voltammograms at a dme for the reduction $Cd(II) + 2e \rightleftharpoons Cd(Hg)$.
The reverse sweeps are larger than the forward sweeps in the case of
amalgam formation at spherical electrodes [51] (anodic stripping
phenomena). For a reversible system involving two soluble species
$(A + ne \rightleftharpoons B)$, the forward and reverse scans completely overlap
(equal diffusion coefficients assumed). Obviously, ac cyclic voltammetry should be able to provide a powerful means for investigating
electrode processes, because it substantially extends the capabilities of the very popular dc version described in Chap. 5. This
technique is presently in its infancy only, but results are extremely
encouraging [45].

7.5.6 Pulsed DC Potentials
       in AC Polarography

In differential pulse polarography (Chap. 6), a periodic potential
pulse of fixed amplitude E_p is applied to the normal dc voltage ramp
just prior to the end of the drop life. The difference between these
two current values is electronically stored and presented to a suitable readout device. If a similar pulse is now applied to the voltage
ramp used in ac polarography, then the resultant voltage-time curve
will be as shown in Fig. 7.37. Using similar current sampling as in
differential pulse dc polarography, a differential pulse ac polarogram
results from plotting the difference in alternating current

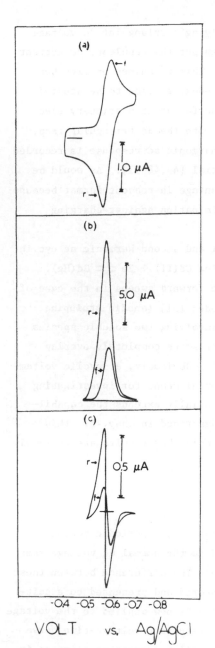

FIGURE 7.36 Cyclic voltammograms at
a dme of 1×10^{-4} M Cd(II) in
1 M HCl. (a) Direct current mode.
Presweep delay = 1 sec.
v = 100 mV sec^{-1}. (b) Fundamental
harmonic ac mode. Presweep
delay = 1 sec. v = 100 mV sec^{-1}.
ΔE = 10 mV p-p; f = 400 Hz. In-
phase component. (c) Second harmonic
ac mode. Presweep delay = 1 sec.
v = 50 mV sec^{-1}. ΔE = 10 mV p-p;
f = 400 Hz. In-phase component.
f = forward scan; r = reverse scan.
[Reproduced from Anal. Chem. 44,
1934 (1974). © American Chemical
Society.]

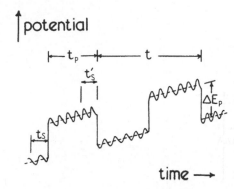

↑potential

├─t_p─┼──── t ────┤

time ⟶

FIGURE 7.37 Representation of E-t curve in differential pulse ac polarography: (---) differential pulse ramp, (─) pulsed ac ramp. ΔE_p = pulse height, t_p = pulse duration, t = drop time, t_s and t'_s are current sampling times before and during the pulse, respectively. [Reproduced from Anal. Chem. 47, 1906 (1975). © American Chemical Society.]

$\Delta I(\omega L)$ versus dc potential [52]. Since it results from a difference, the shape of the $\Delta I(\omega t)$-vs.-E curve is akin to that of a derivative fundamental harmonic ac polarogram. The difference measurement leads to a decrease in the charging current component of the experiment and, when coupled with phase-sensitive readout, a dual approach to minimizing the charging current is in operation. Figure 7.38 shows a range of differential pulse ac polarograms. The total current differential pulse ac voltammograms show very flat baselines compared to the normal ac polarograms. With phase-selective readout the quadrature component measurement also shows a marked superiority in the differential pulse ac mode because of the subtraction of charging current terms. At high frequencies (for example, 1000 Hz) where the charging current is extremely high, even the in-phase component of conventional ac polarograms contains a substantial charging current component due to nonideality. However, as demonstrated in Note 52, the combination of phase-selective detection (in-phase component) and differential pulse ac polarography provides virtually complete rejection of charging current, even at high frequencies and low concentrations. In summary, the method under discussion is a linear combination of differential pulse and ac polarography and combines the attributes of both approaches in discriminating against charging current.

In the section on rapid ac polarography, it was noted that any modification which improves the dc aspect of the ac polarographic experiment will be advantageous. On this basis, it may well be that

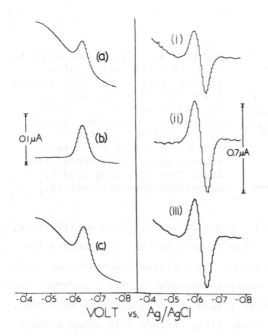

VOLT vs. Ag/AgCl

FIGURE 7.38 Phase-selective and total ac differential pulse polarograms of 1×10^{-4} M Cd(II) in 1 M KNO_3. Pulse height = -50 mV.
Pulse duration = 57 msec. ΔE = 30 mV p-p; f = 957 Hz. Drop time
= 0.5 sec. Current sample time = 17 msec. (a) Quadrature component;
ac polarogram. (b) In-phase component; ac polarogram. (c) Total
current; ac polarogram. (i) Quadrature component; differential
pulse ac polarogram. (ii) In-phase component; differential pulse
ac polarogram. (iii) Total current; differential pulse ac polarogram. [Reproduced from Anal. Chem. 47, 1906 (1975). © American
Chemical Society.]

the conventionally used linear dc potential ramp is not the optimum
one for ac polarography, but relatively little work to evaluate this
hypothesis has been undertaken. In Chap. 6, the method of normal
pulse polarography was shown to be very attractive as a means of
eliminating undesirable phenomena arising from mercury film formation.
Bond and Grabarić [53] have now demonstrated that using the normal
pulse potential ramp to provide the dc component of the ac polarographic experiment is highly advantageous in the same situation, or
indeed whenever the normal pulse polarograms are superior to their
dc counterparts.

Virtually all of the developments and innovations of the last
decade have been confined to the ac aspects of ac polarography.
However, from the limited information now available real, scope
for further improvements may now lie in the alternative, but neglected,
direction of the dc component of the experiment [53].

7.5.7 Analytical Applications
of High-Frequency AC Polarography

Logically, high frequency ac polarography is a simple extension of
previous discussion and most of the details have already been covered.
However, some important but not widely recognized facets of the
theory can be shown to provide unique possibilities in analytical
work [54-56]. Consequently, a brief discussion relevant to the high-
frequency domain is presented in a separate section to highlight
the specificity which can be gained in ac polarography.

In analytical applications of ac polarography, as was seen
earlier, low frequencies are usually recommended because electrode
processes are more reversible and waves are therefore often better
defined at lower frequencies.

Figure 7.39 illustrates the ability of the high-frequency ac
method to discriminate against the irreversible reduction wave for
hydrogen ion, thus allowing zinc to be determined in a medium in
which it could not be determined by differential pulse or dc polarography. The concept of discriminating against the irreversibly reduced hydrogen ion species is also demonstrated clearly in fast sweep
ac voltammetry (Fig. 7.33) and in data contained in Notes 29 and 56.
Figure 3.40 shows how the determination of uranium and lead in a
complex uranium mineral, Samarskite, can be achieved directly via
high-frequency ac polarography [55].

One disadvantage of polarographic methods of analysis, compared
with others, is the frequent need for removal of oxygen. Dissolution
of many samples requires the end solutions to be acidic. In acidic
media, oxygen is often irreversibly reduced and the use of the high-
frequency ac technique successfully discriminates against the waves
arising from the oxygen electrode processes [29,54]. Further, in

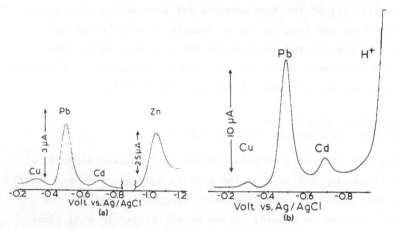

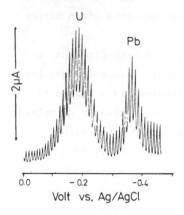

FIGURE 7.39   (a) Phase-selective fundamental harmonic ac polarogram
of a zinc concentrate in 2 M HCl. ΔE = 10 mV p-p at 400 Hz.
(b) Differential pulse polarogram of the same solution as in (a).
Note how the irreversibly reduced hydrogen wave masks the zinc wave
under differential pulse conditions. [Reproduced from Anal. Chim.
Acta 75, 409 (1975).]

FIGURE 7.40   High-frequency phase-
selective ac polarography for the direct
determination of uranium and lead in
the mineral Samarskite. ΔE = 10 mV p-p
at 900 Hz. [Reproduced from Anal. Chem.
46, 1551 (1974). © American Chemical
Society.]

acidic media, coupled chemical reactions associated with the oxygen
electrode process in neutral or alkaline media are usually absent.
Considerable gains in convenience and time saving are therefore
sometimes made possible by the use of high-frequency ac polarography
[54] when the removal of oxygen can be avoided.

The above rather specialized use of ac polarography in the high-
frequency domain clearly illustrates the need to be able to

characterize the nature of electrode processes. It also illustrates
that ac and pulse methods of polarography, for example, are by no
means equivalent. AC methods tend to be more specific than pulse
or dc methods and this feature can be exploited on occasions such
as those discussed above.

Of course, the other side of the coin reveals that differential
pulse and dc methods are generally more useful in determining a wider
range of species. In this context, the often complementary rather
than competitive nature of the various branches of polarography is
well illustrated. Instrumentation and other aspects of low-frequency
ac polarography are also simplified relative to the high-frequency
case, and to date the 10 to 100 Hz range has been used almost exclusively in analytical work. However, in geochemical and other
complicated areas of analysis, for example, literally almost any
element in the periodic table can be simultaneously present with
the species to be determined, and if chemical separations are to
be avoided on a large percentage of analyses the analytical method
needs to have a high degree of specificity [54,55].

High-frequency ac polarography can provide specificity because
of its ability to discriminate against irreversible electrode processes. The magnitude of the alternating current for reversible
electrode processes increases in proportion to the square root of
frequency. However, the ac technique is not particularly sensitive
to nonreversible electrode processes (i.e., low current per unit
concentration) as noted elsewhere. Indeed, for nonreversible electrode processes, the magnitude of the alternating current is generally
independent of frequency. Thus if a medium can be chosen in which
the ac electrode process for the species to be determined remains
close to reversible, even at high frequencies, and simultaneously
in which potentially interfering species are irreversibly reduced,
high-frequency ac polarography can be used to provide considerable
selectivity toward the species being determined. In complex analytical problems, the somewhat unusual approach of recommending high
frequencies over low frequencies is therefore likely to have much in
its favor.

The use of high frequencies necessitates the use of high-quality
instrumentation. The contribution of the charging current is prohibitive at the 100-Hz region and needs to be eliminated or discriminated against if useful measurements are to be made. Phase-sensitive
detection provides a means for doing this. The systematic use of
phase-sensitive detection in which only the in-phase component of
the alternating current is measured, depends on the ohmic (iR)
losses being minimal. High frequency ac polarography suffers severely
from iR drop effects and a three-electrode, iR compensating, potentiostat system is essential.

7.6 SECOND HARMONIC AC POLAROGRAPHY

At various stages in Sec. 7.2, the consequences of the electrochemical cell behaving as a nonlinear network are reported to give rise
to second-order techniques. The first of these to be discussed
in any detail is second harmonic ac polarography, in which the
response of the cell at twice the fundamental frequency is examined.
Figures 7.4, 7.6, 7.32, 7.33, and 7.36 already provide representative
examples of both total and phase-selective second harmonic polarograms or voltammograms (2f techniques), and it is clear that all the
options available to the fundamental mode (1f technique) are retained.
Indeed, almost all the discussion given for the fundamental harmonic
is readily extended to the 2f readout as soon as the realization is
made that the time scale is shorter (2f versus f) and that while the
magnitude of the current is low, the charging current which behaves
as a linear network (see Sec. 7.2) is extremely small. The net consequence of these two observations is that the method is extremely
sensitive for the determination of reversibly reduced or oxidized
species but gives low currents per unit concentration for species
exhibiting irreversible electrode reactions. With this in mind,
only features unique to the 2f mode or those not immediately obvious
are discussed, as virtually all the information required by the
analytical chemist has been conveyed in the fundamental harmonic
section.

Figure 7.41(a) shows a comparison of phase-selective and total alternating current 2f polarograms for a reversible system. In this figure, ψ is the alternating current at 2f, [I(2ωt)], and ψ_p the maximum or minimum value of ψ. ψ_p is the parameter proportional to concentration in second harmonic polarography. The third harmonic, with Φ and Φ_p as the analogous parameters, is shown in Fig. 7.41(b). Since charging current has been stated to be very low in second harmonic polarography, the need to use phase-sensitive detection may be queried. There are, however, still some advantages in doing so. The calibration curve in the phase-sensitive mode, for example, can be constructed from a peak-to-peak value, i_{p-p} [58], as shown in Fig. 7.41. A peak-to-peak value is a uniquely defined parameter measured independently of a base line, as is required for obtaining ψ_p in the non-phase-selective version. Furthermore, the charging current is not exactly zero and does in fact have a finite value in second harmonic polarography [9,11]. The use of phase-selective detection, therefore, still offers the possibility of discriminating against charging current. In general, employment of phase-sensitive detection in the analytical use of second harmonic ac polarography is a prudent experiment on many occasions, although it is certainly true that it is not nearly as important as with the fundamental harmonic method.

From Fig. 7.41, it can be seen that location of $E_{1/2}^r$ (reversible case) is simple and accurate in the 2f mode [59]. Either a zero crossing point (phase-selective) or minimum (non-phase-selective) is involved. The wave shape in second harmonic or other second-order techniques is generally conducive to very accurate measurement of the required parameters.

Examination of the theory for the reversible second harmonic ac response readily leads to the understanding of a few "tricks of the trade" in the 2f method. Devay et al. [60] have described the optimum conditions for measurement of the second harmonic.

Equations presented by Smith [9], using notation already employed, gives the magnitude of the alternating current, I(2ωt), as

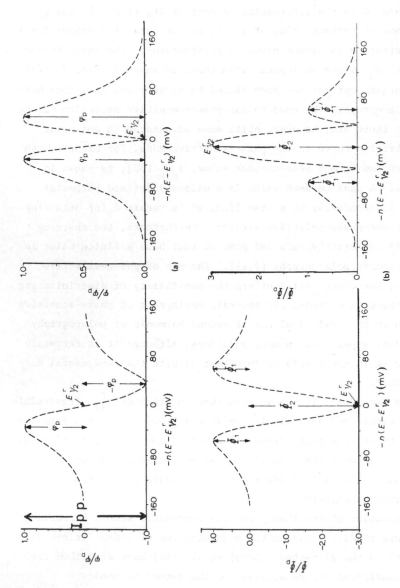

FIGURE 7.41 (a) Second harmonic and (b) third harmonic ac polarograms for a reversible process: (left) phase-sensitive readout; (right) non-phase-sensitive readout. [Reproduced from J. Electroanal. Chem. 35, 343 (1972).]

$$I(2\omega t) = \frac{2^{1/2} n^3 F^3 AC_A (\omega D_A)^{1/2} \Delta E^2 \sinh (j/2)}{16R^2 T^2 \cosh^3 (j/2)} \sin (\omega t - \frac{\pi}{4}) \quad (7.25)$$

This equation is only strictly valid at small amplitudes ($\leq 16/n$ mV p-p), but nevertheless it suffices for the present discussion.

One of the possible difficulties in undertaking second harmonic measurements is the fact that $I(2\omega t)$ is small in magnitude and that instrumental noise, in fact may limit the sensitivity rather than the charging current. The ΔE^2 dependence predicted by Eq. (7.25) therefore makes it attractive to use larger amplitudes of around 50 mV p-p. The broadening of the wave as a result of exceeding the $16/n$ mV p-p limit noted above is still not severe at this amplitude, and in all other respects the departure from Eq. (7.25) is not substantial. Since the charging current is extremely small, higher frequencies may be considered than in the 1f mode. However, it must be remembered that most electrode processes rapidly become irreversible at 2f because of the short time scale, and the villains of uncompensated resistance and the small but finite charging current can readily assert most undesirable influences at high frequency. Consequently a frequency range similar to that in the 1f experiment still is used normally in second harmonic ac polarography. The n^3 term in Eq. (7.25) also means that the limit of detection for a reversible three-electron reduction [such as for Sb(III) in HCl] will be extremely low relative to a reversible one-electron reduction [such as for Tl(I) in the same medium].

Since the 2f technique is at an even shorter time scale than the 1f method, classification of the reversibility or otherwise of an electrode process with respect to this technique is very important. Certain criteria for doing this are tabulated by Smith [9], but the easiest one for the analytical chemist to use is the peak-to-peak separation which should be $68/n$ mV at 25°C. This parameter is akin to the half width of $90/n$ mV in the fundamental harmonic. More detailed analyses of the reversibility might include the frequency dependence of $I(2\omega t)$ (linear $\omega^{1/2}$ plot), phase angle measurement

(45° and 225° relative to fundamental harmonic on positive and negative peaks, respectively) or shape of the wave (equivalent to the
second derivative of the dc polarogram).

Armed with the above similarities to and differences from the
fundamental harmonic, one could readily work one's way through all
the obvious variations of the second harmonic method, e.g., short
controlled drop time (rapid) 2f polarography, current-sampled 2f
polarography, subtractive mode 2f polarography, etc., but all this
can be summarized by stating that the same conclusions given with
respect to the fundamental mode, apply. For example, the second
harmonic response in the fast sweep and cyclic ac voltammetry section was included in Figs. 7.32, 7.33, and 7.36, and the relative
performances of the f and 2f responses do not provide any surprising
results. For reversible or close to reversible processes, there is
no doubt that the claimed advantages of the 2f method can be realized
readily with modern instrumentation. The second harmonic method,
as predicted theoretically, is indeed one of the most sensitive
methods available for this class of electrode process, with a limit
of detection in the 10^{-7} to 10^{-8} M concentration range in aqueous
media. The low detection limits are a direct consequence of the
highly favorable faradaic-to-charging current ratio. Unfortunately,
however, as with several of the other polarographic methods regarded
as rather exotic by nonspecialists in electroanalytical chemistry,
a real paucity of actual analytical examples exists to support such
claims. However, as most of the newer commercially available ac
polarographs now provide phase-selective second harmonic polarography
as a standard feature or as an optional accessory, this situation
could be expected to change over the next few years. For excellent
and most readable accounts of the second harmonic method in all its
aspects, two excellent reviews by Smith [9,11] are essential background reading, as they convey the feeling that there is a considerable, but as yet untapped, scope for using the second harmonic
method.

While it may be true that third and higher harmonics can readily
be measured with modern instrumental methods [Figs. 7.6 and 7.41(b)],
this author can see no likely analytical advantage to be gained by
doing so and, thus apart from acknowledging their existence, no
further mention of these techniques is given. Devay et al. [60]
have discussed the optimum conditions for measurement of the third
harmonic and this work can be consulted by the more academically
inclined.

7.7 INTERMODULAR AND FARADAIC RECTIFICATION
AC POLAROGRAPHY BASED ON SINE WAVES

Reinmuth [61] has presented an elegant review of the various facets
of second order electroanalytical methods. Included in this article
is a very definitive account of the nomenclature and an assessment
of the various approaches in analytical studies, as well as electrode
kinetics. According to Reinmuth [61], the second-order techniques
account for three well-known forms of distortion of electrical signals, namely, second harmonic distortion production of second harmonic currents on application of pure fundamental frequency voltages,
or vice versa; intermodulation distortion-production of sum and
difference frequency components when signals of two or more frequencies are applied simultaneously; rectification-production of
direct current voltages on application or alternating currents and
vice versa.

The second harmonic method already considered is a readily implemented extension of the fundamental mode and the question of concern to an analytical chemist is probably whether intermodular or
rectification methods have any advantages to offer over the second
harmonic method, because presumably this will be the criterion by
which the alternative techniques need to be judged. The theory for
each of the second order methods predicts that for a reversible
process, the shape of the wave will be similar to a second-
derivative ac polarogram, so that improved resolution or readout
mode is unlikely to provide an advantages over the second harmonic.

Indeed, by considering the intermodular technique, the close relationship of all methods is revealed clearly. On applying two sine
waves of frequencies f_1 and f_2 to an electrochemical cell, the output will consist of signals at f_1, f_2, $2f_1$, $2f_2$, $f_1 + f_2$, $f_2 - f_1$,
etc., as well as the normal dc component and the faradaic rectification signals. The signals of interest to the intermodular technique
are at $f_1 + f_2$ and $f_2 - f_1$, and it can be noted that if $f_1 = f_2$,
then the frequency sum is equivalent to a second harmonic and the
frequency difference to the faradaic rectification component [61].

In developing instrumentation for the intermodular technique,
it is clear that there is the disadvantage of having two sine wave
signal sources instead of one. Furthermore, it is obvious that
additional operations must be performed in the setting up procedures
compared with the second harmonic technique, because two oscillators
instead of one are involved. As the literature relating to analytical applications of intermodular techniques [42,61-64] fails to
convince this author that adequate advantages over the second
harmonic method exist it is recommended that if the analytical chemist desires to use a second order technique based on a sinusoidal
wave form, it may as well be the second harmonic one, since it is
the simplest to implement and use.

Most of the methods of faradaic rectification [61,65-68]
(usually based on high-frequency sinusoidal techniques) need not be
discussed in this book as they are generally too inconvenient for
analytical work [9], even though excellent applicability to electrode kinetics [68] has been demonstrated convincingly. Related
second order techniques have also been proposed. Those employing
a dc potential ramp would appear to be the most promising. For
example, Brocke [69] has reported a method he calls "demodulation
polarography," in which a high-frequency sinusoidal signal of, say,
100 kHz is amplitude modulated by a low-frequency sinusoidal signal
of 37 Hz. The demodulation of the sinusoidal signal (37 Hz) by a
reversible electrode process such as that for cadmium was reported
to give a detection limit of 5×10^{-7} M. An improved form of this
technique using a triangularly modulated polarizing voltage has also

been proposed [70]. However, even with this approach there are no
data to suggest any improvement or advantage over the second harmonic
technique. Zheleztsov [71,72] has also described the theory and
implementation of an ac polarographic technique using an amplitude-
modulated sinusoidal voltage with a reported sensitivity of less than
5×10^{-9} M for cadmium. Despite the fact that Zheleztsov suggests
that the method is an order of magnitude improvement over the second
harmonic method, it is still somewhat premature to expect this technique to be used in routine analytical applications of polarography.
Further comments on ac techniques using one very high frequency
signal and another low-frequency signal (with detection of the response at the lower frequency), appear in Sec. 8.2. In general,
while many more variations of second-order techniques can be envisaged,
they are likely to be regarded as being rather esoteric. Consequently,
very convincing data on a far wider scope than simply the limit of
detection of cadmium will need to appear in the literature before
an analytical chemist is likely to be induced to use them in his
laboratory.

7.8 TENSAMMETRY

One of the most rapidly expanding areas in the analytical use of ac
polarography is the use of so-called tensammetric waves. These nonfaradaic processes occur at very positive or negative potentials,
where an adsorbed species is displaced from the electrode by virtue
of an increased affinity of the supporting electrolyte. In Chap. 5
such nonfaradaic processes resulting from an adsorption-desorption
mechanism were observed also to give peaks in linear sweep and cyclic
voltammetry at stationary electrodes. At the dc potential region
where the adsorption-desorption double-layer transformations occur,
a large differential capacity exists which produces large nonfaradaic
waves.

Figure 7.42 is a schematic representation of a tensammetric
curve [7]. The curve is similar to a capacity versus potential curve,
as expected from discussion in Chap. 5, and the current is

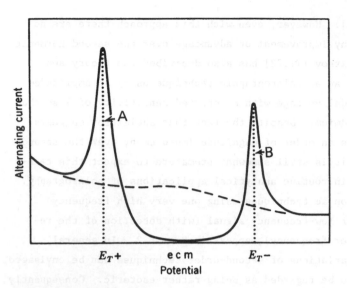

FIGURE 7.42 Tensammetric waves (-) and base current (---).
(A) Positive tensammetric wave; (B) Negative tensammetric wave.
E_{T+} and E_{T-} are their respective peak potentials. (Reproduced from
B. Breyer and H. H. Bauer, *Alternating Current Polarography and
Tensammetry*, Interscience, New York, 1963.)

considerably depressed between the two waves, which are situated on
either side of the electrocapillary maximum (ecm). At sufficiently
negative or positive potentials, where no adsorption occurs, the
tensammetric curve coincides with that of the supporting electrolyte.
Since the adsorption is at its maximum value around the ecm, the
depression of the base current is greatest around this potential
region. The large changes in alternating current at the adsorption-
desorption potentials arise because of the nature of the alternating
potential. Typical fundamental and second harmonic tensammetric
ac waves are illustrated in Figs. 7.43 and 7.44.

Bauer et al. [73] have presented a review of adsorption measurements at electrodes, which can be consulted by those interested
in some of the theoretical aspects. Jehring [74] has published a
survey of developments in this area. His work is extremely valuable
to the analytical chemist because the detailed characteristics of
the responses of tensammetric waves [74-77] have been reported under

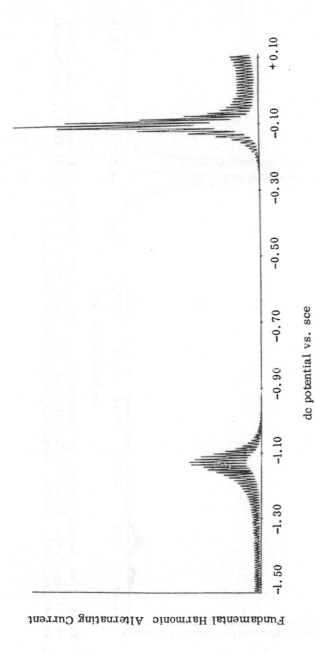

FIGURE 7.43 Fundamental harmonic ac polarogram showing tensammetric waves of cyclohexanol. Solution is 0.050 M cyclohexanol in 0.5 M KNO$_3$ at 25°C. ΔE = 20 mV p-p at 38.6 Hz. [From C.R.C. Crit. Rev. Anal. Chem., D. E. Smith, 2, 247 (1971), The Chemical Rubber Co., 1971. Used by permission of The Chemical Rubber Co.]

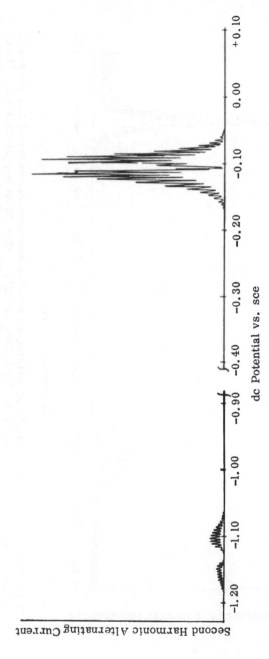

FIGURE 7.44 Second harmonic polarogram showing tensammetric waves of cyclohexanol. Conditions as for Fig. 7.43. [From C.R.C. Crit. Rev. Anal. Chem., D. E. Smith, 2, 247 (1971), The Chemical Rubber Co., 1971. Used by permission of The Chemical Rubber Co.]

a variety of conditions for both fundamental and second harmonic
ac polarography. The square wave response (see Chap. 8) is also
encompassed in Jehring's work. With regard to sensitivity, Jehring
and Stolle [76], in a study of the characterization of approximate
molecular weight, report the relative sensitivities of fundamental
harmonic measurements at a dme, fundamental harmonic measurements at
a stationary mercury drop, and second harmonic measurements with a
stationary mercury drop, to be in the ratio 1:25:250, for the determination of polyethylene glycol. The most sensitive measurement is,
as anticipated, the second harmonic one, with a reported detection
limit of 0.02 mg liter^{-1} of polyethylene glycol (average molecular
weight is 1000).

The usual word of warning pertaining to the use of electrode
processes including adsorption must be reiterated in discussing tensammetric processes. A wide range of compounds, particularly organic
compounds, gives rise to a tensammetric response; however, the reader
who is interested in potential applications should be very familiar
with their variability toward medium effects before undertaking
too many experiments [74]. Furthermore, such waves frequently behave rather differently from faradaic processes. For example, peak
heights are unlikely to be a linear function of concentration, and
the peak position is usually a function of concentration. Electrolyte type and concentration also may influence significantly the
wave shape and position far more than for faradaic processes. Extreme care and common sense must be employed therefore in using
tensammetric waves in analytical work. The work of Bauer et al.
[78] provides data in support of these remarks.

In such a potentially tricky area of analysis, as tensammetry
undoubtedly is, the best didactic approach possibly may be to report
comprehensively on an actual application from the literature; thus
Note 79 is cited extensively to conclude this section.

A variety of methylcarbamates are used in agriculture as insec-
ticides. Among the most common are Carbaryl (1-naphthyl
methylcarbamate), Aldecarb (formerly temik, [2-methyl-2-
methylthiopropionaldehyde O-methylcarbamoyl oxime) and Butacarb

(3,5-di-*t*-butylphenyl N-methylcarbamate). The work of Booth and
Fleet [79] describes a method for the direct determination of
methylcarbamates, based on their adsorption-desorption phenomena.
Both ac polarographic and cyclic voltammetric techniques have been
used.

None of the compounds studied show conventional dc polarographic
waves. However, with ac polarography, all compounds show current
peaks at potentials in the region from -1.0 to -1.5 V vs. sce.
That these peaks were tensammetric in character was elucidated by the
depression of the base line of the supporting electrolyte (Fig. 7.45)
and the absence of a significant dc polarographic wave. The electrocapillary curves (Fig. 7.46) for these compounds show that they are
adsorbed over a large voltage range on either side of the electrocapillary maximum, where maximum adsorption occurred. At high negative potentials, the capacitive current coincides with that of the
supporting electrolyte, since at these potentials the surfacant is
not present at the mercury-solution interface. These sharp capacitive peaks therefore would appear to be due to a desorption process,
and as such reflect periodic changes in the capacity of the double
layer. Contributory evidence of the tensammetric character of these
peaks is the rectilinear dependence of the peak potential on the
log of the concentration of insecticide. It is also worthy of note
that small steps are observed in the dc curve for these compounds
at potentials where tensammetric peaks occur. These adsorption steps
are characteristically much smaller than conventional dc polarographic
waves.

The optimum pH was found to be 8.4 (aqueous 0.5 M sodium di-
hydrogen phosphate/0.5 M sodium hydroxide buffer). The effect of
variation of buffer concentration indicated that maximum peak heights
for Aldecarb, Methiocarb and Carbaryl were obtained in 0.5 M buffer.
Owing to the surface-active properties of the alcohols, a maximum of
10% methanol could be tolerated. A 1 to 2% methanol solution was
desirable. Because of the solubility of Butacarb, 10% methanol
solutions, 0.2 M in buffer, were necessary. A rectilinear dependence

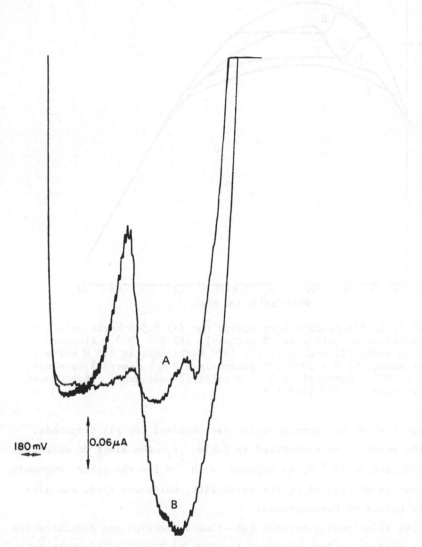

FIGURE 7.45 AC polarograms for (A) 0; (B) 5×10^{-5} M. Butacarb
in 10% methanol at pH 8.98 (0.2 M boric acid/0.2 M sodium
hydroxide). [Reproduced from Talanta *17*, 491 (1970).]

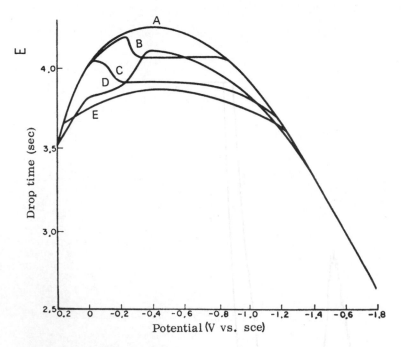

FIGURE 7.46 Electrocapillary curves for (A) 0.5 M boric acid/0.5 M
sodium hydroxide buffer in 2% methanol; (B) 5×10^{-4} M Aldecarb
in 0.5 M buffer/2% methanol; (C) 10^{-4} Methiocarb in 0.5 M buffer/
2% methanol; (D) 5×10^{-5} M Butacarb in 0.2 M buffer/10% methanol;
(E) 2×10^{-4} M Carbaryl in 0.05 M buffer/2% methanol. [Reproduced
from Talanta *17*, 491 (1970).]

of peak current vs. concentration was obtained for all compounds,
and the results are summarized in Table 7.6. The limit of detection
for Aldecarb is 10^{-4} M, as opposed to 10^{-5} M for the other compounds.
This may be attributed to the solubility, molecular size, and ali-
phatic nature of the compound.

The relationship between base-line depression and concentration
of the surface-active carbamate is shown in Fig. 7.47 for Carbaryl
and Aldecarb. A rectilinear relationship occurs over a limited
concentration range, above which the form of a Langmuir isotherm is
followed. This would appear to signify that the mercury surface is
now completely covered with the adsorbed species.

The effect of various cations (0.1 M, as chlorides) and various anions (0.1 M, as sodium salt) on the tensammetric peaks was investigated.

TABLE 7.6 Concentration Dependences for Methylcarbamates

Concentration of methylcarbamate (M)	Peak current[a] (μA)							
	AC polarography				DC cyclic voltammetry			
	B	C	M	A	B	C	M	A
2×10^{-6}	--	--	--	--	--	--	--	--
4×10^{-6}	--	--	--	--	0.40	--	0.40	--
6×10^{-6}	--	--	--	--	0.60	0.45	0.60	--
8×10^{-6}	--	--	--	--	0.87	0.60	0.75	--
1×10^{-5}	0.05	--	0.05	--	1.10	0.70	1.00	--
2×10^{-5}	0.07	0.05	0.09	--	2.00	1.30	2.40	--
4×10^{-5}	0.17	0.09	0.22	--	4.40	2.40	4.70	--
6×10^{-5}	0.28	0.14	0.29	--	6.60	3.20	7.60	0.30
8×10^{-5}	0.37	0.18	0.41	--	9.00	4.40	9.20	0.40
1×10^{-4}	0.49	0.24	0.52	0.06	11.00	4.80	11.40	0.50
2×10^{-4}	L.S.	0.50	L.S.	0.12	--	--	--	0.95
4×10^{-4}	--	1.00	--	0.28	--	--	--	1.70
6×10^{-4}	--	1.41	--	0.43	--	--	--	2.80
8×10^{-4}	--	1.73	--	0.59	--	--	--	3.75
1×10^{-3}	--	2.07	--	0.71	--	--	--	4.40

[a] B, Butacarb; C, Carbaryl; M, Methiocarb; A, Aldecarb. L.S., limited by solubility. Data taken from Note 79.

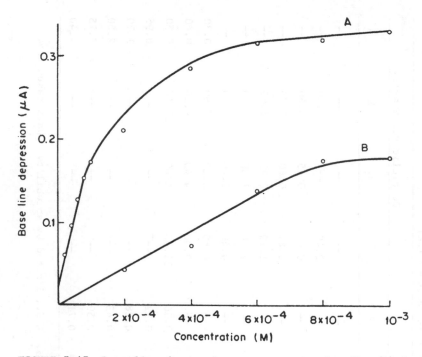

FIGURE 7.47 Base-line depression vs. concentration for (A) Carbaryl
and (B) Aldecarb in 0.05 M boric acid, 0.05 M sodium hydroxide
buffer (pH 8.4), 2% methanol. [Reproduced from Talanta 17, 491
(1970).]

In the case of cations, the desorption potentials are increasingly
negative in the sequence $La^{3+} < Mg^{2+} < Li^{+}$. The observed effect
will depend to some degree on the displacement capability of the
cation and also on the solubility of the species in solution. In
the case of the anions NO_2^{-}, NO_3^{-}, ClO_4^{-}, CO_3^{2}, CN^{-}, and SO_4^{2-}, the
effect on peak height and peak potential follows no specific pattern.
This is primarily because, at negative potentials, it is the cation
that determines the capacity and structure of the double layer.

The extent of adsorption of the organic molecule therefore is
not governed entirely by its own solubility. The concentration of the
supporting electrolyte and any extraneous ions in solution must also
be considered. This is an inherent drawback in the analytical utility of capacitive (tensammetric) peaks. The conditions under which

the experiment is to be conducted must be clearly defined and closely
adhered to.

The cyclic voltammogram for Butacarb is shown in Fig. 7.48.
A single cathodic-anodic peak is observed at the sweep rates used.
The cathodic peak corresponds to the tensammetric peak caused by the
desorption of the surface-active carbamate compound. This peak was
utilized for analytical purposes. The anodic peak observed on the
reverse scan is due to adsorption of the carbamate onto the electrode
surface, replacing the cations of the supporting electrolyte.
Methiocarb and Carbaryl show similar behavior. Aldecarb shows, in
addition, the capacitive peaks corresponding to adsorption of the
surfacant on the cathodic scan and desorption on the reverse scan.

A rectilinear dependence of peak current against concentration
was observed for all compounds (Table 7.6). Concentrations of the
order of 5×10^{-6} M (1 ppm) for Butacarb, Carbaryl, and Methiocarb
must be taken as the lower limit attainable with the apparatus
used. The limit for Aldecarb is 5×10^{-5} M.

The above almost verbatim report from Booth and Fleet's paper
testifies to the difficulties and restrictions imposed on using

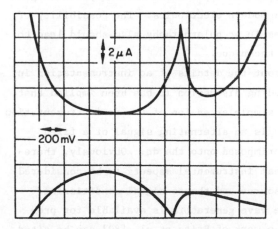

FIGURE 7.48 DC cyclic voltammogram of 5×10^{-5} M Butacarb in 0.2 M
boric acid/0.2 M sodium hydroxide buffer, 10% methanol. Scan rate
= 4 V sec^{-1}. [Reproduced from Talanta *17*, 491 (1970).]

tensammetric waves. From the above, it may appear that dc cyclic
voltammetry is more sensitive than ac polarography. However, had
a phase-selective fundamental or second harmonic ac polarograph been
used, the converse probably would have been observed.

7.9 EXAMPLES AND TRENDS IN THE USE
OF SINUSOIDAL AC POLAROGRAPHY

Sinusoidal ac polarography constitutes one of the most widely used
non-dc polarographic method. The monograph by Breyer and Bauer [7]
provides a survey of the earlier applications in analytical chemistry
and the extent of work undertaken by the early 1960s can be seen to
be substantial. Unlike the work reviewed in their monograph, recent
literature is dominated by the use of phase-selective fundamental
harmonic ac polarography. The phase-selective versions of both the
fundamental and second harmonic method are now standard features of
the new commercially available ac polarographs, and applications of
the second-order 2f mode, while rare now, may become more prevalent
in the future. Widespread use of other second-order methods seems
less likely. Trends toward the phase-selective readout form undoubtedly will be continued, as the approach is obviously advantageous
and readily implemented with modern electronics. The possibility of
exploiting tensammetric waves in ac polarography also should lead to
further applications of the technique.

Little has been said about the details of ac instrumentation in
this chapter, but from preceding discussion it has been implied that
an ac polarograph is only a simple extension of the dc instrumentation
in that the only new feature is an alternating signal of a fixed
amplitude and frequency superimposed onto the dc. Obviously, therefore, the important additional instrumental aspects to be considered
are the alternating signal source and the ac-signal conditioning
network. Any number of sine wave generators is available for providing the ac signal, and the work of Britz et al. [80] can be cited
as a source of information in this area. Importantly, however, it
is probably the advent of high-performance, inexpensive phase-
sensitive detectors (also known as *lock-in amplifiers*) into ac

polarography [7,11,81-85] that has stimulated manufacturers of commercially available instruments to feature phase-sensitive detection
as a standard accessory, and which has made the phase-selective version the routine mode for undertaking ac polarographic measurements.

Phase-sensitive detectors are readily constructed with the aid
of operational amplifiers. Figure 7.49 gives an example of a signal
conditioning network for phase-sensitive ac detection. All components
are now very inexpensive, and so the possibility of using several
phase-sensitive detectors in the one instrument to perform a number
of functions is now a reality. Smith and his group [86,87], for
example, use this approach when simultaneously recording dc, fundamental and second harmonic polarograms (ac polarography in the multi-
plex mode), and Blutstein et al. [64] employed this approach for
developing a multifunctional ac detection system which included
phase-selective intermodular, fundamental and second harmonic ac
polarography, as well as simultaneous in-phase and quadrature component measurements. Thus, development of the second-order techniques,
as for the fundamental harmonic method, has been facilitated greatly
by the use of phase-sensitive detectors.

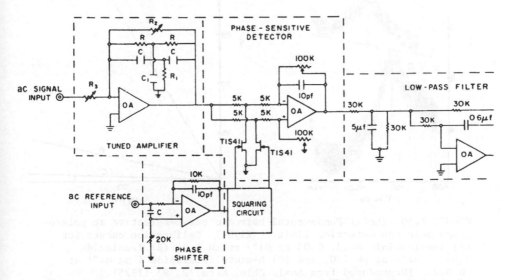

FIGURE 7.49 Signal conditioning network for phase-sensitive ac signal detection. [Reproduced from Computers in Chemistry and Instrumentation 2, 369 (1972).]

   Several practical applications of ac polarography have been
mentioned so far, to illustrate particular concepts. To provide a
bibliography of applications of ac polarography, while no doubt a
most desirable exercise, is not the purpose of the present book and
is not provided. Rather, several additional and pertinent applications demonstrating general principles and trends are discussed to
conclude this chapter.

   Figure 7.50 shows phase-selective polarograms of three organic
species at low concentrations and their calibration curves [88].
Several typical features emerge from examining the polarograms.
First, despite the use of phase-selective detection, the sensitivity

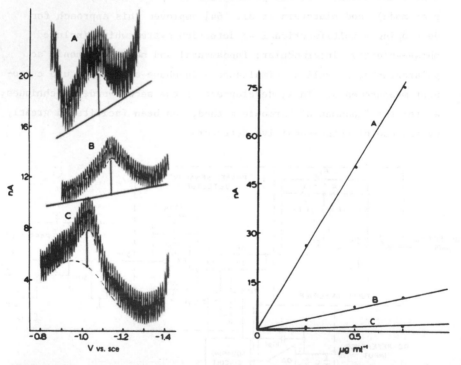

FIGURE 7.50   (Left) Fundamental harmonic phase-selective ac polarograms near the detection limit.   (Right)   Calibration curves for
(A) isonicotinic acid, 0.03 µg ml^{-1} at pH 5.0, (B) isoniazide
0.02 µg ml^{-1} at pH 7.0, and (C) N-acetylisoniazide 1 µg ml^{-1} at
pH 6.0.   [Reproduced from Anal. Chim. Acta 78, 93 (1975).]

in ac polarography is still limited ultimately by the background or
charging current, and thus the model of an ideal capacitor and instrumentation, which theoretically should enable complete rejection of
the charging current, is not realized experimentally. Second, the
variety of wave shapes and sensitivities possible in ac polarography
is revealed in this figure. Despite the probable mechanistic differences in the electrode processes for the three species, linear
calibration curves are obtained in each case. (Linearity can be
obtained for many mechanistically complicated processes.) Third, it
was mentioned that for nonreversible electrode processes, solution
conditions can be very critical and must be controlled strictly.
For many organic electrode processes in aqueous media, protonation
or deprotonation of the reactant or products is involved, and buf-
fering of solutions is required (see also Chap. 2). This is likely
to be even more critical in ac polarography and has been undertaken
in the example considered above.

Woodson and Smith [89] show that, since the use of nonaqueous
solvents eliminates or decreases the influence of coupled chemical
reactions for many organic systems, the ac polarography of a wide
range of pharmaceuticals is often substantially more suitable in
such media. Table 7.7 summarizes some of the data contained in
their paper. For those systems where the fundamental harmonic technique gives a lower limit of detection than the dc method, the electrode processes are reversible, or nearly so.

Mechanistic complications, probably involving coupled chemical
reactions, decrease the sensitivity of the ac method for some of
the pharmaceuticals. In such cases, the dc method can have the
lower detection limit. Had phase-selective fundamental harmonic
measurements been used, an order of magnitude or more improvement
presumably would have resulted. This assessment is indicated by the
second harmonic results tabulated alongside those from dc and fundamental harmonic ac polarography. Figures 7.51 and 7.52 provide
examples of dc and ac polarograms of some of the pharmaceuticals.
Any serious student of the use of modern polarographic methods in
analytical chemistry would be well advised to read this particular

TABLE 7.7 Compilation of DC and AC Polarographic Responses of Pharmaceuticals in Acetonitrile (0.1 M Tetrabutylammonium Perchlorate)[a]

Compound	$E_{1/2}$[b]	E_p[c]	E_{min}[d]	n_{dc}[e]	$n_{\omega t}$[f]	$n_{2\omega t}$[g]	Approximate detection limits		
							I_{dc}[h]	$I_{(\omega t)}$[i]	$I_{(2\omega t)}$[j]
Acetylsalicylic acid	-1.64	-1.76	-1.87	0.44	0.45	0.40	5×10^{-5}	1×10^{-4}	1×10^{-4}
Atropine	-2.14	N.R.[k]	N.R.	0.60	N.R.	N.R.	5×10^{-5}	N.R.	N.R.
Atropine methyl nitrate	-2.04	-2.10	-2.13	0.50	N.A.[l]	0.60	1×10^{-4}	5×10^{-4}	5×10^{-4}
Colchiceine	-0.96	-1.00	-1.02	0.80	0.58	0.55	2×10^{-5}	3×10^{-5}	2×10^{-4}
Colchicine	-1.47	-1.47	-1.47	1.00	1.00	1.00	3×10^{-5}	1×10^{-5}	5×10^{-4}
Deserpidine	-2.14	-2.14	-2.12	1.00	1.00	1.00	3×10^{-5}	1×10^{-5}	5×10^{-4}
Estradiol	-0.64	N.R.	N.R.	0.30	N.R.	N.R.	3×10^{-5}	N.R.	N.R.
Hydrochlorothiazide	-1.56	01.65	-1.62[m]	0.80	0.70	N.A.	1×10^{-5}	3×10^{-5}	3×10^{-4}
Hydrocortisone	-1.58	-1.63	-1.68	0.90	0.56	0.54	1×10^{-5}	3×10^{-5}	2×10^{-4}
Methylclothiazide	-1.58	-1.60	1.69[m]	0.58	0.45	N.A.	3×10^{-5}	3×10^{-4}	6×10^{-4}
Nitroglycerine	-0.59	N.R.	N.R.	0.50	N.R.	N.R.	2×10^{-5}	N.R.	N.R.
Phenobarbital	-2.07	-2.14	-2.18[m]	0.83	0.60	N.A.	5×10^{-5}	1×10^{-4}	2×10^{-4}
Prednisolone	-1.64	-1.80	-1.85	0.93	0.50	0.38	5×10^{-5}	8×10^{-5}	8×10^{-4}
Prednisone	-1.65	-1.70	-1.80	0.32	0.35	0.26	5×10^{-5}	5×10^{-4}	5×10^{-5}
Progesterone	-1.85	-1.91	-1.90	0.93	0.84	0.85	5×10^{-5}	1×10^{-4}	5×10^{-5}
Reserpine	-2.14	-2.15	-2.14	1.00	1.00	1.00	5×10^{-5}	5×10^{-6}	1×10^{-6}
Salicylic acid	-1.54	-1.70	-1.66	0.50	0.60	0.60	8×10^{-4}	2×10^{-4}	2×10^{-4}

Sulfacetimide	-1.85	-1.98	-1.98^m	0.35	0.27	N.A.	5×10^{-5}	5×10^{-5}	5×10^{-5}	1×10^{-5}
Sulfamethazine	-1.45	-1.57	-1.51^m	0.33	0.37	N.A.	5×10^{-5}	3×10^{-5}	3×10^{-5}	5×10^{-6}
Testosterone	-2.02	-2.03	-2.02	1.10	0.95	1.00	3×10^{-5}	2×10^{-5}	2×10^{-5}	2×10^{-6}

^{a} AC data refer to total alternating current measurements.

^{b} $E_{1/2}$ = dc polarographic half-wave potential (volt vs. silver reference).

^{c} E_p = fundamental harmonic ac polarographic peak potential (volt vs. silver reference).

^{d} E_{min} = potential of second harmonic ac polarographic minimum (volt vs. silver reference).

^{e} n_{dc} = apparent n value calculated from dc polarographic data using the relationship;
$n = 56/(E_{3/4} - E_{1/4})$, where $E_{1/4}$ is the quarter-wave potential and $E_{3/4}$ is the three-quarter-wave potential in millivolt.

^{f} $n_{\omega t}$ = apparent n value calculated from fundamental harmonic ac polarograms using the relationship
$n = 90/W_{1/2}$ where $W_{1/2}$ = full width of wave at half height in millivolt.

^{g} $n_{2\omega t}$ = apparent n value calculated from second harmonic ac polarograms using the relationship
$n = 68/\Delta E_p$, where ΔE_p is the separation of second harmonic peak potentials in millivolt.

^{h} I_{dc} = detection limit based on the dc polarographic current (mole liter^{-1}).

^{i} $I(\omega t)$ = detection limit based on fundamental harmonic ac polarographic current (mole liter^{-1}).

^{j} $I(2\omega t)$ = detection limit based on second harmonic ac polarographic current (mole liter^{-1}).

TABLE 7.7 (Continued)

[k] N.R. designates no detectable response, i.e., no wave.

[l] N.A. designates either that measurements were not performed, or nature of data precluded essential calculation.

[m] Designates that second harmonic ac polarogram does not exhibit a minimum; number represents peak potential.

Data taken from Note 89.

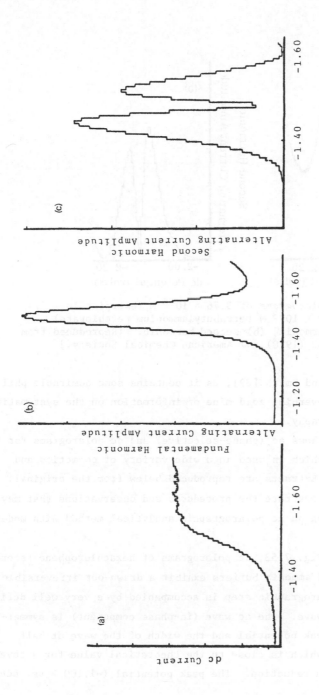

FIGURE 7.51 Polarograms of 3.27×10^{-4} M colchicine in acetonitrile (0.1 M tetrabutylammonium perchlorate): (a) dc polarogram, (b) fundamental harmonic ac polarogram, and (c) second harmonic ac polarogram. [Reproduced from Anal. Chem. 42, 242 (1970). © American Chemical Society.]

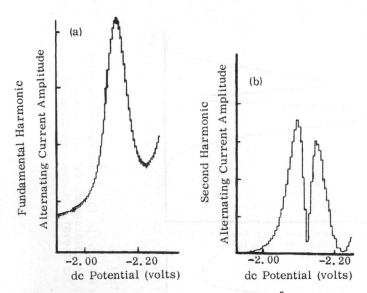

FIGURE 7.52 AC polarograms of 3.76×10^{-5} M reserpine in
acetonitrile (1.00×10^{-3} M tetrabutylammonium perchlorate).
(a) fundamental harmonic, (b) second harmonic. [Reproduced from
Anal. Chem. *42*, 242 (1970). © American Chemical Society.]

paper by Woodson and Smith [89], as it contains some admirable philosophy and the proverbial gold mine of information on the systematic
use of ac polarography.

Figure 7.53 shows ac (phase-selective) and dc polarograms for
hexachlorophene, which is used in a wide variety of cosmetics and
pharmaceuticals. Extracts are reproduced below from the original
paper [90], to demonstrate the procedures and observations that may
occur in developing an ac polarographic analytical method with modern
instrumentation.

As shown in Fig. 7.53, dc polarograms of hexachlorophene recorded
from phosphate and ammonia buffers exhibit a drawn-out irreversible
wave. The dc polarographic step is accompanied by a very well defined
ac polarographic wave. The ac wave (in-phase component) is symmetrical about the peak potential and the width of the wave at half
height is 90 mV, which is close to the theoretical value for a reversible one-electron reduction. The peak potential (-1.100 V vs. sce)

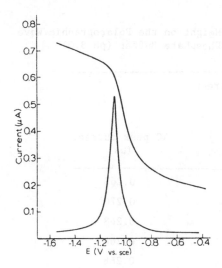

FIGURE 7.53 DC and ac polarograms of 0.1 mM hexachlorophene in phosphate buffer pH 8.0. The ac current is the in-phase component. [Reproduced from Anal. Chim. Acta *61*, 320 (1973).]

and the half-wave potential ($E_{1/2}$ = -1.020 V vs. sce) are independent of pH in the range 6 to 11, indicating that hydrogen ions are not involved in the rate-determining step.

The effect of drop time was investigated by recording polarograms of 0.1 and 0.05 mM hexachlorophene in phosphate buffer (pH 8) at various heights of the mercury column. As indicated in Table 7.8, the height of the ac wave is independent of the drop time, whereas the limiting current of the dc wave increases linearly with the height of the mercury column, indicating that the current is controlled by an adsorption process. AC polarograms of the drug, recorded with non-phase-sensitive detection, showed that the base current is depressed at potentials more positive than the peak potential, which implies that it is the oxidized form of the electroactive species which is adsorbed on the electrode surface.

Polarograms recorded from 0.1 M phosphate buffer (pH 8.0) with various amounts of hexachlorophene present, showed that the height of the ac wave increases linearly with the bulk of concentration of the drug, in the range 10^{-6} to 5.10^{-5} M (Table 7.9). As a result of the strong adsorption of the depolarizer, the value I_p/C decreases at higher bulk concentrations, but the polarograms are still perfectly reproducible. When the concentration of hexachlorophene is decreased

TABLE 7.8   Effect of Mercury Column Height on the Polarographic Wave
of 0.05 mM Hexachlorophene in 0.1 M Phosphate Buffer (pH 8.0)

h_{corr}[a] (cm)	DC limiting current		AC peak current (μA)
	μA	$1 \times h_{corr}^{-1}$ ($\mu A\ cm^{-1}$)	
39.9	0.265	0.0066	0.268
44.9	0.280	0.0062	0.270
49.9	0.310	0.0062	0.268
54.9	0.350	0.0064	0.266
59.9	0.370	0.0062	0.268
64.9	0.440	0.0067	0.270

[a]h_{corr} = mercury column height corrected for back pressure.   Data
taken from Note 90.

TABLE 7.9   AC Polarographic Data for the Reduction
of Various Amounts of Hexachlorophene in 0.1 M
Phosphate Buffer (pH 8.0)

Concentration (mM)	I_p[a] (nA)	I_p/C ($\mu A\ mM^{-1}$)
0.100	415	4.15
0.050	270	5.40
0.025	138	5.52
0.010	56.0	5.60
0.0075	41.5	5.53
0.0050	28.0	5.60
0.0010	5.6	5.60

[a]I_p = ac peak height.   Data taken from Note 90.

below 10^{-5} M, reproducible waves are obtained only at relatively long
drop times. The data in Table 7.9 were obtained at a mercury height
of 45 cm, which corresponds to a drop time of 6 sec. The dc polarographic step is ill defined at low concentration of the drug, and
the whole wave disappears at concentrations below 5.10^{-5} M. Consequently, only phase-selective polarography is applicable for the
determination of small amounts of hexachlorophene.

Based on the experimental results, the following electrode
reaction is proposed

$$RCl + e \xrightarrow{\text{slow}} R_{(ads)} + Cl^-$$

$$R_{(ads)} + e \overset{\text{fast}}{\rightleftharpoons} R^-$$

$$R^- + H^+ \rightarrow RH$$

where RCl denotes hexachlorophene.

In summary, hexachlorophene produces a very well defined ac
polarographic wave at the dropping mercury electrode. By means of
a phase-sensitive ac polarograph the drug can be determined easily
in the concentration range 0.4 to 40 µg ml^{-1}. The limit of detection is about 100 ng ml^{-1}. The peak height is independent of pH
in the entire range 6 to 11. The proposed method for determination
of the drug is very rapid and selective and is more sensitive than
the spectrophotometric methods. Unfortunately, the ac wave is de-
formed in the presence of surface-active substances like proteins.
Hence, hexachlorophene must be separated from biological samples
before the polarographic determination.

The preceding examples were deliberately chosen because they
are of the kind involving considerable mechanistic complications.
For further examples that can be examined to show the methodology
and extent of the use of ac polarography, the reader is referred
directly to the literature [91-96]. These references include mention
of determining explosives [91], nitrilotriacetic acid [92], tin in

alloys [93], dissolved oxygen in gasoline [94], and lead and zinc in lubricating oil [95].

The breadth and depth of ac polarography, however, is nowhere near completely encompassed by these selected references, and only by further detailed examination of the original literature and by undertaking ac polarographic experiments for oneself can all the possibilities of this technique be appreciated.

NOTES

1. B. Fleet and R. D. Jee, *Electrochemistry, Vol. 3*, Specialist Periodical Reports, The Chemical Society, London, 1973.

2. G. C. Barker and R. L. Faircloth, in *Advances in Polarography* (I. S. Longmuir, ed.), Pergamon, Oxford, 1960, p. 313.

3. G. C. Barker and I. L. Jenkins, Analyst 77, 685 (1952).

4. J. H. Sluyters, J. S. M. C. Breukel, and M. Sluyters-Rehbach, J. Electroanal. Chem. *31*, 201 (1971).

5. I. Ruzic, J. Electroanal. Chem. *39*, 111 (1972).

6. G. C. Barker, J. Electroanal. Chem. *41*, 95 (1973).

7. B. Breyer and H. H. Bauer, in *Alternating Current Polarography and Tensammetry*, Chemical Analysis Series (P. J. Elving and I. M. Kolthoff, eds.), Interscience, New York, 1963, vol. 13.

8. D. E. Smith, *Applications of Minicomputers to Measurement of Faradaic Admittance*. Topics in Pure and Applied Electrochemistry SAEST, Karaikudi, India, 1976.

9. D. E. Smith, in *Electroanalytical Chemistry* (A. J. Bard, ed.) Dekker, New York, 1966, vol. I, chap. 1.

10. M. Sluyters-Rehbach and J. H. Sluyters, in *Electroanalytical Chemistry* (A. J. Bard, ed.), Dekker, New York, 1970, vol. IV.

11. D. E. Smith, C.R.C. Crit. Rev. Anal. Chem. *2*, 247 (1971).

12. D. E. Smith, Computers in Chemistry and Instrumentation *2*, 369 (1972).

13. A. M. Bond, Anal. Chem. *44*, 315 (1972).

14. B. Timmer, M. Sluyters-Rehbach, and J. H. Sluyters, J. Electroanal. Chem. *14*, 169 (1967); *14*, 181 (1967).

15. D. E. Smith and T. G. McCord, Anal. Chem. *40*, 474 (1968).

16. I. Ruzic and D. E. Smith, Anal. Chem. *47*, 530 (1975).

17. H. H. Bauer and P. J. Elving, Electrochim. Acta *2*, 240 (1960).

18. A. M. Bond and A. B. Waugh, Electrochim. Acta *15*, 1471 (1970).

19. J. E. B. Randles and K. W. Somerton, Trans. Faraday Soc. *48*, 951 (1952).

20. A. A. Moussa and H. M. Sammour, J. Chem. Soc. 1960, 2151.

21. A. M. Bond and R. J. Taylor, J. Electroanal. Chem. *28*, 209 (1970).

22. B. J. Huebert and D. E. Smith, J. Electroanal. Chem. *31*, 333 (1971).

23. A. M. Bond and D. R. Canterford, Anal. Chem. *44*, 721 (1972).

24. A. M. Bond, J. Electrochem. Soc. *118*, 1588 (1971).

25. A. M. Bond, Anal. Chem. *45*, 2026 (1973), and references cited therein.

26. A. M. Bond, Talanta *20*, 1139 (1973).

27. A. M. Bond and D. R. Canterford, Anal. Chem. *44*, 1803 (1972).

28. A. M. Bond and R. J. O'Halloran, J. Phys. Chem. *77*, 915 (1973).

29. A. M. Bond, Talanta, *21*, 591 (1974).

30. G. S. Smith, Trans. Faraday Soc. *47*, 63 (1951).

31. L. Namec, J. Electroanal. Chem. *8*, 166 (1964).

32. I. M. Kolthoff, J. C. Marshall, and S. L. Gupta, J. Electroanal. Chem. *3*, 209 (1962).

33. W. B. Schaap and P. S. McKinney, Anal. Chem. *36*, 1251 (1964).

34. G. L. Booman and W. B. Holbrook, Anal. Chem. *35*, 1793 (1963).

35. D. R. Canterford, Anal. Chem. *46*, 763 (1974).

36. E. R. Brown, T. G. McCord, D. E. Smith, and D. D. DeFord, Anal. Chem. *38*, 1119 (1966).

37. G. C. Barker and R. L. Faircloth, J. Polarogr. Soc. *4*, 11 (1958).

38. W. L. Underkofler and I. Shain, Anal. Chem. *37*, 218 (1965).

39. H. Blutstein and A. M. Bond, Anal. Chem. *46*, 1531 (1974).

40. C. I. Mooring, Polarogr. Ber. *6*, 63 (1958).

41. R. Neeb, Z. Anal. Chem. *186*, 53 (1962).

42. R. D. Jee, Z. Anal. Chem. *264*, 143 (1973).

43. B. Fleet, R. D. Jee, and C. J. Little, J. Electroanal. Chem. *43*, 349 (1973).

44. H. Blutstein and A. M. Bond, Anal. Chem. *46*, 1934 (1974).

45. A. M. Bond, R. J. O'Halloran, I. Ruzic, and D. E. Smith, Anal. Chem. *48*, 872 (1976).

46. A. M. Bond, Anal. Chim. Acta *74*, 163 (1975).

47. R. D. Jee, Proc. Analyt. Div. Chem. Soc. *12*, 184 (1975).

48. A. L. Juliard, Nature *183*, 1040 (1959).

49. A. L. Juliard, J. Electroanal. Chem. *1*, 101 (1959).

50. A. M. Bond, J. Electroanal. Chem. *50*, 285 (1974).

51. A. M. Bond, R. J. O'Halloran, I. Ruzic, and D. E. Smith, Anal. Chem. *50*, 216 (1978).

52. A. M. Bond and R. J. O'Halloran, Anal. Chem. *47*, 1906 (1975).

53. A. M. Bond and B. S. Grabarić, J. Electroanal. Chem. *87*, 251 (1978).

54. A. M. Bond, Anal. Chem. *45*, 2027 (1973).

55. A. M. Bond, V. S. Biskupsy, and D. A. Wark, Anal. Chem. *44*, 1551 (1974).

56. A. M. Bond and M. E. Beyer, Anal. Chim. Acta *75*, 409 (1975).

57. A. M. Bond and J. H. Canterford, Anal. Chem. *43*, 228 (1971).

58. H. Blutstein and A. M. Bond, Anal. Chem. *46*, 1754 (1974).

59. A. M. Bond and D. E. Smith, Anal. Chem. *46*, 1946 (1974).

60. J. Devay, T. Garai, L. Meszáros, and B. Palagyi-Fenyes, Magyar Kém, Folyóirat *75*, 460 (1969).

61. W. H. Reinmuth, Anal. Chem. *36*, 211R (1964).

62. R. Neeb, Z. Anal. Chem. *188*, 401 (1962); *208*, 168 (1965); Naturewiss *49*, 1 (1962); *49*, 477 (1962).

63. D. Sauer and R. Neeb, J. Electroanal. Chem. *75*, 171 (1977).

64. H. Blutstein, A. M. Bond, and A. Norris, Anal. Chem. *48*, 1975 (1976).

65. P. Delahay, in *Advances in Electrochemistry and Electrochemical Engineering* (P. Delahay and C. Tobias, eds.), Interscience, New York, 1961, vol. I, chap. 5.

66. W. A. Brocke and H. W. Nürnberg, Z. Instrumentenk *75*, 291, 315, 355 (1967).

67. G. C. Barker, *Transactions Symposium Electrode Processes, Philadelphia 1959*, Wiley, New York, 1961, chap. 18.

68. H. P. Agarwal, in *Electroanalytical Chemistry* (A. J. Bard, ed.), Dekker, New York, 1974, vol. 7, pp. 161-271.

69. W. A. Brocke, J. Electroanal. Chem. *30*, 237 (1971).

70. W. A. Brocke, J. Electroanal. Chem. *33*, App 1 (1971).

71. A. V. Zheleztsov, Zhur. Anal. Khim, *26*, 644, 650 (1971); *27*, 1461 (1972).

72. A. V. Zheleztsov and R. K. Rafikov, Zh. Anal. Khim. *28*, 867 (1972).

73. H. H. Bauer, H. B. Herman, and P. J. Elving, in *Modern Aspects of Electrochemistry* (J. O. M. Bockris and B. E. Conway, eds.), Plenum Press, New York, 1971.

74. H. Jehring, J. Electroanal. Chem. *21*, 77 (1969).

75. H. Jehring, E. Horn, A. Reklat, and W. Stolle, Collect. Czech. Chem. Commun. *33*, 1038 (1968).

76. H. Jehring and W. Stolle, Collect. Czech. Chem. Commun. *33*, 1670 (1968).

77. H. Jehring, J. Electroanal. Chem. *20*, 33 (1969).

78. H. H. Bauer, H. R. Campbell, and A. K. Shallal, J. Electroanal. Chem. *21*, 45 (1969).

79. M. D. Booth and B. Fleet, Talanta *17*, 491 (1970).

80. D. Britz, J. S. Jackson, and H. H. Bauer, Chem. Instrum. *3*, 229 (1971).

81. R. F. Evilia and A. J. Diefenderfer, Anal. Chem. *39*, 1885 (1967).

82. T. F. Retajczyk and D. K. Roe, J. Electroanal. Chem. *17*, 21 (1968).

83. R. de Levie and A. A. Husovsky, J. Electroanal. Chem. *20*, 181 (1969).

84. H. H. Bauer and D. Britz, Chem. Instrum. *2*, 361 (1970).

85. D. E. Glover and D. E. Smith, Anal. Chem. *43*, 775 (1971).

86. D. E. Glover and D. E. Smith, Anal. Chem. *44*, 1140 (1972).

87. B. J. Huebert and D. E. Smith, Anal. Chem. *44*, 1179 (1972).

88. J. J. Vallon, A. Badinand, and C. Bichon, Anal. Chim. Acta *78*, 93 (1975).

89. A. L. Woodson and D. E. Smith, Anal. Chem. *42*, 242 (1970).

90. E. Jacobsen and T. Rojahn, Anal. Chim. Acta *61*, 320 (1972).

91. J. S. Hetman, Z. Anal. Chem. *264*, 159 (1973).

92. G. den Boef, R. Oostervink, and F. Freese, Z. Anal. Chem. *264*, 147 (1973).

93. T. Mukoyama, T. Yemane, N. Kiba, and M. Tanaka, Anal. Chim. Acta *61*, 83 (1972).

94. I. Ishii and S. Musha, Rev. Polarogr. *16*, 61 (1969).

95. T. Ishii and S. Musha, Bunseki Kagaku *19*, 938 (1970).

96. H. K. Hoff and E. Jacobsen, Anal. Chim. Acta *54*, 511 (1971).

Chapter 8

MISCELLANEOUS POLAROGRAPHIC
METHODS USED IN ANALYTICAL CHEMISTRY

In previous chapters, individual techniques that are widely used in
analysis were described in considerable detail. There are obviously
an extraordinarily large number of other polarographic methods that
have been proposed for analytical use in an initial publication, but
have not been heard of since, hopefully only because they were not
all that good. Such methods need not be discussed in this book.
Then again, there are other techniques which have found quite wide
acceptance but have yet to be referred to. The range of techniques
considered in the preceding chapters represents, in this author's
opinion, the basic selection that an analytical chemist wishing to
undertake routine polarographic analysis should have access to. All
of these methods have been discussed widely in the literature, are
constructed from relatively simple instrumentation, and are based on
well-established theoretical principles which enable them to be used
in a systematic rather than empirical fashion. Furthermore, the
philosophy behind modern polarographic analysis is adequately conveyed
by reference to the methods already considered. However, so as not
to offend too many of my colleagues, several other of the well-
established polarographic techniques with distinct analytical use
or claimed analytical advantage are discussed briefly in this
chapter.

8.1 SQUARE-WAVE POLAROGRAPHY

As the name implies, *square-wave polarography* [1] involves the use
of a small-amplitude square-wave voltage in place of the sinusoidal
one used in (sinusoidal) ac polarography. Alternating current detection is effected by sampling the current near the end of each
square-wave half cycle, and the i-E curve consists of a plot of the
alternating current (or difference in current) sampled during the
positive and negative square-wave half cycles versus the dc potential.

Detailed discussion of the technique of square-wave polarography
is available in the literature [1-6]. By its very nature, the square-
wave polarographic experiment is obviously akin to a high-frequency
pulse polarographic experiment. The relationship to sinusoidal ac
polarography is also obvious. Only a brief résumé of its operation
therefore need be given.

Discrimination against charging current is achieved by several
measurement "tricks" encountered previously in other techniques.
The charging current resulting from the growth of the mercury drop
is minimized by measuring the current only at the end of the drop
life, where the rate of area change is quite small. The small
charging current contribution still remaining from area growth during
the current measurement period is sometimes compensated for by using
a tilted square-wave (as shown in Fig. 8.1), analogous to the method
Ilkovic and Semerano [7] used in dc polarography.

Figure 8.2 demonstrates that the method of eliminating the
charging current resulting from the periodic nature of the signal
is based on the same principle as that in pulse polarography
(Chap. 6). If square-wave pulses are superimposed onto the dc potential, the faradaic (alternating) current changes as a function
of $t^{-1/2}$ in the case of a diffusion-controlled process. The charging current changes exponentially as a function of time, with the
decay being proportional to $e^{-t/RC_{dl}}$ (where R is the ohmic resistance
of the polarographic circuit, t is time, and C_{dl} is the capacity
of the double layer). Consequently, the decay of the (alternating)
charging current is faster than that for the faradaic current and

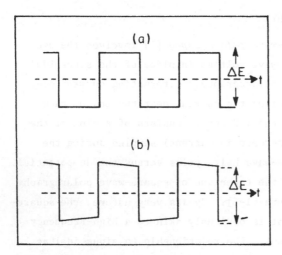

FIGURE 8.1   (a) Normal square wave and (b) tilted square wave used
to decrease the charging current resulting from area growth of the
dme.   The square wave is superimposed onto the dc ramp in square-
wave polarography.

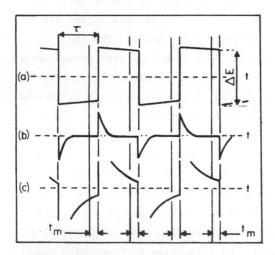

FIGURE 8.2   Schematic representation of the periodic behavior of
(a) the applied square-wave voltage, (b) the charging current, and
(c) the faradaic current.

measurements are made at the end of the square wave. The charging
current value after the decay time of 5RC is practically zero in
aqueous media, so that measurements made at any time, t_m, after the
5RC decay period and before the end of the square-wave pulse can be
investigated to determine the most favorable ratio of signal to
charge current or signal to noise. The other condition for minimization of the charging current is that the value of R should be kept
as low as possible, since the capacitance of the double layer is
fixed.

Figure 8.3 shows a square-wave polarogram of a solution containing cadmium and indium and its appearance is similar to a current-
sampled ac polarogram. According to Barker [5], in the small amplitude limit $\Delta E \ll RT/nF$, the peak height I_p for a reversible process
A + ne $\rightleftharpoons$ B, is given approximately by

$$I_p = \frac{k\ n^2 F^2\ \Delta E\ D_A^{1/2} C_A}{RT} \tag{8.1}$$

where k is a constant for a square wave of amplitude ΔE and fixed
frequency and other symbols are those used conventionally. The sim-
ilarity of the response to the sinusoidal case is obvious. Equations
for the irreversible case are much more complicated, but as is the
case in sinusoidal ac polarography, the current per unit concentration is considerably lower (see Chap. 7).

The square-wave ac technique has not been exploited under such
a wide range of operating conditions as has the sinusoidal version,
and, for example, much of the data appears to have been reported
at a fixed frequency of 225 Hz. However, the technique has been
eminently successful in trace analysis, permitting concentrations
as low as 5×10^{-8} M to be determined for some reversibly reduced
metal ions in aqueous solutions [8,9]. Figure 8.4 shows the determination of cadmium near the detection limit. The lower limit of
detection for irreversibly reduced species would appear to be only
about 10^{-6} M. Square wave polarography operates on essentially the
same basic principles as phase-selective ac polarography. Both

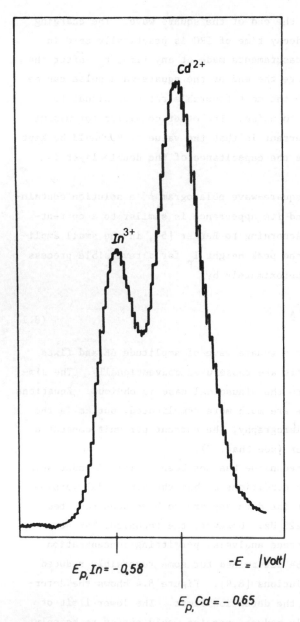

FIGURE 8.3   Square-wave polarogram of 5×10^{-5} indium(III) and 1×10^{-4} M cadmium.   (Reproduced from Note 6.)

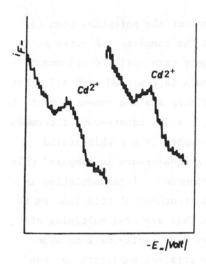

FIGURE 8.4 Determination of
2×10^{-7} M cadmium by square-wave
polarography. (Reproduced from
Note 6.)

techniques discriminate predominantly against charging current on
the basis of the smaller RC time constant associated with the
double-layer charging process relative to the faradaic process. A
phase angle of nearly 90° in a sinusoidal technique and a decay of
current early in the half cycle of a square wave are essentially
both equivalent to a small RC time constant, and on this basis both
techniques should be essentially equivalent. Using three-electrode
instrumentation, this has been the author's experience, although
earlier results in the literature indicated greater sensitivity for
square-wave polarography.

The analytical application of square-wave polarography has recently been reviewed by Geissler [10]. Examples are available in
almost every field, and for the analytical chemist interested in the
kind of determinations possible, Notes 8 to 20 can be consulted.
Further aspects of the theory can also be found in the work of
Ruzić [21].

Since the technique is akin to the sinusoidal version of ac
polarography described in detail in Chap. 7, it comes as no surprise
to find advances also follow the same pathways.

Synchronization of the commencement of the potential scan to
the growth of the mercury drop so that the complete i-E curve is
recorded on a single drop is an obviously very useful development
[22,23] as is the case when this approach is employed with all other
polarographic methods. Also not surprising are the recent reports of
second-order square-wave polarograms [23-27]. Square-wave intermodulation polarography [23,25,26], for example, is a sophisticated
technique that exploits the fact that a square-wave is composed only
of odd harmonics of the square-wave frequency. Intermodulation is
a result of the nonlinearity of the electrochemical cell leading to
small faradaic currents at frequencies that are even multiples of
the fundamental frequency of the square-wave (unity mark to space
ratio). Although not mentioned in the original publications; one
could expect a second harmonic contribution also to be present.
Figure 8.5 shows a comparison of fast sweep square-wave and intermodular square-wave polarograms for the reduction of Pb(II) and

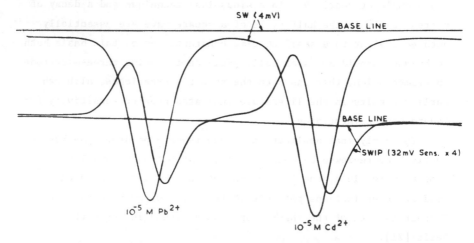

FIGURE 8.5 Fast sweep square-wave and square-wave intermodular
voltammograms for lead (II) and cadmium(II) in 1 M HClO$_4$. Scan
commenced at −0.25 V vs. sce, 4 sec after drop detachment, and lasted
6 sec. Square-wave (SW) curve refers to a square-wave of frequency
225 Hz, while square-wave intermodular (SWIP) results were obtained
with a square-wave frequency of 112.5 Hz. (Reproduced from Note 25.)

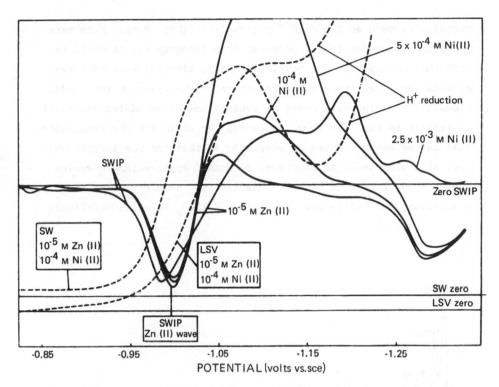

FIGURE 8.6 Square-wave intermodular (SWIP), square-wave (SW), and
linear sweep (LSV) voltammogram of 0.2 M HCl containing 10^{-5} M
Zn(II) and various amounts of Ni(II), showing the superiority of the
SWIP mode for the detection of Zn(II) (quasi-reversible electrode
process) in the presence of two highly irreversible electrode reactions [H^+, Ni(II)]: SWIP, full curve; SW and LSV. broken curve.
(Reproduced from Proc. Anal. Div. Chem. Soc., 1975, 179.)

Cd(II) in 1 M $HClO_4$. Without an exact theoretical proof, it has been
stated that the sensitivity is very low both for totally irreversible
reductions and for very reversible reductions, the maximum sensitivity
being observed for quasi-reversible electrode processes. This very
unusual type of selectivity can make the square-wave intermodular
polarograph ideal for the detection of a minor constituent in the
presence of either a reversibly or irreversibly reduced species [26].
Unfortunately, only the latter example has been experimentally verified. For example, Barker has shown that Zn(II) can be determined
easily in 0.2 M HCl at levels below 10^{-5} M, even when the solution

contains as much as 2.5×10^{-3} M nickel(II) (Fig. 8.6). With many
polarographic modes (e.g., pulse or dc polarography), it would be
difficult to separate the quasi-reversible zinc(II) wave from that
arising from the irreversible reduction of the hydrogen ion. Additionally, the zinc wave would be rapidly masked on adding nickel(II),
as results in Fig. 8.6 show. Assuming the claim for the reversible
case can be substantiated as elegantly as that for the irreversible
one, this mode seems to have many potential applications stemming
from its novel type of kinetic selectivity. However, since the
sensitivity for Pb(II) and Cd(II) in 1 M HClO$_4$ is still excellent,

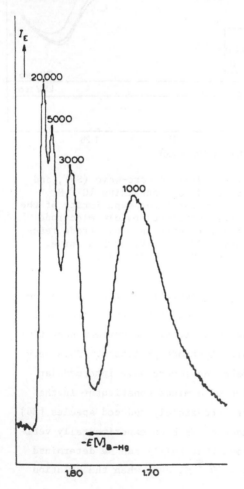

FIGURE 8.7 Square-wave polarogram showing tensammetric waves
from a mixture of polyethylene
glycols of different average
molecular weights, in 1 M LiCl.
Solutions contain four polyethylene glycol samples, whose
average molecular weights are
1000, 3000, 5000, and 20,000.
The concentrations of each of
the samples are 20.0, 7.5, 3.0,
and 15.0 mg liter^{-1}, respectively.
(Reproduced from Note 28.)

the query remains in the author's mind as to what k_s is in fact needed to achieve the low sensitivity claimed for the "very reversible" reduction. Indeed, one wonders whether this suggestion would be substantiated if a complete theoretical treatment, taking into account all of the second-order terms, were undertaken.

Based on a knowledge of sinusoidal ac polarography, tensammetric waves are expected in the square-wave version. A particularly novel application [28-30] of a square-wave tensammetric response is the characterization of approximate average molecular weight of poly- mers. In Fig. 8.7, the tensammetric response of a mixture of poly- ethylene glycols, characterized by four different average molecular weights, is shown. A resolved peak is obtained for each of these average molecular weights, and the peak width is a function of molec- ular weight.

8.2 RADIO-FREQUENCY POLAROGRAPHY

Radio-frequency polarography is another variety of ac polarography developed by Barker [5,31,32]. To adapt the measurement of the faradaic rectification effect to analytical work, he devised a polar- ographic technique in which a sinusoidal radio-frequency (100 kHz to 6.4 MHz) signal, ω_1, square-wave modulated at 225 Hz, ω_2, is superimposed onto the dc potential ramp. The response at 225 Hz is measured as in square-wave polarography. Although Barker pre- sented this technique as a method of measuring the dc rectification component in the presence of the normal dc polarographic current, it can be considered as a form of intermodulation polarography, as indicated by Reinmuth [33,34]. The method therefore is yet another second-order technique resulting from the nonlinearity of the elec- trochemical cell. The applied waveform obviously includes Fourier components of frequency ω_1, $\omega_1 - \omega_2$, and $\omega_1 + \omega_2$. Thus, the current observed at frequency ω_2 could be interpreted as faradaic intermodu- lation of the component of frequency ω_1 with the two side-band fre- quencies. Theoretical treatments using this approach give the same expressions as those presented by Barker, who showed that the current

is the same as that produced under ordinary ac polarographic conditions when using a signal of peak-to-peak amplitude equal to minus
the faradaic rectification potential at a frequency ω_2.

Until recently this long-known mode of polarography has been
largely ignored by analysts, probably because of a lack of instrumentation of sufficiently low noise level to realize the potential
gain in performance that results when the normal square-wave signal
is replaced by a square-wave modulated high-frequency one [26].
However, Barker and colleagues have now developed a multimode polarograph [23,26] which meets the necessary requirements. They use a
modulated (225 Hz) signal of frequency 72 kHz, and the normal square-
wave circuits are employed to measure the response produced by the
nonlinearity of the cell. Although the measured signal may be
smaller than that produced in the corresponding square-wave experiment, there is often a reduced noise level and a gain in precision
which results in considerable capabilities for low-level determinations. The radio-frequency technique, while very complicated from
the theoretical point of view, is quite simple in concept [26].
Schmidt and Von Stackelberg [4] have reviewed the basic principles
of radio-frequency polarography, and their article can be consulted
for further details. Figure 8.8 gives an indication of the sensitivity and base line reproducibility when this technique is used
with high-quality instrumentation under fast sweep conditions at
a dme, enabling concentrations below 10^{-8} M to be determined for
reversible systems in aqueous media. As with all ac methods, the
sensitivity decreases with decreasing reversibility. For irreversibly reduced species, a sensitivity of 10^{-7} M has been claimed [4].

Activity in the field of radio-frequency polarography has recently been more extensive and substantial, which suggests that
analytical chemists have become alerted to the potentially very high
sensitivity obtainable via this method. The major difficulty experienced by the author in accurately summarizing the current status
of this technique is one of nomenclature. Apart from the dilemma
of considering it as a method based on faradaic rectification or as

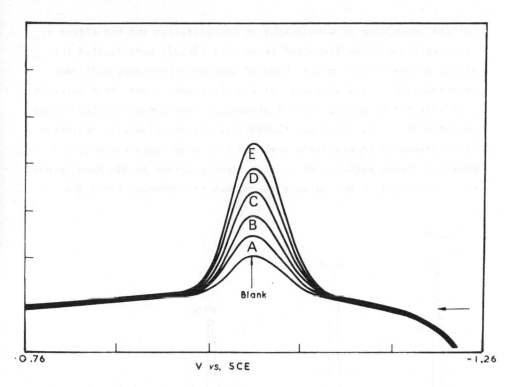

FIGURE 8.8   Fast sweep radio-frequency voltammograms at a dme for addition of Zn(II) to 0.2 M KCl/2 × 10^{-4} M HCl: (A) 10^{-8} M, (B) 2 × 10^{-8} M, (C) 3 × 10^{-8} M, (D) 4 × 10^{-8} M, (E) 5 × 10^{-8} M. (Reproduced from Note 23.)

a class of intermodular polarography (as noted when introducing the method), the situation is further confused by the fact that radio-frequency, high-frequency, and detector polarography all appear to refer to the same technique.   Literal translations of Russian ter-minology cause added confusion.   However, assuming the author has correctly assessed the situation with respect to the technique under question in individual papers, the following statements are appro-priate.

Kambara et al. [35], based on assessment of data from a number of experiments, have concluded that radio frequency is one of the most elegant polarographic methods of analysis.   These workers have formulated theoretical expressions and experimentally verified them

for the dependence of wave height on concentration and the effect of
cell resistance. Vasileva and co-workers [36,37] have studied the
effect of sweep rate on the shape of the radio-frequency voltammogram obtained at a stationary mercury electrode. Apart from Barker's
work referred to above, other instrumental developments include those
in Notes 38 to 40. From the theoretical and experimental considerations presented in Kambara's work and from other data elsewhere, a
three-electrode system such as that used by Barker in his more recent
work is probably required to give optimum performance, since the

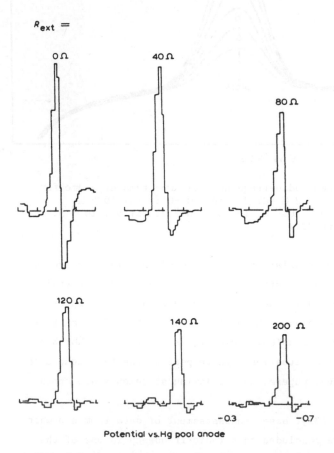

FIGURE 8.9 Variation of radio-frequency polarograms with external
resistance. Solution is 1×10^{-5} M Pb(II) in 0.5 M KNO_3. (Reproduced from Note 35.)

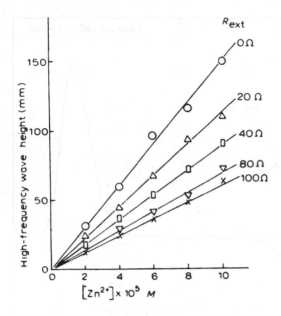

FIGURE 8.10   Calibration curves for Zn(II) in 1 M KCl with various external resistances.   Note that the linearity, but not the sensitivity, is independent of resistance.   (Reproduced from Note 35.)

sensitivity is inversely proportional to uncompensated resistance [35].   Most of the work to date has yet to employ the three-electrode format, so substantial distortion in shape and other aspects of many of the published curves has probably occurred.

For actual analytical determinations of various species, the reader is referred to Notes 39 to 46.   From the more recent work, and indeed even Barker's early studies [31], it is apparent that the shapes of radio-frequency polarograms are extremely dependent on the kinetics of the electrode process, as well as uncompensated resistance.   Figures 8.9 to 8.13 provide some representative examples of radio frequency polarograms which verify these features.   The theory taking into acocunt all the kinetic nuances and other subtleties has still not been developed to the degree of sophistication of techniques described in earlier chapters, and the analytical chemist frequently will have to adopt an empirical approach in using this technique for some time yet.   However, at the very high frequencies

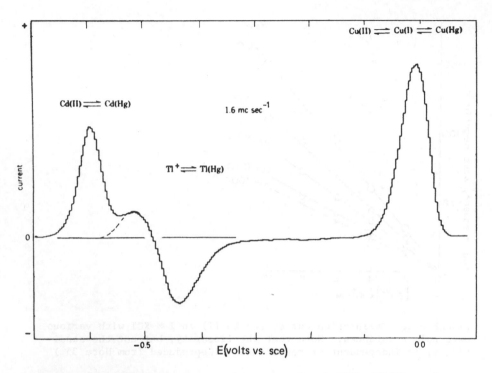

FIGURE 8.11   Radio-frequency polarograms for Cu(II), Tl(I), and
Cd(II) in 0.5 M K_2SO_4.   (Reproduced from Note 31 by permission of
the publisher, The Electrochemical Society, Inc.)

used in this technique, it is obvious that few electrode processes
will be reversible or diffusion controlled.   This substantial dependence of electrode processes on kinetics was shown to be capable
of exploitation in sinusoidal ac polarography.   A sometimes important
feature of the radio-frequency method is that it enables distinc-
tions to be made between species reduced at similar potentials, as
wave symmetry is very dependent on the kinetics of the reaction.
This may lead to more certain identification of an organic species,
for example [26], since characteristic wave shapes are often found.
However, a great deal more practical work on actual analytical examples is still needed to establish the intrinsic merits and disadvantages of radio-frequency polarography to the satisfaction of the
practising analytical chemist.   Hopefully, such information will be

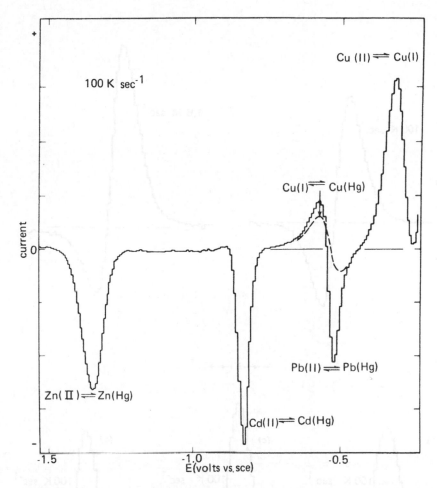

FIGURE 8.12 Radio-frequency polarograms for Cu(II), Pb(II), Cd(II), and Zn(II) in 2 M NH₄OH/2 M NH₄Cl. (Reproduced from Note 31 by permission of the publisher, The Electrochemical Society, Inc.)

forthcoming in the not too distant future, as the prognosis for radio-frequency polarography has been excellent ever since its development about 20 years ago. The assumed hindrance to progress, viz., inadequate instrumentation, should have been overcome by recent developments [26], and the next few years should see a substantial increase in the use of this technique, if it lives up to expectations.

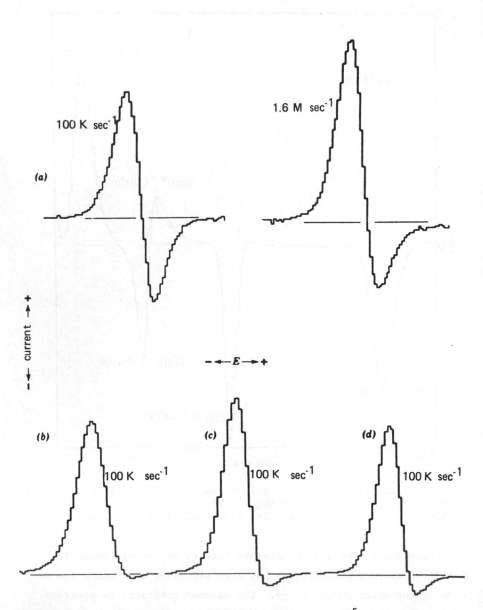

FIGURE 8.13 Radio-frequency polarograms for 2×10^{-5} M Cd in various media: (a) 1 M KNO_3, (b) 0.5 M K_2SO_4, (c) 1 M KF, (d) 1 M $HClO_4$. (Reproduced from Note 31 by permission of the publisher, The Electrochemical Society, Inc.)

8.3 CHRONOPOTENTIOMETRY

As the name implies, *chronopotentiometry* utilizes the measurement
and interpretation of E-t curves. The experiment is carried out by
applying a controlled current between the working and auxiliary
electrodes with a galvanostatic system, as described in Chap. 2.
The resulting chronopotentiogram showing the variation of potential
with time can be used, in principle, for a variety of analytical
applications or investigations of electrode kinetics. Figure 8.14
gives an example of a chronopotentiogram for a simple electrode
process, A + ne $\rightleftharpoons$ B. Before the constant current electrolysis
begins, the solution contains only A and it has a relatively oxidizing (positive) potential. As soon as the electrolysis is initiated
by application of the current, some A is reduced to B and the working
electrode potential becomes more reducing (negative). If the electrode process is reversible, the potential of the working electrode
obeys the Nernst equation

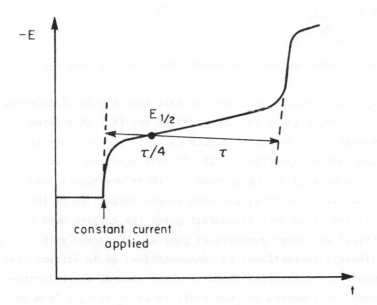

FIGURE 8.14 Chronopotentiogram for a simple electrode process.

$$E \simeq E° + \frac{RT}{nF} \ln \frac{C_{A(x=0)}}{C_{B(x=0)}} \tag{8.2}$$

The potential changes only slowly with time during the period that the ratio $C_{A(x=0)}/C_{B(x=0)}$ is close to 1. As the electrolysis continues, the concentration of A decreases to the point where there is insufficient A to accommodate all of the constant current, despite the fact that A diffuses toward the electrode surface (unstirred solution, and sufficient supporting electrolyte present to prevent migration assumed). Ultimately, the concentration of A near the electrode surface drops toward a negligible value, and the current forces the electrode to a potential at which a different reaction can occur. At this stage, a rapid change in potential is observed. The time from the start of the electrolysis until the rapid potential change is designated the transition time τ in Fig. 8.14.

Under conditions of linear diffusion in an unstirred solution, the transition time τ (seconds) is given by the Sand's equation

$$\tau^{1/2} = \frac{\pi^{1/2} nFAD_A^{1/2} C_A}{2i} \tag{8.3}$$

where i is the constant current in amperes and other symbols are as used previously.

For analytical chemists, the two features that may be of interest are that $\tau^{1/2}$ is proportional to concentration and that this parameter is independent of electrode kinetics (k_s), in the same way as is the limiting current specified by the Ilkovic equation in dc polarography. In principle, the parameter τ therefore should have considerable analytical utility, but even casual inspection of the E-t curve is likely to be most discouraging for the chemist used to seeing symmetrical or almost symmetrical peak-shaped curves with peak height linearly proportional to concentration, as in differential pulse polarography, for example. Even if there were no other limitations in chronopotentiometry, the necessity of calculating τ from an E-t curve with an analytically disadvantageous shape would mitigate against its widespread use (remember that Fig. 8.14 is an ideal case and real curves are usually much harder to evaluate).

In the early days of chronopotentiometry, this method of analysis was optimistically considered as fairly accurate and sensitive
[47-49], with the possibility of determining concentrations as small
as 10^{-6} M being espoused (Note 48, p. 214). The early euphoria,
however, was soon replaced by pessimistic assessments: Lingane [50]
mentioned 10^{-3} M as the lower limit, Reinmuth's opinion was, to say
the least, unfavorable [51], and Davis [52] stated, "As a strictly
analytical tool, chronopotentiometry definitely is not as useful as
(dc) polarography. The concentration range that can be measured
does not extend below about 10^{-4} M if accuracy better than a few
percent is desired." By the 1970s Bos and Van Dalen [53] were able
to really put the nail in the coffin on the early data:

> Chronopotentiometry suffers from two disadvantages which can
> easily be verified but have never been clearly admitted in the
> literature: (a) the results are incorrect except in favourable
> circumstances; (b) the reproducibility of measurements is poor.
> The causes of their disadvantages are rather complex, but in
> extensive preliminary experimental investigations it became
> clear to us, that there are two main factors: (1) the extra-
> ordinarily large influence of contaminations and (2) the dis-
> turbing effect of the capacity current.

For any analyst contemplating using chronopotentiometry to solve
his or her problem, the complete paper by Bos and Van Dalen should
be compulsory reading. In addition to placing the method in perspective, they developed far improved instrumentation compared with
that used in the earlier work, and still were only able to accurately
determine concentrations of thallium and lead down to 10^{-5} M.
Sturrock et al. [54,55] have very recently applied more of the
"electronic and other trickery" each electroanalytical method seems
ultimately to be subjected to. However, in both these studies, work
still is being undertaken only on model rather than "real life"
problems, on systems that would be handled almost trivially by many
of the controlled-potential techniques described earlier.

From the above it should be clear to most readers that the
controlled-current method of chronopotentiometry is not the author's

preferred analytical technique, and certainly would not appear to
be competitive with many of the controlled-potential methods.
Rather than discuss all the variations of chronopotentiometry contained in the extensive literature on the subject, it would probably
be more profitable to conclude this discussion of chronopotentiometry
by briefly comparing the more important features of the controlled-
current and controlled-potential techniques and then giving an example
of the application of chronopotentiometry to electrode kinetics.

The general aspects of the controlled-potential and controlled-
current methods, when performed with modern instruments, are super-
ficially very similar. The construction of either a potentiostat
or a galvanostat is closely related, as shown in Chap. 2, and indeed
interconversion of one mode to the other is usually included in
commercially available instruments. The construction of a simple
galvanostat is probably simpler, but instruments allowing for correction of capacitive currents end up more complicated than the
equivalent potentiostat-based instrument. However, the mathematical
treatment of both techniques is fundamentally different. In controlled-
current experiments, the boundary condition for solution of the
integral equations is based on the known current function or the
known flux at the electrode surface, while in controlled-potential
experiments, it is the concentration itself as a function of potential,
rather than the concentration gradient, which is used in the boundary
condition. It turns out that the mathematical treatments in
controlled-current methods are frequently much simpler and closed-
form solutions are almost always obtained. The relative popularity
of chronopotentiometry as a method for studying solution and electrode kinetics is probably a direct consequence of this phenomenon.
An example of the theoretical treatment of an electrode process
under conditions of chronopotentiometry is presented in subsequent
discussion.

A disadvantage of controlled-current methods is the larger and
continuous effect of double-layer charging. This, as mentioned
above, is one of the main reasons for the present very limited use

of chronopotentiometry in analytical work where low concentrations
and, consequently, charging current are of considerable interest.
Electrode kinetic investigations using chronopotentiometry can usually
be undertaken at concentrations above the 10^{-3} M level, where the
faradaic-to-charging current ratio is relatively favorable. In a
controlled-potential experiment, charging of the double layer occurs
at the very first instant of the experiment. In a galvanostatic
experiment, the controlled current is divided between the faradaic
process and the charging process, and since the potential is changing
throughout the experiment, double-layer charging occurs continuously
and in a fashion that is difficult to compensate for [53]. For
example, if the constant current applied is I, then the current going
to the faradaic reaction, i_f, is

$$i_f = I - i_c \qquad (8.4)$$

During the electrolysis, a varying part of the applied current, depending on the varying slope of the curve, is therefore consumed in
charging the double layer, the decrease in current actually used for
electrolysis results in lengthening of the transition time
(Fig. 8.15). Furthermore, the potential change at the transition
time is less pronounced, the rise being flattened by the charging
process continuing to consume a portion of the current (Fig. 8.15),
and τ becomes more difficult to calculate. Similarly, impurities

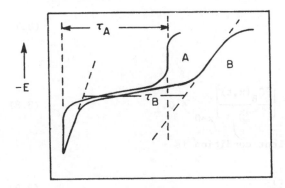

FIGURE 8.15 The distortion of chronopotentiograms by capacity
effects: (a) theoretical shape; (b) curve considerably distorted
by capacity effects.

in the solution will of course cause complications as the diminished
current available for the electroactive species of interest results
in a lengthening of the transition time. That is, waves are not
additive as in controlled-potential methods such as dc polarography,
and determinations of a mixture of species is concomitantly complex.
The reasons for the preference of controlled-potential methods in
analytical work are quite clear from the preceding discussion.

As expected, the shape of the E-t curve in the controlled-
current method is dependent on the electrode kinetics, as is the
case with the i-E curve in polarography.

Under linear diffusion conditions to a planar electrode with
only A initially present at concentration C_A, the equations governing
the system A + ne → B are

$$\frac{\partial C_A(x,t)}{\partial t} = D_A \frac{\partial^2 C_A(x,t)}{\partial x^2} \tag{8.5a}$$

$$\frac{\partial C_B(x,t)}{\partial t} = D_B \frac{\partial^2 C_B(x,t)}{\partial x^2} \tag{8.5b}$$

At $t = 0$, $C_A(x,t) = C_A$,

$$C_B(x,t) = 0 \tag{8.6}$$

At $x \to \infty$, $C_A(x,t) \to C_A$,

$$C_B(x,t) \to 0 \tag{8.7}$$

The flux condition is

$$D_A \left(\frac{\partial C_A(x,t)}{\partial x} \right)_{x=0} + D_B \left(\frac{\partial C_B(x,t)}{\partial x} \right)_{x=0} = 0 \tag{8.8}$$

and the concentration gradient condition is

$$D_A \left(\frac{\partial C_A(x,t)}{\partial x} \right)_{x=0} = \frac{i(t)}{nFA} \tag{8.9}$$

Equations (8.5) to (8.8) are the same as used previously, but Eq. (8.9), a concentration gradient condition, replaces the concentration condition of the controlled potential methods.

If

$$i(t) = i \text{ (constant)} \tag{8.10}$$

as is the case in chronopotentiometry, then

$$C_A(0,t) = C_A - \frac{2it^{1/2}}{nFAD_A^{1/2}\pi^{1/2}} \tag{8.11}$$

and

$$C_B(0,t) = \frac{2it^{1/2}}{nFAD_B^{1/2}\pi^{1/2}} \tag{8.12}$$

For reversible behavior, the Nernst equation holds and substituting Eqs. (8.11) and (8.12) into it gives

$$E = E^\circ + \frac{RT}{nF} \ln \frac{C_A - t^{1/2}(2i/nFAD_A^{1/2}\pi^{1/2})}{t^{1/2}(2i/nFAD_B^{1/2}\pi^{1/2})} \tag{8.13}$$

For the particular case when the electrolysis is complete, $C_A(0,t) = 0$ and $t = \tau$. Substituting this condition into Eq. (8.11) leads to the Sand equation presented in Eq. (8.3). Using the Sand equation to obtain an expression for C_A gives

$$E = E_{\tau/4} + \frac{RT}{nF} \ln \frac{\tau^{1/2} - t^{1/2}}{t^{1/2}} \tag{8.14}$$

where

$$E_{\tau/4} = E^\circ + \frac{RT}{nF} \ln \left(\frac{D_A}{D_B}\right)^{1/2} \tag{8.15}$$

Note that when $t = \tau/4$, $E =$ the reversible polarographic half-wave potential $E_{1/2}^r$.

Figure 8.14 shows a reversible chronopotentiogram. From Eq. (8.14), a test for reversibility is that a plot of E vs. $\log [(\tau^{1/2} - t^{1/2})/t^{1/2}]$ should give a straight line of slope $2.303RT/nF$ or $|E_{1/4} - E_{3/4}| = 47.9/n$ mV at 25°C, and analogies with polarographic equations are quite obvious.

For a totally irreversible electrode process, the rate expression is introduced in the normal way to give the end result

$$E = \frac{RT}{\alpha nF} \ln \frac{nFAC_A k_f}{i} + \frac{RT}{\alpha nF} \ln \left[\frac{\tau^{1/2} - t^{1/2}}{\tau^{1/2}} \right] \tag{8.16}$$

or

$$E = E° + \frac{RT}{\alpha nF} \ln \frac{nFAC_A k_s}{i} + \frac{RT}{\alpha nF} \ln \left[\frac{\tau^{1/2} - t^{1/2}}{\tau^{1/2}} \right] \tag{8.17}$$

The origin of an irreversible curve at $t = 0$, therefore, depends on the quantities αn and k_s, and compared with a reversible process, the curve shifts toward more negative potentials. For an irreversible process, a linear dependence is obtained for a plot of E vs. $\log [1 - (t/\tau)^{1/2}]$ with slope $2.303RT/\alpha nF$. The value of αn and k_s can be obtained from this plot. For an irreversible wave at 25°C,

$$|E_{1/4} - E_{3/4}| = \frac{33.6}{\alpha n} \text{ mV}$$

If the reversible electrode process is followed by a chemical reaction, the entire E-t curve is displaced to more positive potentials, as expected by analogy with results from dc polarography. The quantitative relationship for this EC mechanism, the CE case, disproportionation, dimerization, adsorption, etc., have all been reported for chronopotentiometry and related constant current techniques (see Notes 52, 56-71 for examples).

A great many papers, including some of those referenced previously, have been devoted also to chronopotentiometry with alternative controlled-current formats [52,66,67]. For example, in current reversal chronopotentiometry [62,66,68,71], the direction of the current is changed at the instant when the transition time ends.

The electroactive species is thus subject to the reverse electrode process, provided it is able to undergo this reaction, i.e., after reduction it is re-oxidized or vice-versa, and similar to the controlled-potential technique of cyclic voltammetry (Chap. 5). A great deal of important information on the electrode process is gained by using current reversal or cyclic chronopotentiometry (current direction reversal several times).

In conclusion, it should be emphasized again that despite the fact that some authors have recommended chronopotentiometry as an analytical tool, its greatest importance, however, is likely to be found in the investigation of electrode kinetics. In this context it has all the capabilities and applications associated with the controlled-potential techniques such as linear sweep and cyclic voltammetry described in Chap. 5. For example, since theoretical relationships in chronopotentiometry hold for short transition times (from tenths to tens of seconds), the technique can be applied at a dme using a long drop time of, say, 10 sec. Late in the drop life, a constant current impulse lasting a fraction of a second is applied, and the E-t curve recorded at constant electrode area [66]. Thus, advantages associated with mercury electrodes can be enjoyed also with chronopotentiometry. Application of chronopotentiometry to solid electrodes or stationary mercury electrodes is, of course, also exceedingly simple. Measurements of rates of electron transfer and coupled chemical reactions several orders faster than by conventional dc polarography can be made via chronopotentiometry. However, since the majority of these measurements can be achieved also by many of the controlled-potential methods discussed in earlier chapters, and even more importantly, since these methods are of considerable value in both analytical chemistry and interpretation of electrode processes, the use of chronopotentiometry is not likely to be considered by the nonelectrochemical specialist whose main aim is an analytical determination of an electroactive species. Thus, even in determining the mechanism of analytically used electrode processes, e.g., assessing reversibility or otherwise as is usually essential, resort to the technique of chronopotentiometry is unlikely to be the most expedient approach [72,73].

8.4 DC POLAROGRAPHY
WITH CONTROLLED CURRENT

Conventionally, dc polarography is performed under controlled-
potential conditions. However, with chronopotentiometry the first
formal introduction to the genuinely controlled current techniques
was made and this concept can be applied in most areas of polarography and voltammetry. Quite obviously with dc polarography, instead
of having the potential as the independent variable, the current
could be controlled and the potential would be the dependent variable in this instance.

Adams et al. [74] introduced a manual method for obtaining
controlled-current voltammograms at a platinum electrode. Ishibashi
and Fujinaga [75-77] extended this work so that controlled-current
polarograms could be recorded at a dme in much the same way as the
well-known controlled-potential version. Figure 8.16 shows a comparison of controlled-potential and controlled-current dc polarograms,
and it is no surprise to see that the major difference with the
controlled-current method from the other is the anticipated one that
the oscillations arising from the growth and fall problem of the
dropping mercury electrode lie along the current axis. Contrary to
conventional polarographic behavior where only small oscillations
are obtained when the current oscillates at a given applied potential,
a constant current applied to the dme produces wide oscillations in
potential, which increase in magnitude as the applied current approaches the limiting current. Early in the drop life, when the
instantaneous diffusion current is small, a constant current, higher
than the instantaneous diffusion current, drives the potential of
the dme to that where reduction (oxidation) of another electroactive
species, the supporting electrolyte or the solvent, occurs. Thus,
as shown in Fig. 8.16, simultaneous reduction of lead(II) and hydrogen ion occurs in the 1 mM $Pb(NO_3)_2$/0.1 M HCl system, and simultaneous
reduction of lead(II) and cadmium(II) occurs in the
1 mM $Pb(NO_3)_2$/0.05 M $CdCl_2$ system at the higher current values.

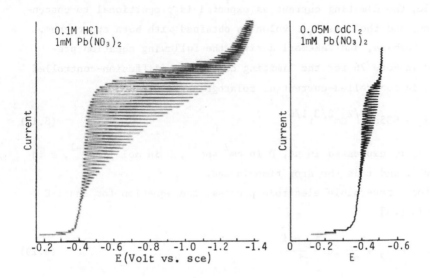

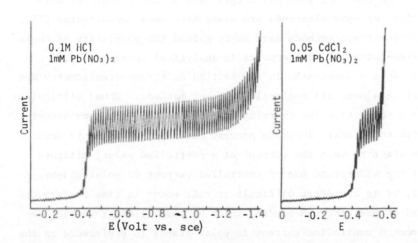

FIGURE 8.16 Comparison of controlled-current (upper curves) and controlled-potential (lower curves) dc polarograms. (Reproduced from Note 76.)

Theoretical treatment of controlled-current dc polarography is entirely analogous to the controlled-potential case. Thus, for example, the limiting current as expected is proportional to concentration, and the same $E_{1/2}$ value is obtained with both approaches. Senda, Kambara, and Takemori derived the following equation presented in Note 76 for the limiting current i_1 (diffusion-controlled case), in controlled-current dc polarography:

$$i_1 = 635nD^{1/2}Cm^{2/3}t^{1/6} \tag{8.18}$$

where i_1 is expressed in μA, D in $cm^2\ sec^{-1}$, C in mol $liter^{-1}$, m in mg sec^{-1}, and t is the drop time in sec.

For a reversible electrode process, the equation for the i-E curve is [76]

$$E = E_{1/2}^r + \frac{RT}{nF} \ln \frac{i_1 - i}{i} \tag{8.19}$$

where E being measured at the end of the drop life.

Despite the fact that advantages such as the absence of maxima and minima for some electrode processes have been demonstrated [78], controlled-current methods have never gained the popularity of their controlled-potential counterparts in analytical chemistry. The disadvantages demonstrated in the section on chronopotentiometry are inherent in almost all controlled-current methods. Thus, difficulties with correction for charging current, and the problems associated with additional electrode process necessarily having to occur at some stage to keep the current at a controlled value, mitigate against the widespread use of controlled current dc polarography. Overall, it is therefore difficult at this point in time to foresee the kind of breakthrough needed, which ultimately might enable one to recommend controlled-current dc polarography in preference to the well-established, well-understood, and widely used controlled-potential method. However, it should be noted at least that Kies and co-workers [77-79] in their contributions involving the use of controlled-current density, have presented a significant advance which makes the controlled-current method more competitive. At controlled-current density rather than controlled current, complications due to drop growth at the dme are eliminated and the technique therefore becomes essentially equivalent in ease of use to the controlled-potential method.

Instead of working with a controlled but varied current, as is usually done, measurement at constant current can be performed at a dme [80]. In this case, the potential of the dme depends on the concentration of the electroactive species in solution when a constant current, lower than the diffusion current, is maintained. The theory for this method has been developed, and it has been suggested for analytical work within limited concentration ranges [80].

8.5 CHRONOPOTENTIOMETRY WITH CONTROLLED ALTERNATING CURRENT

It is not easy to find an unambiguous name for this electroanalytical technique which, like dc polarography, was introduced by Heyrovský [81]. Heyrovský himself called it "alternating current oscillographic polarography" because it makes use of an alternating current and in his version he used an oscilloscope to record the E-t curves. Schmidt and von Stackelberg [4] refer to the method as "Oscillographic Polarography According to Heyrovský and Forejt," a name which provides recognition of the developers of the method. A most comprehensive account of the technique is a book by Kalvoda [82], which is entitled "Techniques of Oscillographic Polarography." As stated in Chap. 5 on linear sweep voltammetry, also called *oscillographic polarography* by some workers, this author regards the nature of the readout device as being incidental to the technique. For example, there is no doubt that nowadays, alternatives to oscilloscopes exist for measuring the E-t curve in Heyrovský's method. Nomenclature based on "oscilloscope" is therefore considered inappropriate and, for lack of a better alternative, the name chronopotentiometry with controlled alternating current" is suggested as being at least partially informative, even though still not entirely satisfactory. Certainly it

has distinct similarities to cyclic chronopotentiometry, mentioned
briefly in Sec. 8.3 of this chapter.

The technique of chronopotentiometry with controlled alternating
current is analogous to electrolysis with an alternating potential.
The time dependence of the alternating current may be triangular,
square wave, or sinusoidal, and the E-t curve consists of two
branches. In one half-period of the alternating current waveform,
the working electrode functions as a cathode, and in the next half-
period as an anode. Thus, like dc cyclic voltammetry and current
reversal techniques in chronopotentiometry, reduction and oxidation
aspects of the electrode process may be observed.

At a mercury electrode and in aqueous media, the potential range
that can be used without oxidizing the mercury electrode or reducing
the cation of the supporting electrolyte is about 0 to -2 V vs. sce.
When applied at a dme, all measurements are of course made late in
the drop life to ensure constant area conditions apply. Naturally,
if electrodes other than mercury are used, different potential limits
apply. In order to restrict consumption of faradaic current to
alternate oxidation and reduction of mercury during the application
of the alternating current, a direct current is superimposed onto
the alternating current. The direct current is cathodic and amounts
to a few percent of the amplitude of the alternating current.

Provided that the current is so small that the supporting electrolyte or solvent is not reduced, the resulting E-t curve depicts
the charging current. Assuming the capacitance of the double layer
C_{dl} is independent of potential, the usual relationship

$$i_c = -C_{dl} A \frac{dE}{dt} \tag{8.20}$$

applies. Since no electrolysis occurs, the charging current i_c is
equal to the applied current i, which is in turn defined by the
amplitude of the applied voltage ($V(t) = V \sin \omega t$ if a sine wave is
used) and resistance R of the circuit. Thus,

$$i_c = i(t) = i \sin \omega t = \frac{V}{R} \sin \omega t \tag{8.21}$$

From Eq. (8.20),

$$\frac{dE}{dt} = - \frac{V}{C_{d1}AR} \sin \omega t \tag{8.22}$$

On integrating,

$$E = \frac{V}{C_{d1}A\omega R} \cos \omega t = \frac{i_c}{\omega C_{d1}A} \sin (\omega t - \frac{\Pi}{2}) \tag{8.23}$$

From Eq. (8.23), it follows that the potential lags behind the current by $\Pi/2$ and the potential is proportional to the applied current and inversely proportional to the frequency of the applied signal and the differential capacity of the double layer.

On the device for recording the E-t curve, the curve depicted in Fig. 8.17(a) would be seen. The lower horizontal portions correspond to oxidation of mercury at about 0 V vs. sce. The upper horizontal portion corresponds to reduction of the cation of the supporting electrolyte [for example, $Na^+ + e \rightleftharpoons Na(Hg)$] at about -2 V vs. sce. The rising and descending portions of the E-t curve correspond to the charging and discharging of the electrical double layer. In the absence of the faradaic currents associated with oxidation of mercury and reduction of, say, Na^+ the broken line shown in Fig. 8.17(Ia) would be observed. (The same remark applies to all broken lines associated with curves in Fig. 8.17.)

Only the rising and descending portions of curve I are of interest to the analytical chemist. On adding an electroactive species to the solution, a curve of the shape shown in Fig. 8.17(Ib) will be seen. An inflection point appears on the E-t curve; the duration of this "time-lag" is analogous to the transition time discussed in the section on chronopotentiometry. For a reversible electrode process, the time-lags on the cathodic and anodic branches are formed at the same potentials; for irreversible processes, the anodic time-lag is either shifted to more positive potentials and the cathodic one to more negative potentials, or one of them is absent. For a reversible process, the potential at the inflection point is close to $E_{1/2}$. The time-lags last from the time the deposition potential

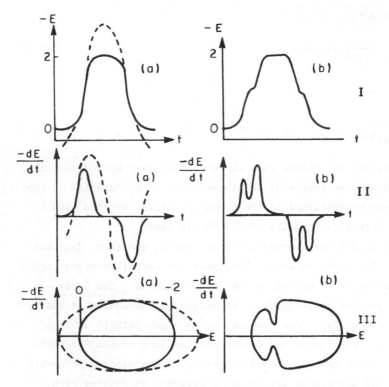

FIGURE 8.17 Various types of curves used in chronopotentiometry with controlled alternating current. I: E vs. t curves; II: dE/dt vs. t curves; III: dE/dt vs. E curves. Supporting electrolyte (a) with and (b) without reducible species.

of the electroactive species has been reached until its concentration at the electrode surface becomes zero, and they are therefore proportional to concentration. Equations for reversible reduction have been derived by Micka [83], Kambara [84], and Matsuda [85]. After reduction of the first electroactive species is complete, the potential increases again until another electrode process occurs and a time-lag is observed for each process.

The origin of the time-lags is as follows. The total current i is equal to the sum of the charging current i_c and the faradaic current i_f. In the absence of an electroactive species, $i = i_c$. If a species is reduced, the faradaic current flows at the expense

of the charging current, which is equivalent to saying that the potential remains almost constant, while the charging current and,
hence, dE/dt approach zero. The complete curve therefore provides
information on the nature of the electroactive species (potential
of time-lags), its concentration (length of time-lag), and reversibility of the electrode process (separation in potential of time-
lags). However, the quantitative evaluation of these curves is quite
inaccurate, as can be seen from the nature of the E-t curve in
Fig. 8.17. In order to obtain improved curves for evaluation, the
derivative of the E-t curve can be obtained as shown in Fig. 8.17
(IIa) and (IIb) or alternatively, one can combine the functions
E vs. t and dE/dt vs. t to give a new one, dE/dt vs. E (Fig. 8.17
(IIIa) and (IIIb), which is the function most widely used in analytical work.

If the electrode capacity of the double layer suddenly changes
at a certain potential because of the adsorption or desorption of a
surface active substance, a time-lag will also appear on the E-t
curve, as shown in Fig. 8.18. The method of alternating current
polarography at controlled current therefore also has the usual
characteristic of being sensitive toward adsorption processes as is
usually associated with ac and fast sweep potential methods.

Despite the fact that chronopotentiometry with controlled
alternating current has been widely used in analytical work in
Eastern European countries, this author is completely unenthusiastic

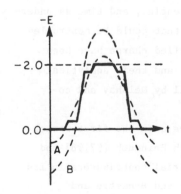

FIGURE 8.18 E-t curve in the presence
of adsorption: (A) normal E-t curve,
(B) E-t curve for an electrode with an
adsorbed layer. The full black line
depicts the E-t curve obtained in a
supporting electrolyte containing a
surface-active species adsorbed and
desorbed at particular potentials.

about it as a competitive analytical technique. The shapes of the
E-t curves are a sufficient reason on their own, to avoid the technique. Furthermore, the controlled-potential method of fast scan
linear sweep voltammetry can be used to solve the same problems in
most instances with better results, in the author's experience. If
the reader requires a more optimistic view of this technique, several
reviews are available [4,82,86,87]. However, rarely in these reports
is the method compared with anything other than dc polarography,
where it is undoubtedly true that some relative advantages are
offered. Nowadays, the method must not only be superior to dc
polarography but also offer real advantages over all the other commonly used modern polarographic methods described in earlier chapters,
and evidence that this criterion can be met has not been presented
to this author's satisfaction. Heyrovský and Micka, for example, in
their review [87] state that "in quantitative analysis the main advantage over the classical (dc) polarographic method is in its
speed.... On the other hand, its sensitivity is on the average by
one order of magnitude lower. For analytical purposes, the most
suitable concentration range is between 5×10^{-5} and 5×10^{-4} M;
within these limits the relative error does not exceed 5%."

8.6 CHARGE-STEP POLAROGRAPHY

In previously considered electroanalytical techniques, controlled
current or controlled potential was employed as the independent
variable. In measurements of current, potential, and time as undertaken in polarography, a related variable that could be controlled
is the charge. Techniques based on controlled charge have been
termed *coulostatic* or *charge-step* methods, and their analytical
implementation has been described in detail by Delahay and co-
workers [89-95].

Once again Barker would appear to have been the first worker
to describe the technique [31,96], although Reinmuth [97,98] and
Delahay [95] were undoubtedly the main initial contributors in its
application to electrode kinetics. Astruc and Bonastre and

co-workers [99-105] further improved the technique with respect to instrumentation, and other workers have also made further contributions [106-108] in the analytical area.

In coulostatic or charge-step polarography, the potential of the electrode is set initially at a value, E_i, at the foot of the usual i-E curve. That is, where the faradaic current for reduction (or oxidation) is practically zero. The charge on the electrode is then rapidly changed in a few microseconds or less with a coulostat (cf. potentiostat and galvanostat). The time for charging is sufficiently short, compared to the rate constant for the electrode reaction, that essentially no faradaic current can flow during the charge pulse and all of the charge goes to the double layer. The charge supplied is sufficient to change the potential E of the electrode to a value, E_c, which should lie on the limiting current or plateau region of the conventional dc polarogram. After charging, the cell is essentially at open circuit, and so the faradaic current can be supplied only by discharge of the double layer.

The charge is governed by the potential of the electrode relative to the electrocapillary capillary maximum or point of zero charge, as described in Chap. 2. Thus,

$$C_{dl}(E - E_m) - C_{dl}(E_c - E_m) = \int_0^t I_d \, dt \qquad (8.24)$$

where I_d is the diffusion-controlled limiting current per unit area, t is the time elapsed since commencement of electrolysis, and C_{dl} is assumed to be independent of potential. As for most polarographic methods, measurements are made near the end of the drop life if a dme is being used, so that conditions of constant area can be assumed.

For (linear) diffusion control

$$I_d = \pm nFC_A \left(\frac{D_A}{\pi t}\right)^{1/2} \qquad (8.25)$$

the $\pm$ sign being for reduction and oxidation, respectively. Combining Eqs. (8.24) and (8.25) gives

$$\Delta E = E - E_c = \pm \left[\frac{2nFC_A (D_A)^{1/2}}{\pi^{1/2} C_{dl}} \right] t^{1/2} \tag{8.26}$$

This equation is a general one describing the open circuit E-t variation after the termination of charge injection. The potential therefore varies with $t^{1/2}$ and the concentration, C_A, can be determined from the slope of the ΔE-vs.-$t^{1/2}$ plot.

In controlled-charge methods, the problem of the double layer charging current is avoided. Also, measurements are made at times during which no appreciable net current flows through the cell, so that no corrections for ohmic potential losses are needed for interpreting the results. Thus, the experimental difficulties associated with using high-resistance nonaqueous solvents are greatly reduced, and much lower concentrations of potentially contaminating supporting electrolytes can be used in trace analysis. Concentrations down to at least 10^{-7} M can be determined by charge-step polarography, so the method is very sensitive. Despite these demonstrated advantages, reported practical applications of the method have been rare. This is due perhaps to the lack of selectivity (resolvability of the method) and the need to replot real time data on a root time scale for accurate measurements in the method, as originally reported.

The advantage of computerized readout techniques in charge-step polarography is clearly revealed in the work of Kudirka et al. [107], who have described an instrumental method combining the advantages of the coulostatic technique with the simplicity of operation and selectivity of dc polarography. In their approach, a dme is held at a potential where the faradaic current is essentially zero. Late in the drop life, a small charge is rapidly applied to the dme. Data points on the resulting E-t curve are acquired on a small digital computer which automatically calculates the slope and intercept (E at t = 0) for the first portion of the E vs. $t^{1/2}$ curve. The experiment is then repeated on the next mercury drop using a somewhat larger charge. When the charge is sufficient to bring the electrode to a potential approaching the polarographic $E_{1/2}$ value, the slope will increase. Further increases in charge will bring the electrode potential to the polarographic limiting current region, at which point the electroactive species should yield a potential-independent slope value. In other words, a plot of slope versus intercept values for incrementally advancing charge-steps applied to successive drops at a dme should resemble a current-sampled dc polarogram and permit both quantitative and qualitative determinations or mixtures of electroactive species. A curve for a mixture of 5×10^{-6} M cadmium and 5×10^{-6} M zinc is shown in Fig. 8.19, and a block diagram of the components of the measurement system and their interconnections is shown in Fig. 8.20. Figure 8.21 shows the relationship between concentration and measured slope for a range of inorganic species. The slopes were obtained in potential regions where the reductions of the particular ions were diffusion controlled and the expected linear responses were obtained. Astruc and Bonastre [105] have reported a similar approach but employing simpler instrumentation.

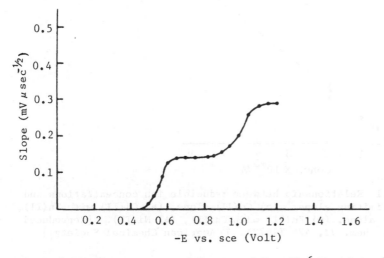

FIGURE 8.19 Charge-step polarogram of 5×10^{-6} M cadmium(II) and 5×10^{-6} M zinc(II). [Reproduced from Anal. Chem. *44*, 425 (1972). ©American Chemical Society.]

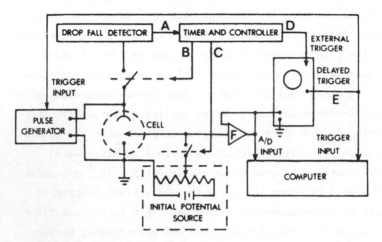

FIGURE 8.20 Block diagram of the instrumentation used to record
the charge-step polarograms presented in Fig. 8.19. [Reproduced
from Anal. Chem. 44, 425 (1972). © American Chemical Society.]

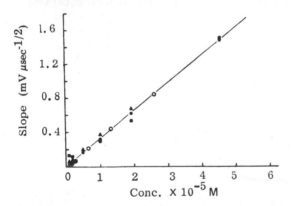

FIGURE 8.21 Relationship between reducible ion concentrations and
measured limiting slope: (o) Cd(II) alone, (•) Cd(II) with Zn(II),
(⊕) Zn(II) alone, (■) Zn(II) with Cd(II), (▲) Ni(II). [Reproduced
from Anal. Chem. 44, 425 (1972). © American Chemical Society.]

Katzenberger and Daum [108] have developed a related differential method which also appears to be highly successful. Experimental results with this approach indicate a detection limit below 10^{-7} M and show that supporting electrolyte concentrations as low as 10^{-4} M can be tolerated.

The charge-step method is obviously one which could become widely used as the use of minicomputers and/or microprocessors becomes widespread (see Chap. 10). Despite the claimed and almost certainly obtainable advantages of controlled-charge methods, unfortunately the methods have apparently yet to be used on many real analytical problems. At the risk of becoming very repetitious, it must therefore be stated once again, that a highly promising technique can be evaluated adequately only by tackling real problems rather than the determination of cadmium in distilled water/supporting electrolyte. This task has yet to be satisfactorily undertaken with respect to coulostatic methods, and its use in analytical chemistry is therefore still essentially in hibernation compared with the widely used technique of, say, differential pulse polarography. Future very valuable work could include a direct comparison of charge step and, say, differential pulse polarography on a particular analytical problem. Then and only then would the assessment of the charge-step method as a modern polarographic method be complete.

NOTES

1. G. C. Barker and I. L. Jenkins, Analyst 77, 685 (1952).

2. G. C. Barker, *Modern Electroanalytical Methods* (G. C. Charlot, ed.), Elsevier, Amsterdam, 1958, pp. 118-131.

3. B. Breyer and H. H. Bauer, in *Alternating Current Polarography and Tensammetry*, Chemical Analysis Series (P. J. Elving and I. M. Kolthoff, eds.), Interscience, New York, 1963, vol. 13.

4. H. Schmidt and M. von Stackelberg, *Modern Polarographic Methods*, Academic Press, New York, 1963.

5. G. C. Barker, Anal. Chim. Acta *18*, 118 (1958).

6. J. Havas, B. Juhasz, and B. Koszegi, Hung. Sci. Instrum. *15*(15) (1968).

7. D. Ilkovic and G. Semerano, Collect. Czech. Chem. Commun. 4, 176 (1932).

8. D. J. Ferret and G. W. C. Milner, Analyst 80, 132 (1955); 81 (1956).

9. D. J. Ferret, G. W. C. Milner, and A. A. Smales, Analyst 79, 731 (1954).

10. M. Geissler, Hung. Sci. Instrum. 1973, 9.

11. E. B. Buchanan, Jr., and J. R. Bacon, Anal. Chem. 39, 615 (1967).

12. M. Noshiro and M. Sugisaki, Bunseki Kagaku 15, 356 (1966).

13. A. Mizuike, T. Miwa, and S. Oki, Anal. Chim. Acta 44, 425 (1969).

14. R. G. Pac, Zavod Lab. 28, 18 (1962).

15. E. Sudo and H. Okochi, Bunseki Kagaku 17, 338 (1968).

16. P. Bersier and F. Z. von Sturm, Z. Anal. Chem. 224, 317 (1967).

17. O. Gürtler and Chu-Xuan-Anh, Mikrochim. Acta 1970, 941.

18. A. Mizuike and T. Kono, Microchim. Acta 1970, 665.

19. N. Yamata, Kuji Seijo 9, 55 (1971).

20. E. B. Buchanan, Jr., T. D. Schroeder, and B. Novosel, Anal. Chem. 42, 370 (1970).

21. I. Ruzić, J. Electroanal. Chem. 39, 111 (1972).

22. G. C. Barker, G. W. C. Milner, and H. I. Shalgosky, Polarography, Proc. Congr. Mod. Anal. Chem. Ind., St. Andrews, 1957, p. 199.

23. G. C. Barker, A. W. Gardner, and M. J. Williams, J. Electroanal. Chem. 42, App. 21 (1973).

24. G. C. Barker, in Polarography 1964 (G. J. Hills, ed.), Macmillan, London, 1966, vol. I, p. 25.

25. G. C. Barker, A. W. Gardner, and M. J. Williams, J. Electroanal. Chem. 41, App. 1 (1973).

26. G. C. Barker, Proc. Anal. Div. Chem. Soc. 1975, 171.

27. A. M. Bond and U. S. Flego, Anal. Chem. 42, 2321 (1975).

28. H. Jehring, J. Electroanal. Chem. 21, 77 (1969).

29. H. Jehring, E. Horn, A. Reklat, and W. Stolle, Collect. Czech. Chem. Commun. 33, 1038 (1968).

30. H. Jehring and W. Stolle, Collect. Czech. Chem. Commun. 33, 1670 (1968).

31. G. C. Barker, in Transactions on the Symposium on Electrode Processes (E. Yeager, ed.), Wiley, New York, 1961, p. 325.

32. G. C. Barker, R. L. Faircloth, and A. W. Gardner, Nature 181, 247 (1958).

33. W. H. Reinmuth, Anal. Chem. *36*, 211R (1964).

34. D. E. Smith, in *Electroanalytical Chemistry* (A. J. Bard, ed.), Dekker, New York, 1966, chap. 1.

35. T. Kambara, S. Tanaka, and K. Hasebe, J. Electroanal. Chem. *21*, 49 (1969).

36. L. N. Vasileva and H. V. Lukashenkova, Zhur. Anal. Khim *25*, 412 (1970).

37. L. N. Vasileva and N. B. Kogan, Zhur. Anal. Khim. *26*, 1932 (1971).

38. C. L. Roughton, M. Harrison, and B. Surfleet, Analyst *95*, 894 (1970).

39. B. S. Bruk and B. M. Sternberg, Zavod. Lab. *36*, 365 (1974).

40. R. M. Salikhdzhanova and I. E. Bryskin, Zavod. Lab. *37*, 765 (1971).

41. Y. Osajima, K. Matsumoto, M. Nakashima, F. Hashinaga, and S. Furutani, Bunseki Kagaku *20*, 1291 (1971).

42. S. Furutani, Bunseki Kagaku *16*, 103 (1967).

43. L. P. Chernega, V. I. Bodya, and Yu. S. Syalikov, Zhur. Anal. Khim. *26*, 1686 (1971).

44. Y. Osajima, M. Nakashima, and S. Furutani, Bunseki Kagaku *16*, 1297 (1967).

45. T. Kambara, S. Tanaka, and K. Hasebe, J. Chem. Soc. Jap., Pure Chem. Sec. *88*, 644 (1967).

46. G. Wolff and H. W. Nürnberg, Z. Anal. Chem. *224*, 332 (1967).

47. P. Delahay and C. Mamantov, Anal. Chem. *27*, 478 (1958).

48. P. Delahay, *New Instrumental Methods in Electrochemistry*, Interscience, New York, 1954, chap. 8.

49. C. N. Reilley, G. W. Everett, and R. H. Johns, Anal. Chem. *27*, 483 (1955).

50. J. J. Lingane, *Electroanalytical Chemistry*, 2nd ed., Interscience, New York, 1958, p. 633.

51. W. H. Reinmuth, Anal. Chem. *38*, 270R (1966).

52. D. G. Davis in *Electroanalytical Chemistry* (A. J. Bard, ed.), Dekker, New York, 1966, pp. 157-196.

53. P. Bos and E. Van Dalen, J. Electroanal. Chem. *45*, 165 (1973).

54. P. E. Sturrock, J. L. Hughey, B. Vandreuil, and G. E. O'Brien, J. Electrochem. Soc. *122*, 1195 (1975).

55. P. E. Sturrock and B. Vandreuil, J. Electrochem. Soc. *122*, 1311 (1975).

56. P. Delahay and T. Berzins, J. Amer. Chem. Soc. *75*, 2486 (1953); , 4205 (1953).

57. P. Delahay, C. C. Mattax, and T. Berzins, J. Amer. Chem. Soc. *76*, 5319 (1954).

58. O. Fischer and O. Dracka, Collect. Czech. Chem. Commun. *24*, 3046 (1959); *27*, 2727 (1962).

59. O. Fischer, O. Dracka, and E. Fischerová, Collect. Czech. Chem. Commun. *25*, 323 (1960); *26*, 1505 (1961).

60. O. Dracka, Collect. Czech. Chem. Commun. *24*, 3523 (1959).

61. O. Fischer, Collect. Czech. Chem. Commun. *27*, 1119 (1962).

62. O. Dracka, Collect. Czech. Chem. Commun. *25*, 338 (1960); *26*, 2144 (1961).

63. R. M. King and C. N. Reilley, J. Electroanal. Chem. *1*, 434 (1959-60).

64. W. Jaenicke and H. Hoffman, Z. Elektrochem. *66*, 809 (1962); *66*, 814 (1962).

65. L. B. Anderson and D. J. Macero, Anal. Chem. *37*, 322 (1965).

66. J. Heyrovský and J. Kuta, *Principles of Polarography*, Academic Press, New York, 1966, pp. 512-516.

67. R. W. Murray, *Physical Methods of Chemistry. Part IIa, Electrochemical Methods*, Interscience, New York, 1971.

68. H. B. Herman and A. J. Bard, Anal. Chem. *35*, 1121 (1963); *36*, 510 (1964); *36*, 971 (1964).

69. A. C. Testa and W. H. Reinmuth, Anal. Chem. *32*, 1512 (1960).

70. H. B. Herman and H. N. Blount, Anal. Chem. *25*, 165 (1970).

71. K. W. Hanck and M. L. Deanhardt, Anal. Chem. *45*, 179 (1973), and references cited therein.

72. P. J. Lingane, C.R.C. Crit. Rev. Anal. Chem. *1*, 587 (1970).

73. R. S. Nicholson, Anal. Chem. *44*, 478R (1972).

74. R. N. Adams, C. N. Reilly, and N. H. Furman, Anal. Chem. *25*, 1160 (1953).

75. M. Ishibashi and T. Fujinaga, Denki Kagaku *24*, 375 (1956); *24*, 525 (1956); Anal. Chim Acta *18*, 112 (1956).

76. M. Ishibashi and T. Fujinaga, in *Modern Electroanalytical Methods, Proc. Int. Symp. Mod. Electrochem. Methods Anal., Paris, 1957* (G. Charlot, ed.), Elsevier, Amsterdam, 1958.

77. T. Fujinaga, in *Progress in Polarography* (P. Zuman and I. M. Kolthoff, eds.), Interscience, New York, 1962, vol. I, pp. 201-221.

78. H. L. Kies, J. Electroanal. Chem. *16*, 279 (1968); *76*, 1 (1977).

79. H. L. Kies and H. C. Van Dam, J. Electroanal. Chem. *48*, 391 (1973).

80. Y. Israel, Anal. Chem. *31*, 1473 (1959); Talanta *19*, 1067 (1972).

81. J. Heyrovský, Chem. Listy *35*, 155 (1941).

82. R. Kalvoda, *Techniques of Oscillographic Polarography*, 2nd ed., Elsevier, Amsterdam, 1965.

83. K. Micka, Z. Phys. Chem. (Leipzig) *206*, 345 (1957).

84. T. Kambara, Leybold Polarogr. Ber. *2*, 41 (1957).

85. H. Matsuda, Z. Electrochem. *60*, 617 (1956).

86. J. Heyrovský and J. Kuta, *Principles of Polarography*, Academic Press, New York, 166, pp. 516–527.

87. J. Heyrovský and K. Micka, in *Electroanalytical Chemistry* (A. J. Bard, ed.), Dekker, New York, 1966, vol. 2, pp. 193–256.

88. P. Delahay, Anal. Chim. Acta *27*, 90 (1962).

89. P. Delahay, Anal. Chim. Acta *27*, 400 (1962).

90. P. Delahay, Anal. Chem. *34*, 1267 (1962).

91. P. Delahay, Anal. Chem. *34*, 1580 (1962).

92. P. Delahay, Anal. Chem. *34*, 1662 (1962).

93. A. Aramata and P. Delahay, Anal. Chem. *35*, 1117 (1963).

94. P. Delahay and Y. Ide, Anal. Chem. *35*, 1119 (1963).

95. P. Delahay, Anal. Chem. *34*, 1161 (1962).

96. P. Delahay and W. H. Reinmuth, Anal. Chem. *34*, 1344 (1962).

97. W. H. Reinmuth and C. E. Wilson, Anal. Chem. *34*, 1161 (1962).

98. W. H. Reinmuth, Anal. Chem. *34*, 1271 (1962).

99. J. Bonastre, M. Astruc, and J. Leleu, Electron. Ind. *99*, 763 (1966).

100. J. Bonastre, M. Astruc, and J. L. Bentata, Chim. Anal. (Paris) *50*, 113 (1968).

101. M. Astruc and J. Bonastre, J. Electroanal. Chem. *34*, 211 (1972).

102. M. Astruc and J. Bonastre, J. Electroanal. Chem. *36*, 1435 (1972).

103. M. Astruc, F. Del Rey, and J. Bonastre, J. Electroanal. Chem. *43*, 113 (1973).

104. M. Astruc, F. Del Rey, and J. Bonastre, J. Electroanal. Chem. *43*, 125 (1973).

105. M. Astruc and J. Bonastre, Anal. Chem. *45*, 421 (1973).

106. R. W. Sørensen and R. F. Sympson, Anal. Chem. *39*, 1238 (1967).

107. J. M. Kudirka, R. Abel, and C. G. Enke, Anal. Chem. *44*, 425 (1972).

108. J. M. Katzenberger and P. H. Daum, Anal. Chem. *47*, 1887 (1975).

Chapter 9

STRIPPING VOLTAMMETRY

9.1 INTRODUCTION

Typically, the polarographic methods discussed so far offer advantages over direct current (dc) polarography because of the improved
faradaic-to-charging current ratio. In general, the favorable ratio
has been achieved by some "trick" aimed at discriminating against
charging current, although in pulse polarography, for example, larger
faradaic currents are also effective in providing the improved sensitivity.

Stripping voltammetry is an electroanalytical technique which
generates its extremely favorable faradaic-to-charging current ratio
almost solely by its ability to enhance substantially the faradaic
current component response of the experiment, while maintaining the
charging current at values associated with the voltammetric or polarographic techniques described in previous chapters. The extremely
large faradaic currents per unit concentration associated with this
technique give rise to extremely low detection limits, and in the
1950s and 1960s some authors [1-3] felt it to be the most sensitive
technique available. Some idea of the sensitivity attainable can be
gained by the work of Eisner and Mark [4], who showed that stripping
voltammetry was more sensitive than neutron activation analysis for
the determination of silver at the 10^{-10} M level in rain and snow
samples from silver iodide seeded clouds.

In earlier chapters, the technique of cyclic voltammetry [dc, alternating current (ac), and other modes] is described. The dc potential ramp in these experiments is a triangular voltage. In the negative potential direction sweep, reduction of electro- active species in solution may occur. During the second stage of the experiment, when the scan direction is reversed, oxidation of stable products generated at the electrode surface may occur, and the complete cyclic voltammogram may include cathodic and anodic current components. If a reversible process of the A + ne $\rightleftharpoons$ B type occurs, the peak heights of the reduction and oxidation proc- esses are equal (as demonstrated in Chap. 5). However, this is strictly true only when B is soluble in solution. If B forms an amalgam so that the electrode process is of the kind

$$A + ne \rightleftharpoons B \ (Hg) \tag{9.1}$$

and the reduction is carried out at a relatively small volume hanging mercury drop electrode (hmde), for example, cyclic voltammograms of the kind shown in Fig. 9.1 are obtained. It is clear in this figure that on the reverse scan, where the metal is *stripped* from the amal- gam and oxidized back into the solution, that the peak height is larger than the case for the reduction process. The reverse sweep of a cyclic voltammogram for systems involving amalgam formation at a hanging mercury drop electrode, therefore, shows the analytically desired enhancement of the faradaic current and constitutes a special form of *anodic stripping voltammetry*. (Also see further comments in Sec. 9.3.)

Anodic stripping voltammetry in its common form is a combination of a concentration step, in which a metal ion in solution is reduced by controlled-potential electrolysis at a potential more negative than the polarographic half-wave potential, to produce either a metal deposit on a solid electrode (plating) or an amalgam with the mercury of a mercury drop (or mercury film electrode), followed by a stripping process in which the metal is stripped from the electrode (i.e., ox- idized back into solution). The concentration step is carried out for a definite time under reproducible conditions (the solution may

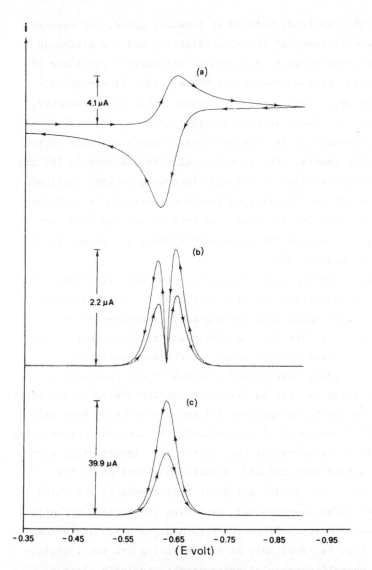

FIGURE 9.1 (a) DC, (b) phase-selective second harmonic ac, and
(c) phase-selective fundamental harmonic ac cyclic voltammograms of
1×10^{-3} M Cd(II) in 1 M KCl at a hmde. Note how the magnitude of
current in reverse sweep is enhanced by amalgam (sphericity) formation [Reproduced from Anal. Chem. *50*, 216 (1978). © American
Chemical Society.]

be stirred or the electrode rotated at constant speed, for example,
to improve the efficiency of the electrolysis), and the stripping
process is performed in most cases via a voltammetric procedure of
the kind discussed in the previous chapters, e.g., linear sweep,
square-wave, pulse, ac voltammetry. Anodic stripping voltammetry,
of course, does not always involve amalgam formation or metal deposition [5]. Obviously it is possible also to use a cathodic variety
of stripping voltammetry. For example, chloride and bromide [6] and
sulfide [7] can be oxidized at mercury electrodes to give insoluble
mercury salts which can be stripped from the electrode by applying a
negatively directed potential scan, and ferrous ion can be determined by anodic deposition and cathodic stripping of hydrous ferric
oxide in acetate medium [8].

The use of stripping voltammetry has been extremely widespread
in environmental investigations, and currently a significant portion
of the literature on analytical polarography (voltammetry) is devoted
to this technique. Reflecting the widespread interest in this branch
of voltammetric analysis is the number of reviews available, and
Notes 9 to 15 provide a most useful coverage of the specialized
aspects of the topic as well as extensive bibliographies of its applications. The review by Barendrecht [9] and the books by Neeb [10]
and Brainina [14] are particularly suitable for workers contemplating
using stripping voltammetry for the first time. Importantly, these
articles incorporate detailed and extensive discussions of the
numerous "tricks of the trade" which are most abundant, and which
must be learned before successfully employing the various techniques
of stripping voltammetry. Since the treatment of stripping voltammetry in this book is aimed only at demonstrating how the technique
fits into the overall pattern of polarographic analysis, little
attention is focused on the many important practical facets. To
avoid any misconceptions (e.g., the method is as easy to use as
polarography at a dme) that such a restricted coverage may convey
it is worth noting at this early point in the chapter that because
it is a technique applied to extremely dilute solutions (10^{-9} M and
sometimes lower), stripping voltammetry requires a considerable

degree of experimental skill and experience to implement successfully. Furthermore, while there is no doubt that this is the most
sensitive polarographic method available and is amenable to an exceedingly straightforward qualitative description in terms of instrumentation and theory, it must be remembered that in reality it is a
very complex technique and there are many more pitfalls for obtaining
erroneous data with this method than with most others. Therefore,
it is most important that rigid data validation procedures be incorporated into the overall analytical procedure when using stripping
voltammetry. For example, even if one does nothing else, one should
always rigorously check, as has been recommended for all polarographic
analyses, that the wave shapes and peak positions of unknown and
standard curves are as expected.

9.2 ELECTRODES USED
IN STRIPPING VOLTAMMETRY

The electrochemical cell and instrumentation used in stripping voltammetry is usually similar to that for polarography. Thus, a three-
electrode potentiostat (galvanostat) system should be employed to
minimize iR drop effects with a suitable working electrode, reference
electrode and auxiliary electrode. The working electrode may be,
however, of many different kinds, as may its operation.

Intuitively, one might expect the working electrode of choice
to be old faithful standby, the dropping mercury electrode. However,
this electrode has been used only sparingly, presumably because
workers have supposed that it is difficult to obtain drop times
sufficiently long to undertake the controlled-potential electrolysis
step required prior to the potential scan. Nevertheless, Velghe
and Claeys [16] have described a slowly growing mercury drop electrode with a drop time of about 18 min. and have used phase-selective
ac techniques to record the stripping curves. Figure 9.2 shows a
stripping curve obtained at the slowly growing dme, and this electrode would appear to have the desired characteristics of being very
reproducible and not susceptible to some of the typical troubles of
stationary drop electrodes (see below), such as solution creeping

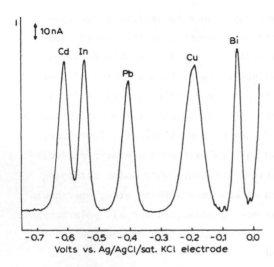

FIGURE 9.2 Phase-selective
ac anodic stripping voltammogram recorded at a slowly
growing dropping mercury
electrode. Solution con-
tains 1.5×10^{-6} M Cd(II),
4.0×10^{-6} M In(III),
8×10^{-7} M Pb(II),
1.3×10^{-6} M Cu(II), and
1.3×10^{-7} M Bi(III) in
1 M HCl. [Reproduced from
J. Electroanal. Chem. 35,
229 (1972).]

up the capillary. Conceivably, more developmental work with this
kind of electrode could lead to its wider use.

The hmde is one of the more popular electrodes for stripping
voltammetry. Figure 9.3(a) shows a commercially available hdme
which basically consists of a microsyringe with a micrometer for
control of drop size. The drop formed at the tip of a capillary
tube via displacement of mercury in the calibrated micrometer syringe
delivery system. Figure 9.3(b) shows an electrochemical cell for
stripping voltammetry using an hmde, whereas Fig. 9.3(c) shows a
schematic diagram of the usual kind of arrangement with any working
electrode. The hmde offers simplicity, economy, and reproducibility.
Figure 9.4 shows curves obtained with linear sweep and differential
pulse stripping readout, and as expected, curves will be similar to
those usually associated with a stationary electrode. Alternatively
[17], mercury drops may be suspended from a tip of platinum wire
imbedded in a glass rod (Fig. 9.5). This version is less easy to
use in routine analysis than the micrometer-controlled system. Two
basic disadvantages have been associated with the hmde [9,11,13].
First, it is a low-surface area-to-volume ratio. The relatively
small surface area reduces the efficiency of the preelectrolysis
step, while the large volume means that a finite time is required

for metals dissolved in the mercury to diffuse from the drop interior.
This latter phenomenon causes a broadening of the stripping peaks,
thereby decreasing the resolution of adjacent peaks. If very long
plating times are used, the metals may even diffuse up into the
mercury in the supporting capillary, causing substantial peak broadening. Second, only low stirring rates of the solution may be employed with a hmde to avoid dislodging the drop. This again places
a restriction on the efficiency of the deposition step and requires
longer preelectrolysis times than would be needed if faster stirring
rates could be employed. In practice, any implied sensitivity problem associated with the hmde is usually relevant only to discussions
related to dc linear sweep methods. Modern techniques of polarography are, of course, recommended for use with the hmde, as they are
with the dme, and limits of detection with ac stripping [18], square-wave stripping [19], derivative techniques of stripping [20], normal
pulse stripping voltammetry [21], and differential pulse voltammetry
[22] are usually far better than purification procedures for reagents
and chemicals, etc., so that sensitivity consideration at a hmde is
actually not a serious problem.

Geometry is obviously an important factor in the sensitivity of
any electrode, and the surface area-to-volume ratio of a mercury
thin-film electrode (mtfe) is extremely high. In addition to having
excellent sensitivity, mercury films give high resolution because
diffusion from within the mercury film to the surface is extremely
fast [23]. This provides a distinct advantage over the hmde, in
some instances when peaks are not completely resolved. The disadvantages of the mtfe are the lack of reproducibility with some types,
and the formation of intermetallic compounds, the latter being enhanced because of the higher concentrations involved.

A great variety of substrates has been used for the mercury
film [9]. Metal wires of materials such as Pt, Ni, or Ag were used
in early work, but generally they were not successful, due to surface
oxide films and poor reprodubility in preparing the mercury film.
Carbon electrodes are now used almost exclusively as the substrate
for mtfe. They are inert, mechanically strong and have good

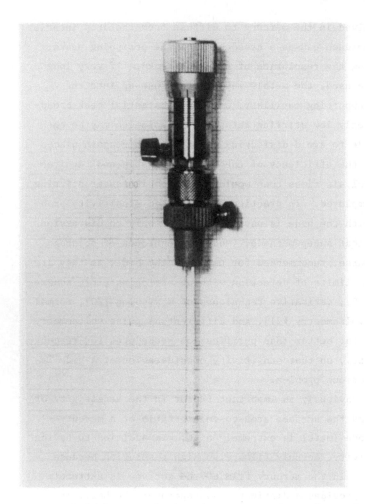

FIGURE 9.3(a) Metrohm hanging mercury drop electrode Model E410
with a micrometer syringe delivery system. (Reproduced by courtesy
of Metrohm Herisau.)

electrical conductivity. Mercury films were originally prepared
via a two-stage process [24], in which mercury first was plated onto
the electrode at a relatively positive potential (e.g., -0.2 V vs.
sce), then the potential was made more negative for the deposition
of metals to be determined. More recently, Florence [25] has
reported a very successful procedure. Mercuric nitrate is added to
the solution to be determined, and a simultaneous deposition is made

FIGURE 9.3(b) Electrode assembly and cell for stripping voltammetry at a hmde. Teflon-coated stirring bar at bottom of cell is rotated by magnetic stirrer (on which cell sits) to stir the solution during the electrolysis period. (Reproduced by courtesy of Metrohm Herisau.)

of both the mercury and the metal in the sample. A mtfe of these types can easily be rotated at high velocity in the region of several thousand revolutions per minute, and therefore high plating efficiency is achieved.

A number of different types of carbon electrode have been reported. Among these are the carbon paste electrode [26], wax-impregnated graphite electrodes [24,27], the pyrolytic graphite electrode [22], and the glassy carbon electrode [25], with the

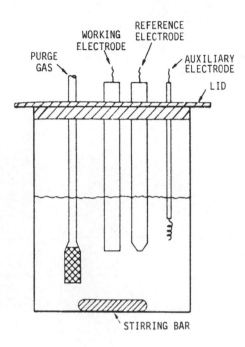

FIGURE 9.3(c) Schematic diagram of electrode assembly and cell for stripping voltammetry. [Reproduced from W. D. Ellis, J. Chem. Educ. *50*, A131 (1973).]

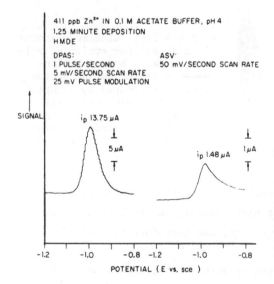

FIGURE 9.4 Stripping curves obtained at a hanging mercury drop electrode. DPAS = differential pulse anodic stripping; ASV = linear sweep anodic stripping. [Reproduced from Amer. Lab. *4*(6), 59 (1972). International Scientific Communications, Inc.]

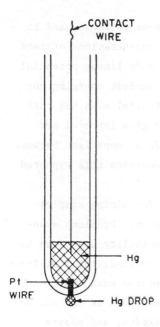

FIGURE 9.5 Hanging mercury drop suspended from a platinum wire. [Reproduced from W. D. Ellis, J. Chem. Educ. *50*, A131 (1973).]

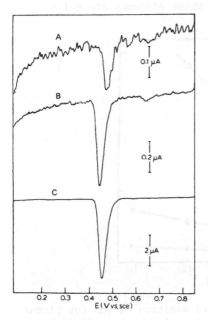

FIGURE 9.6 Determination of lead by anodic stripping voltammetry at a rotating glassy carbon electrode. (A) blank, 20-min deposition, (B) 5×10^{-9} M Pb(II), 10-min deposition, (C) 2×10^{-7} M Pb(II), 7-min deposition. All in supporting electrolyte of 0.1 M KNO_3 plus 2×10^{-5} M Hg(II); scan rate is 3 V min^{-1}; 2000 rpm rotation rate. [Reproduced from J. Electroanal. Chem. *27*, 273 (1970).]

last-mentioned electrode possibly being the best of those used to
date [13]. Figure 9.6 shows curves for low concentrations of lead
at a rotating glassy carbon electrode, using a dc linear potential
sweep. Since other techniques discriminate against charging current, it is not surprising to find the dc potential scan has been
replaced by the differential pulse format, to give improved performance [28,29]. Figures 9.7 and 9.8 provide a comparison between
the dc and differential pulse formats to demonstrate this expected
result.

Generally, poor results are obtained by depositing samples
directly onto solid metal or graphite electrodes. Problems associated with surface contamination and reproducibility are often encountered, and if more than one species is deposited, their stripping
peaks frequently overlap because the metals plated onto the electrode are not interdiffusible. However, metals such as Ag, Au, and
Hg normally must be determined at solid electrodes, and papers
describing the determination of these three elements are quite

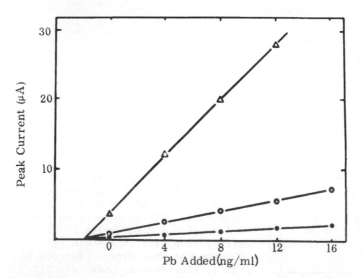

FIGURE 9.7 Comparison of lead standard addition curves for linear
scan and differential pulse stripping techniques at a thin-film wax-
impregnated graphite electrode: (●) linear dc potential, scan rate
of 50 mV sec^{-1}; (o) linear dc potential, scan rate of 200 mV sec^{-1};
(Δ) differential pulse. [Reproduced from Anal. Chem. 45, 2171
(1973). © American Chemical Society.]

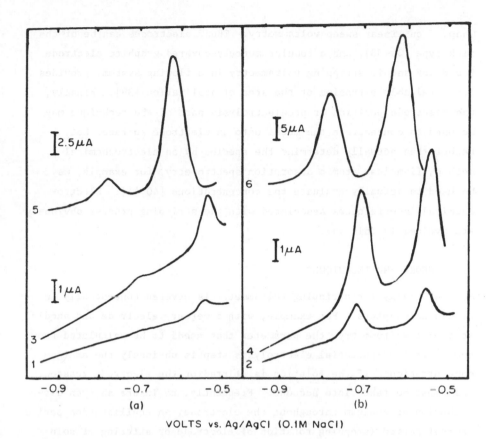

VOLTS vs. Ag/AgCl (0.1M NaCl)

FIGURE 9.8 Comparison of linear sweep and differential pulse stripping curves for a determination of lead and cadmium in a urine sample at a thin film wax-impregnated graphite electrode. Curves 1 and 2, linear dc potential scan rate of 50 mV sec^{-1}, 3 and 4, same as 1 and 2 but 200 mV sec^{-1} scan rate; 5 and 6, differential pulse. (a) no additions; (b) 4 ng/ml Pb(II) and 4 ng/ml Cd(II) added. [Reproduced from Anal. Chem. 45, 2171 (1973). © American Chemical Society.]

widespread [2,4,30-34]. Multiple peaks from oxidation or stripping of a metal are sometimes observed, and peak areas rather than peak heights are recommended for determination of concentration. Pyrolytic graphite [13] or glassy carbon electrodes [13,35] are probably the best choices for determining species having oxidation potentials more positive than mercury itself.

Obviously, the type and arrangement of electrodes for performing stripping voltammetry encompass all the possibilities mentioned in

Chap. 5 on linear sweep voltammetry. Thus, electrodes can be of the
disk type [35-38], and a tubular mercury-covered graphite electrode
for doing anodic stripping voltammetry in a flowing system, provides
for a valuable extension of the area of application [39]. Finally,
the electrode position or preelectrolysis part of the technique may
be used to concentrate the metal onto an electrode surface, but
rather than actually determine the species by an electrochemical
method, flameless atomic absorption spectrometry, for example, may
be used to actually evaluate the concentrations [40,41]. Electrochemical interferences associated with the stripping process obviously
are avoided in this way.

9.3 THEORY AND TECHNIQUES

Much of theory for stripping voltammetry is covered conceptually in
preceding chapters. For example, with a mercury electrode and anodic
stripping voltammetry, the parameter that needs to be calculated from
the controlled-potential electrolysis step is obviously the amalgam
concentration. If the solution is stirred or the electrode rotated,
this must be taken into account. Frequently, to insure an even dis-
tribution of amalgam throughout the electrode, an equilibration period
or rest period (stopping rotation of electrode or stirring of solution) is employed between the deposition and the stripping process.
Again this step must be considered, although the contribution of
this step to the overall electrolysis is relatively small. Thus,
the theory for controlled-potential electrolysis, after taking into
account the nature of the experiment, will enable the concentration
of metal in the amalgam to be calculated. The theory for the faradaic component of the stripping or oxidation step is based on the
same principles described in earlier chapters, with the calculated
amalgam concentration from the electrolysis step being used in the
equations. The charging current terms are essentially the same as
usual; so if the stripping process is undertaken by differential
pulse or phase-selective ac voltammetry instead of dc linear sweep
voltammetry, greater sensitivity is obtained because of the discrimination against the charging current. Of course, in anodic (amalgam)

stripping voltammetry, the sensitivity is also dependent on the concentration of metal in the amalgam and therefore on the geometry of
the electrode and the length and efficiency of the controlled-
potential electrolysis step.

A rigorous theory for the controlled-potential electrolysis
step is difficult to implement because the mass transport processes
are complicated and not always extremely well known or controllable
in the actual analytical situation. However, access to a rigorous
theory is unnecessary because absolute analytical methods are rarely,
if ever, employed in stripping voltammetry. Rather, calibration
curves or even more frequently, standard addition methods, are used
almost invariably to determine concentrations. An elementary theory
is therefore the only one presented for the deposition step to
provide an indication of the parameters which can be adjusted
to optimize the plating efficiency.

Since the controlled-potential electrolysis step in anodic
stripping voltammetry is performed at a potential 300 to 400 mV more
negative than the polarographic half-wave potential, and the solution
is stirred or the electrode rotated, the current flow i(t) at time t
is approximated reasonably by the Levich equation (Chap. 5) (or at
least an expression which is proportional to the Levich equation).
Thus, for the electrode process described by Eq. (9.1),

$$i(t) = k_1 nFAD_A^{2/3} \omega^{1/2} \nu^{1/6} C_A(t) \tag{9.2}$$

where

$\quad k_1$ = a constant appropriate for the electrode under consideration

$\quad \omega$ = rate of electrode rotation or solution stirring

$\quad \nu$ = kinematic viscosity of the solution

$\quad C_A(t)$ = concentration of the metal ion in solution at deposition
time

Other symbols are those conventionally used throughout.

The viscosity of aqueous solutions does not vary greatly. Furthermore, the sixth-root dependency almost ensures, in any case, that
this is unlikely to be a significant term in determining the relative

plating efficiency, unless comparisons in different solvents are
being considered. However, the stirring rate of the solution or the
rotation rate of the electrode are obviously important variables to
be considered, and to improve electrolysis efficiency, they can be
increased to the point where the ability to maintain a drop at a
hanging mercury drop electrode becomes critical, or where unacceptable
solution cavitation occurs with other electrode systems. Increases in
the electrode area also may be used to optimize the plating efficiency,
i.e., the amount of metal deposited per unit time. However, since
the stripping process is usually performed at the same electrode as
the electrolysis step, large-area electrodes are not usually employed in stripping voltammetry, and electrode areas tend to be of
dimensions similar to those used in conventional polarographic or
voltammetric experiments.

Application of Faraday's law enables the concentration of metal
in the amalgam to be calculated. For a simplified treatment, it can
be assumed that a constant current is maintained during the deposition.
This will be a good approximation if the concentration of the metal
ion $C_A(t)$ is not changed appreciably during the course of the electrolysis, and if the solution is stirred or electrode rotated at a constant velocity [see Eq. (9.2)]. Under these conditions, the concentration of the reduced metal in the mercury at either a hmde or mtfe
is given by Faraday's law:

$$C_{B(Hg)} = \frac{it}{nFV} \tag{9.3}$$

where

$C_{B(Hg)}$ = concentration of metal in the amalgam

    i = reduction current

    t = electrolysis time

    V = volume of the mercury in the film or drop

Representing the mass transport terms of stirring, etc., in
Eq. (9.2) by the symbol m (the mass transfer coefficient) and invoking the assumptions listed above, enable Eq. (9.2) to be rewritten
as

$$i = k_1 mnFAC_A \tag{9.4}$$

By substituting Eq. (9.4) into Eq. (9.3) and simplifying the final result obtained for $C_{B(Hg)}$,

$$C_{B(Hg)} = \frac{k_2 mC_A t}{r} \qquad \text{(hmde)} \tag{9.5a}$$

$$C_{B(Hg)} = \frac{k_3 mC_A t}{L} \qquad \text{(mtfe)} \tag{9.5b}$$

where r is the radius of the mercury drop and L is the thickness of the mercury film, and k_2 and k_3 are constants appropriate for the hmde and mtfe, respectively.

Any of the E-t waveforms described in earlier chapters (e.g., linear sweep, sinusoidal alternating current, differential pulse) may be used to strip or oxidize the deposited metal from the amalgam to obtain the peak height (i_p) normally used as the quantization parameter (areas under curves are sometimes used).

The i-E behavior at a hmde can be considered as analogous to that of linear sweep voltammetry (LSV) (Chap. 5). In the case of a reversible electrode process $C_{B(Hg)}$ [from Eq. (9.5a)] is substituted for C_A into the Randles–Sevcik equation. Thus

$$i_p = -k_4 mn^{3/2} D_B^{1/2} rv^{1/2} C_A t \tag{9.6}$$

where k_4 is a numerical constant and v is the scan rate of dc potential. Actually sphericity terms to allow for effects of curvature of the electrode should be added to Eq. (9.6), for a rigorous solution. Such corrections can be quite significant at low scan rates and small values of r [42,43], but the equation is satisfactory for the purposes of this chapter. The enhancement of the reverse sweeps due to amalgam formation in the cyclic voltammograms in Fig. 9.1 is predominantly explicable in terms of the sphericity of the electrode [43,44].

Ignoring sphericity effects and assuming equal diffusion coefficients for oxidized and reduced forms, the peak potential E_p of

the stripping curve at a hmde is given by the usual expression for
oxidation in linear sweep voltammetry, so that

$$E_p = E_{1/2}^r + \frac{1.1RT}{nF} \tag{9.7}$$

where $E_{1/2}^r$ is the reversible polarographic half-wave potential.

With a linear sweep of potential, the rapid change of potential
generates a relatively large nonfaradaic charging current i_c, which
(as we have seen in Chap. 5) is given by the expression

$$i_c = AC_{dl}v$$

where v is the scan rate and C_{dl} is the differential capacity of
the electrode double layer. Thus, while one can increase the absolute magnitude of the faradaic response by increasing A or v, the
charging current is also increased. In fact, in the case of the
hmde, the faradaic signal increases with $v^{1/2}$ and r, whereas the
charging current response increases with v and r^2 ($A \propto r^2$), so that
large area electrodes or fast scan rates in their own right do not
provide for increased sensitivity. Use of other potential waveforms
of course allows for minimization of the residual current.

For mercury thin-film electrodes, the theory developed by deVries
and Van Dalen [45] gives the results for a linear dc sweep of potentials:

$$i_p = -k_5 n^2 AC_{B(Hg)} L\, v \tag{9.9a}$$

or

$$i_p = -k_6 mn^2 AC_A\, vt \tag{9.9b}$$

and

$$E_p = E_{1/2}^r + \frac{2.3RT}{nF} \log \frac{\delta nFLv}{D_A RT} \tag{9.10}$$

where δ is the diffusion layer thickness. Roe and Toni [46] have
also derived equations for the peak position and peak current of the
dc stripping process using a mtfe. $C_{B(Hg)}$ is directly proportional

to the mass of metal dissolved in an ultra-thin film [29], so the
peak current for thin films is determined by the amount of deposited
material present, rather than its concentration in the thin film.
In practical analysis, one does not attempt to construct very thin
mercury film electrodes because they are extremely difficult to reproduce. Rather, thicker films which still exhibit thin-film behavior are employed. Films ranging in thickness from approximately
2 to 10,000 Å show no distortion of any consequence from thin-film
behavior, and offer equivalent sensitivity to very thin films [29,
45].

At a hdme, differential pulse curves are readily understood
via extrapolation from differential pulse and dc linear sweep results. When using reversible electrode processes, the theory for
the ac stripping curve turns out to be analogous to that for ac
voltammetry. Underkofler and Shain's paper on ac stripping
voltammetry [18] demonstrates this conclusion clearly, so new theoretical results do not need to be presented for this situation.
However, such analogies do not hold for thin-film electrode reactions,
and different theoretical expressions from those presented in the
chapter on polarography must be invoked. For pulsed voltammetric
stripping at a mtfe [28,29,47], Osteryoung and Christie [29] have
successfully developed a strategy for solving the theory. They
replace the current-by-time (constant current assumed) term in
Eq. (9.3) and substitute the term Q_M, the total charge passed for
deposition of the metal, to give the more general expression

$$Q_M = nFALC_{B(Hg)} \tag{9.11}$$

for the mtfe. Equation (9.5b) is obviously a special case of
Eq. (9.11) in a rearranged form. Under pulsed conditions for the
stripping process, and using Eq. (9.11) for the controlled-potential
electrolysis step, the approximate result

$$i_p(\text{pulse}) = \frac{-0.138 Q_M}{t_p} \tag{9.12}$$

is obtained at 25°C. In Eq. (9.12), t_p is the pulse width or duration, and the peak current is predicted by the simplified treatment
(as was the case for a linear scans of potential) to be dependent only
on the amount of metal plated onto the electrode and on the pulse
width, provided that the electrode is thin enough and the pulse
width is of sufficient duration so that $L/(D_{B(Hg)}t_p)^{1/2} < 0.2$.

For equally thin films or equivalently slow scan rates under the
same conditions and in the same terminology, the dc linear scan stripping peak current is given by

$$i_p = -11.6nv\,Q_M \tag{9.13}$$

The peak currents for the two techniques, therefore, are equal when
$nvt_p = 1.195 \times 10^{-2}$. If $n = 1$ and $v = 1$ V sec^{-1}, this corresponds
to $t_p = 11.95$ msec, and both the scan rate and pulse duration are
near the usual limits of the two techniques [29]. Therefore, the conclusion can be made that the two techniques are essentially equivalent
in their faradaic responses [29]. However, the dc linear sweep
stripping technique has a substantial double-layer charging current
component, whereas pulsed stripping voltammetry discriminates almost
completely against double-layer charging current, except in the
presence of appreciable uncompensated resistance [28,47], and so the
sensitivity of the pulsed stripping method is likely to be superior.
As will be seen in subsequent discussion, it is a moot point as to
whether or not the increased sensitivity actually can be used, but
theoretically it does exist. Naturally, a staircase waveform could
also be used to discriminate against the charging current [48,49],
and a sensitivity similar to the pulse technique is obtained [49].
However, experiments can be performed much faster with the staircase
waveform; so this offers a minor advantage.

Ac stripping techniques under thin-film conditions (no mercury
film) have been considered by Vydra et al. [36], and data at glassy
carbon electrodes using in situ mercury-deposited films have been
presented by Bond et al. [50] and Batley and Florence [51]. However,
results with ac techniques in general would not appear to be as favorable as pulse methods, because of the relatively high resistance

of a thin-film electrode. At low-resistance hmde's, ac stripping data appear to be equally as sensitive as the pulse methods [18] for reversible systems. Therefore, at this electrode, the ability to discriminate against irreversible processes sometimes provides a decided advantage for the ac stripping method [11,52].

Theories for cathodic stripping voltammetry and other forms of nonamalgam stripping analysis, particularly at nonmercury electrodes, tend to be fiercly complex if they are intended to represent the practical situation, and in general, theory-experiment correlations tend to be poor. Part of the reason for this is that the activity of the solid on an electrode needs to be known, and this is not easily determined when a solid is deposited onto an electrode surface. Furthermore, a multitude of often not well understood surface phenomena are involved in the plating process. Thus, unless a controlled-potential electrolysis step followed by a dc stripping step is involved, in which coulombs are counted and concentrations evaluated via Faraday's laws, completely empirical calibration procedures measured against standard external or internal reference solutions are usually employed. The extreme cases of the reversible and irreversible stripping process at solid electrodes, however, have been discussed [53,54] with a dc linear sweep of potential. For both cases, the peak current is proportional to the scan rate and to the amount of deposit, as expected for a description based on thin-film behavior. The peak potential is proportional to the log of the scan rate, with a slope of $\pm 1.15 RT/nF$ for the reversible case and $-2.3RT/\alpha nF$ and $2.3RT/(1 - \alpha)nF$ for the totally irreversible case, depending on whether the stripping process is reduction or oxidation. For the reversible case, the peak potential is independent of the amount of deposit for coverage of less than a monolayer, and proportional to the logarithm of the amount of material deposited, unless a very thin film is formed. The book and review by Brainina [14,15] should be consulted by any worker contemplating the use of stripping analysis at solid electrodes. To introduce a personal comment, it has been the author's experience that in a complicated solution, such as frequently encountered in environmental analysis, it is very

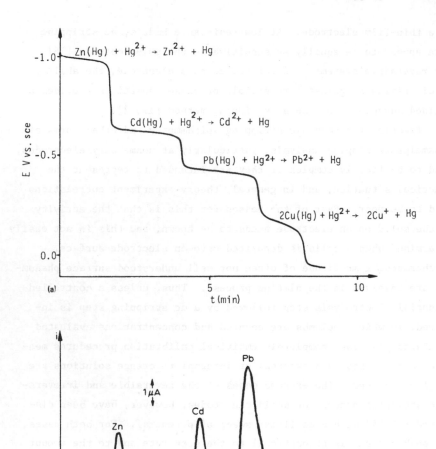

FIGURE 9.9 (a) Time-potential curve at a hmde for a 0.5 M NaCl solution containing 1.5×10^{-6} M Zn(II), Cd(II), Pb(II), and Cu(II) and 5×10^{-4} M Hg(II). A preelectrolysis period of 3 min at −1.25 V vs. sce was used and the t − E curve recorded after a settling period of 30 sec. (b) dc anodic stripping curve of solution in (a). Scan rate is 50 mV sec^{-1}.

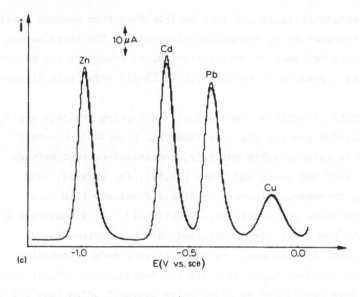

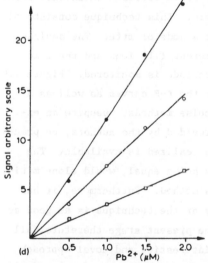

FIGURE 9.9 (c) Differential pulse anodic stripping curve of solution in (a). Scan rate is 5 mV sec^{-1}; pulse amplitude is 50 mV. (d) Calibration curves obtained in the determination of lead(II) by potentiometric stripping analysis (□), dc anodic stripping (●), and differential pulse anodic stripping (o). [Reproduced from Anal. Chim. Acta *83*, 19 (1976).]

difficult to obtain reliable and reproducible data from methods based
on stripping voltammetry at nonmercury electrodes. The introduction
of a plating step followed by nonelectrochemical concentration determination may be a profitable venture [40,41,55-57] under such circumstances.

     The majority of workers concerned with stripping analysis moni-
tor the dissolution process via a voltammetric process (i-E curve).
Naturally, as in polarographic analysis, controlled-current methods
could also be used and Luong and Vydra [58,59], for example, have
used stripping chronopotentiometry on disk and mercury film electrodes (both rotating and stationary). Obviously, the disadvantages
associated with the (i-t) curves obtained chronopotentiometry, as
discussed in Chap. 8, also apply to the stripping mode. Certainly
the sensitivity and reproducibility are not competitive. Jagner and
Graneli [60] have considered an alternative approach which they refer
to as potentiometric stripping analysis. This technique consists of
an initial reduction of metal ions at a hmde or mtfe. The amalga-
mated metals are then oxidized with mercury(II) ions and the time-
potential behavior of the mercury electrode is monitored. Figure 9.9
shows some of the curves and data for the t-E curves as well as a
comparison with dc and differential pulse methods. Despite an extensive list of claimed advantages provided by the authors, no practical demonstration that these can be realized is available. The
shape of the curves, all other things being equal, would alone mitigate against preferential use of this method. Furthermore, it has
yet to be proved that the sensitivity of the technique is as good as
the differential pulse method. At the present stage therefore, all
that can be said is that this is an interesting and novel approach
requiring further assessment to determine whether it will be a valuable contribution to methodology in stripping voltammetry.

9.4 COMPARISON OF SOME
OF THE TECHNIQUES

The practical problems associated with anodic stripping voltammetry
are numerous. As stripping voltammetry is usually applied to concen-
traions in the 10^{-6} M to 10^{-10} M range, many of the problems are
related to difficulties inherent to any form of trace analysis,
namely, contamination of solutions, adsorption of materials onto
container walls, etc. For example, Petrie and Baier [61] have examined the principal lead(II) transport processes occurring in anodic
stripping voltammetric analysis of seawater. From an initial lead(II)
concentration of 6.3×10^{-9} M, the concentration after 95 min is
reduced by 84% at pH 8 to 2, 78% at pH 4 to 6, and 39% at pH 3 to 0.
Primary losses are reported to be due to adsorption onto the cell
and electrode materials; so there can be problems specifically associated with the electrochemistry (i.e., the presence of the electrodes). However, apart from emphasizing the well-recognized fact
that any form of trace analysis is difficult and that stripping
voltammetry is no exception, no direct further reference to the
practical aspects of the methodology itself is covered, because these
problems are not unique to any one particular form of anodic stripping voltammetry. In keeping with the aims of this book, considerable
emphasis is given in Sec. 9.4.6 to comparing the relative virtues
of the more widely used methods. Some comparisons in this direction
have already been offered in previous discussion, but only as passing
comments derived from theoretical considerations. A section concen-
trating on practical data is believed to be potentially more bene-
ficial. This task is facilitated by an article by Batley and
Florence [51] which comprehensively evaluates and compares most of
the important techniques of anodic stripping voltammetry from a
practical viewpoint. Consequently, most of the material cited here
is taken directly from this source and coupled with supplementary
material provided from a less extensive article along similar lines
[52].

The first variable likely to be considered by the analytical
chemist is the choice of electrode. Most authors concur that either
a rotating thin-film electrode (in situ mercury plated if possible)
or a hdme offer the best possibilities if they are applicable to
the situation in question. Remaining discussion will be confined
to these two kinds of electrode. The best stripping techniques will
almost always be selected from the controlled-potential ones (i.e.,
dc linear sweep, ac, or pulse methods, for example), but the choice
from among these is not always obvious.

Actual comparisons with mercury thin-film electrodes at a
glassy carbon electrode and the hmde lead to the following conclusions:

9.4.1 Preformed and In Situ-Deposited Films

Table 9.1 demonstrates that preformed mercury films give results of
inferior precision to those obtained using in situ-deposited mercury.
There is obviously no practical advantage in incorporating the additional step of preforming a mercury film on a substrate into the
analytical method; so subsequent data presented always refer to
mercury films deposited in situ.

9.4.2 Measurement Modes with
Mercury Thin Film Electrodes

The application of differential pulse waveforms to the mtfe with in
situ deposition showed that for the determination of lead and cadmium
in citrate buffer, slightly improved sensitivity could be gained,
compared with the dc linear scan method. This is not surprising in
view of the preceding discussion. However, the differential pulse
waveforms were found to be broader (poorer resolution) and were particularly susceptible to interference from traces of surfacants in
solution [51]. Although not included as a technique for comparison in
Notes 51 and 52, the staircase waveform may be preferable to the differential pulse one under these circumstances [48,49]. Alternating
current methods have also been found to be less suitable than dc ones
with this electrode [51,52]; so there appears to be little advantage

TABLE 9.1 Comparison of Preformed and In Situ Deposited Mercury Films on Glassy Carbon[a]

Film type	Scan number[b]	Average film thickness x 10^6 cm^{-1}	Peak height for Pb(μA)	Peak height for Cd(μA)	Peak width at half-wave height(mV) Pb	Cd
Preformed	1	1.3	18.3	13.7	--	--
	2	1.3	18.3	11.5	32	35
	3	1.3	17.3	10.3	--	--
	4	1.3	21.3	10.3	34	35
	1	11.0	22.0	20.0	--	--
	2	11.0	21.0[c]	20.0	36	35
	3	11.0	20.5	20.3	--	--
	4	11.0	21.5	20.8	35	36
In situ	1	1.3[d]	20.3	16.7	--	--
	2	2.6	20.0	18.5	34	35
	3	3.9	20.7[e]	20.3	--	--
	4	5.2	19.7	19.7	35	35
	1	11.0	21.5	20.7	--	--
	2	22.0	21.5	20.7	34	35
	3	33.0	21.5	20.9	--	--
	4	44.0	21.5	20.7	34	34

[a]1×10^{-7} M Cd^{2+} and Pb^{2+} in 2 M HCl.
[b]Successive scans on same solution at 83 mV sec^{-1} after 5-min depositions.
[c]Results from the analysis of six aliquots (same film) gave a relative standard deviation of 6.5%.
[d]Corresponding to 5×10^{-5} M Hg^{2+} in analytical solution.
[e]Results from the analysis of six aliquots gave a relative standard deviation of 2.5%.
Data taken from Note 51.

in using the instrumentally more complicated techniques with this electrode. (See also Sec. 9.4.4.)

9.4.3 Measurement Modes with
       Hanging Mercury Drop Electrode

Unlike the case with the mfte, dc linear scan methods at the hmde do not have as good resolution or sensitivity as derivative, pulse, or ac methods, at a hmde. Tables 9.2 and 9.3 and Fig. 9.4 give some data for a range of stripping techniques at a hmde which demonstrate that a non-dc linear sweep stripping technique is clearly to be preferred with this electrode. For the determination of a reversible

TABLE 9.2 Limits of Detection Found at a HMDE for Lead(II) and Cadmium(II) in 0.5 M KNO_3[a]

Technique	Limit of detection		Comments
	Pb(II) (M)	Cd(II) (M)	
DC linear sweep	2×10^{-8}	2×10^{-8}	Scan rate 50 mV sec^{-1}
Normal pulse	5×10^{-9}	5×10^{-9}	Nonlinear base line at low concentration in stripping work restricted limit of detection
Pseudoderivative pulse	4×10^{-8}	4×10^{-8}	Signal processing involving taking a difference introduces more noise into signal, thereby limiting sensitivity. Sloping base line, as in normal pulse stripping, also limited sensitivity
Differential pulse	1×10^{-9}	1×10^{-9}	Amplitude 50 mV
Fundamental harmonic (total)	1×10^{-8}	1×10^{-8}	10 mV at 50 Hz
Fundamental harmonic (phase selective)	2×10^{-9}	2×10^{-9}	In-phase component, 10 mV p-p at 80 Hz
Second harmonic (phase selective)	2×10^{-9}	2×10^{-9}	In-phase component, 10 mV p-p at 400 Hz. Magnitude of current small and noise limited sensitivity. Larger amplitude of alternating potential should improve detection limit

[a]Data taken from Note 52. Six minutes of electrolysis used with 5 min of stirring and a 1-min equilibration period.

TABLE 9.3 Comparison of Resolution in Anodic Stripping Voltammetry at a HMDE[a,b]

Technique	More positive process overlaps more negative ($\Delta E_{1/2}$)n(mV)	More negative process overlaps more positive ($\Delta E_{1/2}$)n(mV)
DC linear sweep	168	1840,[c] 1780[d]
First derivative dc linear sweep	148	857,[c] 837[d]
Second derivative dc linear sweep	188, 143[e]	275
Normal pulse voltammetry	236	--
Fundamental harmonic ac, derivative pulse, differential pulse (small amplitude) voltammetry	154	154
Second harmonic ac voltammetry	144[e]	144[e]

[a]DC data are taken from literature cited in Note 52, which should be consulted for further details.
[b]Both electrode processes are reversible reductions, and n1 = n2 = n in the example given. Equal concentrations of both species are present. $\Delta E_{1/2}$ is the minimum separation of $E_{1/2}$ or E_p required for 1% overlap in the two-component system.
[c]Planar diffusion.
[d]Spherical diffusion.
[e]First peak used for analysis.

electrode process such as that of cadmium or lead in 0.5 M KNO_3 considered previously (see Table 9.2), the limits of detection and
reproducibility with phase-selective ac, normal pulse, and differential pulse methods leave little to choose between them. Further, all
give linear calibration curves over wide concentration ranges. The
resolution of the normal pulse method counts against this technique
in multicomponent analysis. In general, the sigmoidal- rather than
peak-shaped curve of the readout of this technique places it at a
disadvantage, as does the asymmetric peak shape of the dc linear
sweep method.

The choice of either an ac or pulse method will usually be
decided by the considerable difference between their behavior toward
the reversibility of the electrode process, as is the case in polarography. AC methods, particularly the second harmonic version, are
remarkably sensitive to departures from reversibility. Thus, the
determination of a particular species giving rise to a reversible
electrode process, in the presence of a complex mixture, is much
more likely to be possible by ac techniques. However, to counter-
balance this, many more of the entities in the complex mixture are
likely to be determinable at the trace level by pulse voltammetry
if the required resolution can be achieved. The complementary
rather than competitive nature of the two stripping electroanalytical
methods at a hmde is therefore seen.

9.4.4 Relative Sensitivities and Limits of Detection at MTFE and HMDE

Both the mtfe and hmde electrodes used in the anodic stripping format have sensitivities which are more than adequate for analysis of
environmental samples. The limit of detection is almost always
governed by the magnitude of the blank value and not by instrumental
sensitivity. Table 9.4 compares the sensitivity of the mtfe and
hdme electrodes using the arbitrary definitions of limit of detection
as three times the standard deviation of the noise level adjacent to
the particular wave considered. The longer deposition times account
for the sensitivity greater than that of the data given in Table 9.2.

TABLE 9.4 Limits of Detection for Cd(II) and Pb(II) in 0.1 M HCl at
the MTFT and HMDE

	Electrode	Stripping method	Limit of[a] detection from noise level(M)	Limit of[b] detection from blank(M)
Cd	MTFE[c]	DC	5×10^{-11}	3×10^{-10}
	HMDE[d]	Differential pulse	5×10^{-11}	6×10^{-10}
	HMDE[d]	Fundamental ac	1×10^{-10}	1.0×10^{-9}
	HMDE[d]	Second harmonic ac	1×10^{-10}	1.2×10^{-9}
Pb	MTFE[c]	DC	5×10^{-11}	6×10^{-10}
	HMDE[d]	Differential pulse	5×10^{-11}	9×10^{-10}
	HMDE[d]	Fundamental ac	1×10^{-10}	1.2×10^{-9}
	HMDE[d]	Second harmonic ac	1×10^{-10}	1.2×10^{-9}

[a]Based on three times the standard deviation of noise level of blank.
[b]Based on three times the standard deviation of blank.
[c]In situ mercury deposition, 15-min deposition, 83 mV sec^{-1} scan rate.
[d]Fifteen-minute deposition.
Data taken from Note 51.

Table 9.5, by providing some data for the determination of cadmium
and lead in a variety of samples, illustrates the remark that adequate sensitivity is obtained by all methods.

9.4.5 Resolution at MTFE and HMDE Electrodes

Data presented in Table 9.6 indicate the greater resolution obtainable at the mtfe compared with the hmde. However, because peak potentials can occur at different positions with the two methods
[Eqs. (9.7) and (9.10)], this is not the entire story. The differences between E_p and $E_{1/2}$ are much larger at the mtfe than at the
hmde, particularly for singly charged ions, and this can result in
different resolution capabilities unrelated to the half width. A
case in point is that of Tl(I) and Cd(II) which in 0.1 M HCl are
unresolved at the mtfe but resolved at the hmde, whereas Tl(I) and
Pb(II) are poorly resolved at this electrode and well separated
at the mtfe.

TABLE 9.5 Analysis of Environmental Samples by Anodic Stripping Voltammetry

Sample	Metal	HMDE[a] Differential pulse	HMDE[a] Fundamental ac	HMDE[a] Second harmonic ac	MTFE[a] dc
0.1 M HCl Electrolyte	$[Cd] \times 10^9$/M	1.6 ± 0.2	1.8 ± 0.3	2.8 ± 0.4	1.28 ± 0.10
	$[Pb] \times 10^9$/M	1.8 ± 0.3	2.0 ± 0.4	2.2 ± 0.4	2.04 ± 0.23
Oyster ash Sample 1[b]	[Cd](ppm in ash)	14.4 ± 0.5	13.6 ± 0.5	15.6 ± 0.9	15.6 ± 1.0
	[Pb](ppm in ash)	8.3 ± 0.7	8.6 ± 1.0	9.3 ± 0.9	10.2 ± 0.9
Oyster ash Sample 2[c]	[Cd](ppm in ash)	36.0 ± 1.5	38.1 ± 3.0	39.8 ± 2.6	43.7 ± 4.6
	[Pb] ppm in ash)	18.6 ± 0.7	20.8 ± 1.3	20.5 ± 1.5	27.0 ± 1.6
NBS Orchard[d] Leaves	[Cd](ppm dry wt.)	0.124 ± 0.011	0.108 ± 0.020	0.114 ± 0.011	0.131 ± 0.009
	[Pb](ppm dry wt.)	42.6 ± 1.1	43.0 ± 1.3	46.3 ± 1.4	45.7 ± 1.0
Seawater	[Cd](μg l^{-1})	0.39 ± 0.11	0.55 ± 0.30	0.69 ± 0.31	0.22 ± 0.06
	[Pb](μg l^{-1})	0.64 ± 0.15	0.98 ± 0.20	0.85 ± 0.51	0.93 ± 0.07
ZnSO$_4$	[Cd](μg ml^{-1})	219 ± 6	214 ± 7	236 ± 10	220 ± 15[e]
	[Pb](μg ml^{-1})	16 ± 1	Unresolvable[e]	Unresolvable[e]	28 ± 3

[a] Results are mean and standard deviation of six determinations.
[b] Ash dissolved in HCl solution.
[c] Ash fumed with HClO$_4$ + HNO$_3$.
[d] NBS certificate values: Cd = 0.11 ± 0.02 ppm, Pb = 45 ± 3 ppm.
[e] Tl interferes.
Data taken from Note 51.

TABLE 9.6 Peak Widths at Half-Wave Height $W_{1/2}$ for Different ASV Techniques

Metal ion in 0.1 M HCl	MTFE $W_{1/2}$(mV)	HMDE		
		Differential pulse $W_{1/2}$(mV)	Fundamental ac $W_{1/2}$(mV)	Second harmonic ac $W_{1/2}$(mV)
Tl(I)	74	86	94	84
Pb(II)	34	44	56	44
In(III)	25	33	44	38

Data taken from Note 49.

9.4.6 Practical Examples of the HMDE and MTFE Techniques

A series of comparative stripping studies with many of the electro-analytical methods described above, based on both thin-film mercury electrodes and hmde's has been undertaken in the author's laboratories. Brief mention of the results of one such study using dc, ac, and pulse methods illustrates the complementary nature of the various approaches.

Zinc sulfate electrolyte is used for the electrolytic production of zinc, and the presence of trace impurities can have considerable influence on the efficiency of the industrial process [62]. Samples consisted of an almost saturated solution of zinc sulfate, and Cd, Tl, Cu, and Sb were to be determined. Samples spiked with known concentrations were also provided, and the concentration range to be covered was approximately 5×10^{-8} to 10^{-5} M. Thus, sensitivity was not a limiting factor with any of the non-dc methods. Cd, Cu, and Tl could be determined in principle directly in the electrolyte. At the hmde, cadmium could be determined equally well by ac or differential pulse methods. For copper, a sloping base line was found in the ac methods, but not with the differential pulse method, which mitigated against the former approach. For thallium, the resolution from cadmium was superior with ac methods, and this provided the preferred approach. Even then, resolution was restricted to favorable concentration ratios, and addition of EDTA was frequently required to separate

the waves. For antimony, the electrolyte was diluted (1:1) with
concentrated hydrochloric acid. Excellent pulse or ac stripping
waves were obtained, and either method could be used.

Thin-film studies were also undertaken at wax-impregnated and
glassy carbon electrodes using both in situ mercury deposition and
preformed electrodeposited mercury films. The dc method itself gave
a peak-shaped curve and extremely high sensitivity. Contrary to
stripping studies at the hmde, little or no advantage was found in

TABLE 9.7 Comparison of the HMDE Using Differential Pulse vs. MTFE
with In Situ Mercury Deposition and Rapid DC Scan

Advantages of hmde	Advantages of mtfe
Control of solution chemistry is much simpler. In general, peak height is less dependent on composition of supporting electrolyte.	Inherently more sensitive, although blank value normally prevents use of full sensitivity.
Interferences are less as a result of greater volume of hanging drop which leads to less interelement compound formation	Results are more reproducible at very low concentrations. Synchronous rotation of electrode provides more reproducible stirring of solution than a magnetic stirring bar.
Multiple voltage scans not required.	Better resolution. Peak widths at half-wave height are smaller, and there is no tailing on anodic side of wave.
Voltammetric cell can be sealed from air more readily than one using a rotating electrode.	Simpler electronic instrumentation. Electrode not affected by vibration. No refilling of mercury electrode and no mercury metal health hazard. Eliminates skill in reproducing size of hanging mercury drop.
Results are generally more reproducible at higher concentrations.	No contamination problems from electrode materials or mercury.

Taken from Note 51.

using non-dc methods. However, in the zinc sulfate electrolyte at the higher concentrations, nonlinear calibration curves and intermetallic interferences were observed. Furthermore, antimony could not be determined satisfactorily. The electrode process used is Sb(III) $\rightleftharpoons$ Sb(amalgam); antimony exists predominantly in the pentavalent oxidation state in the electrolyte, and apparently the reduction to the trivalent state does not proceed readily at thin-film electrodes. Thus, despite the inherently higher sensitivity and greater resolution of thin-film techniques, the hmde was preferred in the above-mentioned example. Generally, the hmde seems to have wider applicability, but the two methods are by no means equivalent and, in a given situation, both might need to be examined.

Batley and Florence [51] have also compared the two methods on a number of samples, and it is abundantly clear that both electrode types may need to be considered. They conclude from their findings that, although a mtfe with in situ mercury deposition gave the most sensitive and precise results for many samples, the hmde method in the differential pulse mode was the most versatile and reliable method investigated. The conclusion is entirely in agreement with the author's experiences. Table 9.7 summarizes some of the comparative data on the two-electrode types provided by Batley and Florence.

To conclude this chapter, a highly selective list of additional references is cited [63-80], which contains reference to the use of stripping voltammetry in its various modern forms. Perusal of these by the interested reader may give at least an intuitive feeling of possible virtues and disadvantages of each of the methods in different situations.

NOTES

1. R. D. De Mars and I. Shain, Anal. Chem. *29*, 1825 (1957).

2. E. S. Jacobs, Anal. Chem. *35*, 2112 (1963).

3. S. P. Perone and J. R. Birk, Anal. Chem. *37*, 9 (1965).

4. U. Eisner and H. B. Mark, Jr., J. Electroanal. Chem. *24*, 345 (1970).

5. M. D. Booth and B. Fleet, Anal. Chem. *42*, 825 (1970).

6. G. Colovos, G. S. Wilson, and J. L. Mogers, Anal. Chem. *46*, 1051 (1974).

7. T. Miwa, Y. Fujii, and A. Mizuike, Anal. Chim. Acta *60*, 475 (1972).

8. C. C. Young and H. A. Laitinen, Anal. Chem. *44*, 457 (1972).

9. E. Barendrecht, in *Electroanalytical Chemistry* (A. J. Bard, ed.) Dekker, New York, 1967, vol. 2, pp. 53–109.

10. R. Neeb, *Inverse Polarographie and Voltammetrie*, Verlag Chemie, Weinheim Bergstr., 1969.

11. T. R. Copeland and R. K. Skogerboe, Anal. Chem. *46*, 1257A (1974).

12. W. Kemula, Pure Appl. Chem. *15*, 283 (1967).

13. W. D. Ellis, J. Chem. Educ. *50*, A131 (1973).

14. Kh. Z. Brainina, *Stripping Voltammetry in Chemical Analysis*, Wiley, Chichester, Sussex, 1975.

15. Kh. Z. Brainina, Talanta *18*, 513 (1971).

16. N. Velghe and A. Claeys, J. Electroanal. Chem. *35*, 229 (1972).

17. W. L. Underkofler and I. Shain, Anal. Chem. *33*, 1966 (1961).

18. W. L. Underkofler and I. Shain, Anal. Chem. *37*, 218 (1965).

19. M. S. Krause, Jr., and L. Ramaley, Anal. Chem. *41*, 1365 (1969).

20. S. P. Perone and J. R. Birk, Anal. Chem. *37*, 9 (1965).

21. G. D. Christian, J. Electroanal. Chem. *23*, 1 (1969).

22. H. Siegerman and G. O'Dom, Amer. Lab. *4*(6), 59 (1972).

23. W. T. deVries, J. Electroanal. Chem. *9*, 448 (1965).

24. W. R. Matson, D. K. Roe, and D. E. Carritt, Anal. Chem. *37*, 1595 (1965).

25. T. M. Florence, J. Electroanal. Chem. *27*, 273 (1970).

26. R. N. Adams, Anal. Chem. *30*, 1576 (1958).

27. J. R. Covington and R. J. Lacoste, Anal. Chem. *37*, 421 (1965).

28. T. R. Copeland, J. H. Christie, R. A. Osteryoung, and R. K. Skogerboe, Anal. Chem. *45*, 2171 (1973).

29. R. A. Osteryoung and J. H. Christie, Anal. Chem. *46*, 351 (1974).

30. S. P. Perone and H. E. Stapelfeldt, Anal. Chem. *38*, 796 (1966).

31. S. P. Perone, Anal. Chem. *35*, 2091 (1963).

32. E. S. Jacobs, Anal. Chem. *35*, 2112 (1963).

33. S. P. Perone and W. J. Kretlow, Anal. Chem. *37*, 968 (1965).

34. G. Raspi and F. Malatesta, J. Electroanal. Chem. *27*, 283 (1970); *27*, 295 (1970), and references cited therein.

35. L. Luong and F. Vydra, Collect. Czech. Chem. Commun. *40*, 1490 (1975).

36. F. Vydra, M. Stulikova, and P. Petak, J. Electroanal. Chem. *40*, 99 (1972).

37. W. R. Seitz, R. Jones, L. Klatt, and W. D. Mason, Anal. Chem. *45*, 840 (1973).

38. G. W. Tindall and S. Bruckenstein, Anal. Chem. *40*, 1637 (1968).

39. D. T. Napp, D. C. Johnson, and S. Bruckenstein, Anal. Chem. *39*, 48 (1967).

40. C. Fairless and A. J. Bard, Anal. Lett. *5*, 433 (1972).

41. C. Fairless and A. J. Bard, Anal. Chem. *45*, 2289 (1973).

42. W. H. Reinmuth, Anal. Chem. *33*, 185 (1961).

43. F. H. Beyerlein and R. S. Nicholson, Anal. Chem. *44*, 1647 (1972).

44. A. M. Bond, R. J. O'Halloran, I. Ruzic, and D. E. Smith, Anal. Chem. *50*, 216 (1978).

45. W. T. deVries and E. van Dalen, J. Electroanal. Chem. *14*, 315 (1967).

46. D. K. Roe and G. E. Toni, Anal. Chem. *37*, 1503 (1965).

47. T. R. Copeland, J. H. Christie, R. K. Skogerboe, and R. A. Osteryoung, Anal. Chem. *45*, 995 (1973).

48. J. H. Christie and R. A. Osteryoung, Anal. Chem. *48*, 869 (1976).

49. U. Eisner, J. A. Turner, and R. A. Osteryoung, Anal. Chem. *48*, 1608 (1976).

50. A. M. Bond, T. A. O'Donnell, and R. J. Taylor, Anal. Chem. *46*, 1063 (1974).

51. G. E. Batley and T. M. Florence, J. Electroanal. Chem. *55*, 23 (1974).

52. A. M. Bond, Anal. Chim. Acta *74*, 163 (1975).

53. M. M. Nicholson, J. Amer. Chem. Soc. *79*, 7 (1957).

54. Kh. Z. Brainina, Sov. Electrochem. *2*, 298 (1966).

55. W. Lund and B. V. Larson, Anal. Chim. Acta *70*, 299 (1974).

56. W. Lund and B. V. Larson, Anal. Chim. Acta *72*, 57 (1974).

57. W. Lund, B. V. Larson, and N. Gunderson, Anal. Chim. Acta *81*, 319 (1976).

58. F. Vydra and L. Luong, J. Electroanal. Chem. *54*, 447 (1974).

59. L. Luong and F. Vydra, Collect. Czech. Chem. Commun. *40*, 1490 (1975).

60. D. Jagner and A. Graneli, Anal. Chim. Acta *83*, 19 (1976).

61. L. M. Petrie and R. W. Baier, Anal. Chim. Acta *82*, 255 (1976).

62. E. S. Pilkington, C. H. Weeks, and A. M. Bond, Anal. Chem. *48*, 1665 (1976), and references cited therein.

63. S. Bubic and M. Branica, Thalassia Jugoslavica *9*, 47 (1973).

64. J. T. Kinard and R. C. Propst, Anal. Chem. *46*, 1106 (1974).

65. H. A. Laitinen and N. H. Watkins, Anal. Chem. *47*, 1352 (1975).

66. J. A. Cox and K. H. Cheng, Anal. Lett. 7, 659 (1974).

67. R. G. Clem, Anal. Chem. *47*, 1778 (1975).

68. R. W. Andrews and D. C. Johnson, Anal. Chem. *47*, 294 (1975).

69. J. P. Roux, O. Vittori, and M. Porthault, Analusus *3*, 411 (1975).

70. E. Ladanyi, U. D. N. Radulescu, and M. Gavan, J. Electroanal. Chem. *24*, 91 (1970).

71. T. M. Florence, J. Electroanal. Chem. *49*, 255 (1974).

72. T. M. Florence, G. E. Batley, and Y. J. Farrar, J. Electroanal. Chem. *56*, 301 (1974).

73. E. D. Moorhead and P. H. Davis, Anal. Chem. *45*, 2178 (1973).

74. L. Duic, S. Szechter, and S. Srinivasan, J. Electroanal. Chem. *41*, 89 (1973).

75. D. I. Levit, Anal. Chem. *45*, 1291 (1973).

76. R. Naumann, Z. Anal. Chem. *270*, 114 (1974).

77. H. Blutstein and A. M. Bond, Anal. Chem. *46*, 1531 (1974).

78. H. Blutstein and A. M. Bond, Anal. Chem. *48*, 759 (1976).

79. E. D. Moorhead and G. A. Forsberg, Anal. Chem. *48*, 751 (1976).

80. T. P. DeAngelis and W. R. Heineman, Anal. Chem. *48*, 2263 (1976).

Chapter 10

COMPUTERS AND DIGITAL DATA
PROCESSING IN POLAROGRAPHY

10.1 MICROPROCESSOR- AND
MINICOMPUTER-CONTROLLED POLAROGRAPHS

Examination of the recent literature on virtually all polarographic
techniques reveals that the use of the laboratory computer in polarographic analysis is becoming extremely common. Advances in electrochemical instrumentation are now closely following the state of the
art in electronic components, and the use of digital circuitry to
perform many functions previously undertaken in the analog format is
becoming widespread. Clearly the most significant advance is the
recent development of the integrated circuit microprocessor, now
widely available at low cost and being incorporated into commercially
available instrumentation. Along with low-cost integrated-circuit
memories and digital-to-analog (D/A) and analog-to-digital (A/D)
converters, the microprocessor makes possible the design of inexpensive instruments which are capable of closed loop control of data
acquisition and reduction. That is, all facets of the experiment
(for example, setting of scan rate, drop time, pulse height, potential increment, measurement of current or peak height, and calculation
of concentration) are performed under computer control and without
operator intervention. Figure 10.1 shows a "smart instrument" which
uses microprocessor control of an analog potentiostat to perform
differential pulse polarography, anodic stripping voltammetry, and

FIGURE 10.1 A microprocessor-controlled polarograph. [Reproduced by courtesy of Princeton Applied Research Corporation (PAR Model 374 Polarographic Analyzer).]

a range of other techniques. Data manipulation procedures, such as rejection of data obtained from a bad drop, averaging of replicate measurements, calculation of peak height and position, background subtraction, and rescaling of the i-E curve, are also performed under microprocessor control. Some of these features are demonstrated in Figs. 10.2 to 10.4. Dessy and co-workers [1-3] have discussed micro-processors from the user's point of view in several lucid articles, and their prognosis [3] is as follows:

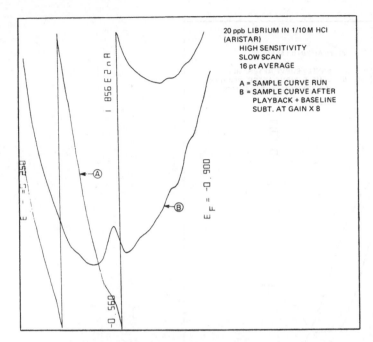

FIGURE 10.2 Microprocessor-controlled polarographs and other "smart"
instruments can average, autorange, and manipulate digital data to
provide optimum presentation of curves. (Some manipulations on the
differential pulse polarogram of librium are shown.) [Reproduced
by courtesy of Princeton Applied Research Corporation (PAR Model 374
Polarographic Analyzer).]

> There is little doubt that microprocessors are going to change
> the way instruments in research laboratories and analytical
> service areas are designed and operated and how they will
> interact with their operators. Within a few years most new
> equipment will be using microprocessors to acquire analytical
> data, perform small manipulations on the data base, and report
> the results.... The promise of complete instrument self-
> calibration and optimization and the potential to change instru-
> ment function drastically by simple changes in the operating
> program has been voiced.

From reading the above, one may be inclined to view micro-

processors as the focal point of an entirely new capability, but

this is far from the truth. Forms of polarographic instrumentation

based on minicomputers and related forms of digital electronics,

which are capable of performing far more complex operations than

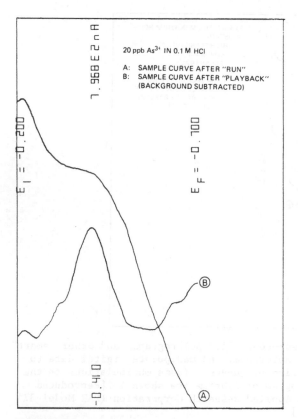

FIGURE 10.3 Background correction is simple when data can be stored digitally. Polarogram of electrolyte is recorded and data are stored in memory. The result is then subtracted from polarogram of species to be determined. Example shown is the determination of arsenic(III) in 0.1 M HCl by differential pulse polarography, using a PAR Model 374 Polarographic Analyzer. With analog equipment, the same kind of experiment would require matched dme's and cells and could be achieved only with great difficulty. (see Chap. 3). (Reproduced by courtesy of Princeton Applied Research Corporation.)

microprocessors, have been in use for many years now. In reality, the microprocessor is only making it possible to perform the same tasks in an electronically different way at a price a manufacturer may consider economically viable for commercial production of the instrumentation. That is, from the manufacturer's point of view, microprocessors have excited much interest, almost solely because of the prospect of minicomputer capabilities at far lower cost.

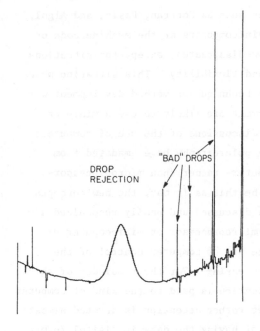

FIGURE 10.4 By developing suitable software, digitally stored raw data can be manipulated in many ways before the final readout is displayed. The above curve may be produced void of bad drops, for example. Such a facility is available on the microprocessor-controlled PAR Model 374 Polarographic Analyzer. (Reproduced by courtesy of Princeton Applied Research Corporation.)

However, most analytical chemists at present are limited in the extent of developmental work that can be undertaken with these devices.

To perform any task, microprocessors must be programmed in a digital environment, using mathematical and logical operations which at the moment must be written in assembly or machine language. Software or program development at this level is tedious, time consuming, and expensive. Analytical application of microprocessor-controlled polarographs in the next few years at least is therefore likely to be dominated by dedicated tasks capable of being undertaken with instrumentation produced by manufacturers of commercially available instrumentation, and much of the research and development work will continue to be based on laboratory minicomputer systems.

Minicomputer systems, in contrast to microprocessor ones, have an electronic architecture which allows them to run sophisticated software systems under the control of "executive" programs. These programs allow the programmer to instruct the minicomputer to perform complex mathematical operations with single-line commands using a variety of high-level languages such as Fortran, Basic, and Algol, and the need for programming minicomputers at the machine code or assembly language level has now disappeared, except for situations requiring considerable speed and flexibility. This situation means that most chemists undertaking technique or method development work necessitating software development are likely to use a mini- or laboratory computer, and most discussions of the use of computers and digital data processing in polarography have emanated from results obtained from minicomputer- rather than microprocessor-controlled systems. However, be this as it may, the new concepts and approaches that need to be discussed and easily recognized are arising not from the use of a microprocessor or minicomputer or equivalent system, but from the use of computer control of the experiment and digital data processing. For the remainder of this chapter, therefore, little attention is paid to the kind of computer used in the work described, but rather attention is devoted almost exclusively to the advantages of having the data in digital format compared with the analog methods on which most data and conclusions presented in earlier chapters are derived.

10.2 THE COMPUTER-CONTROLLED POLAROGRAPH

Figure 10.5 shows one of the simplest representations of computer arrangements that might be used. The potential is applied via the computer to a conventional analog potentiostat. Obviously, the

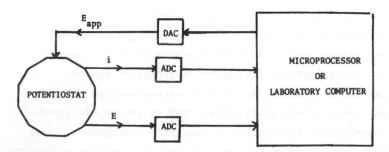

FIGURE 10.5 Schematic diagram of a very simple computerized polarograph.

computer generates a digital voltage which cannot be accepted by the potentiostat; so it is transformed via a digital-to-analog converter to an analog voltage. The measurement or readout of the i-E curve in an analog instrument is made on an X-Y recorder or oscilloscope. With the computer system, the E value and i values, for example, are converted from the analog form provided at the output of the potentiostat to digital data via analog-to-digital converters, and these digitalized signals are fed directly into the computer. With the analog system, once the i-E curve has been recorded, the automatic part of the experiment is finished, and further data manipulations are usually tedious, since they must be done by hand; e.g., the measurement of $E_{1/2}$, i_d, and the correction for iR drop (if necessary) in a dc polarogram are not trivial tasks. Having the data in digital form in a computer is only the start of an automated system, and software instructions allow all kinds of calculations to be performed merely by entering an instruction from a keyboard of a teletype, for example. The results can be printed out, displayed on an oscilloscope, digital display device, or X-Y recorder, and extraordinary simplification, convenience, and time saving, as well as improved accuracy, are possible.

The variations on the preceding instrumentation are numerous and the ratio of digital to analog components in the experimental arrangement can vary enormously. At one extreme, every task may be performed in the usual analog manner, and only at the readout stage is the computer involved; i.e., the computer is really only a fancy recording device in this instrumental arrangement. In other kinds of instrumentation, even the potentiostat itself may be digital, at least in principle, and almost all of the experiment, including potential generation and its application to the potentiostat, can be carried out in the digital domain. At the very least, however, the system should be capable of initiating the experiment and controlling all procedures via an operator keyboard or equivalent input instruction, as well as accepting and storing the data produced by the experiment, to be considered as a computer-controlled experiment.

Let us consider a couple of examples from the literature of fully computerized systems to clearly demonstrate what can be involved in the incorporation and application of a computer in polarography. Bos [4] has described a fully computerized system for current-sampled dc polarography. Figure 10.6 shows the "computer polarograph" described in this paper; it can be seen that it is constructed at a fundamental level using the kind of basic principles and instrumentation likely to be understood in most modern analytical laboratories.

In carrying out the polarographic experiment under digital control, the computer described by Bos performs the following functions:

1. Maintains the potential of the dme versus the reference electrode at a set value by changing the potential of the

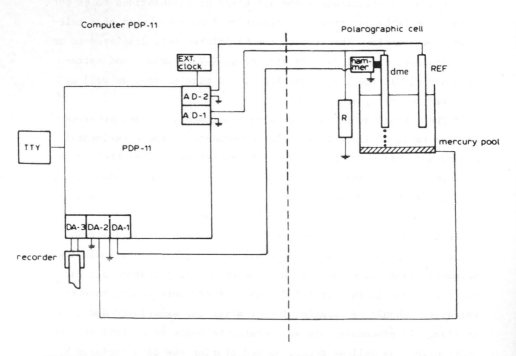

FIGURE 10.6 A computerized polarograph for performing current-sampled dc polarography. [Reproduced from Anal. Chim. Acta *81*, 21 (1976).

auxiliary electrode (mercury pool in Fig. 10.6)

2. Changes the set value of the potential of the dme linearly with time

3. Measures and stores the cell current

4. Synchronizes the measurement of cell current with the mercury drop life

5. Dislodges the mercury drop at fixed intervals

The operator is able to select, via an input instruction, the drop time, the initial potential, the scan direction (positive or negative), the final potential of the scan, the scan rate, the sensitivity of the current measurement, and the amplification of the ohmic cell-resistance compensation circuit. The operator can also display a recorded polarogram on an oscilloscope or on a chart recorder.

A flow chart of the computer program for controlling all these operations is shown in Fig. 10.7. The data processing program is shown in Fig. 10.8. Symbols used in these flow charts are listed in Table 10.1. This program for each wave on the dc polarographic curve produces (a) current-vs.-potential values for all points on the rising part of the wave and (b) values for the dc limiting current, $E_{1/2}$, and the slope of E-vs.-log $[(i_d - i)/i]$ plot. Base-line correction for the wave is done by extrapolating a "least squares" line drawn through 20 points, starting 40 points before the polarographic wave, from the initial potential side. The accuracy of this computerized current-sampled dc polarograph was ±2% for the diffusion current, ±2 mV for $E_{1/2}$, and ±2 mV for the slope of the log plot, all results comparing favorably with data obtained on a conventional analog instrument.

For some applications of polarography, far more sophisticated computer systems and programs may be required. Figure 10.9 provides a schematic diagram for instrumentation capable of performing on-line fast Fourier transform (FFT) data processing in electrochemistry. Use of the digital FFT as an on-line data processing strategy has found widespread acceptance in the areas of nuclear magnetic

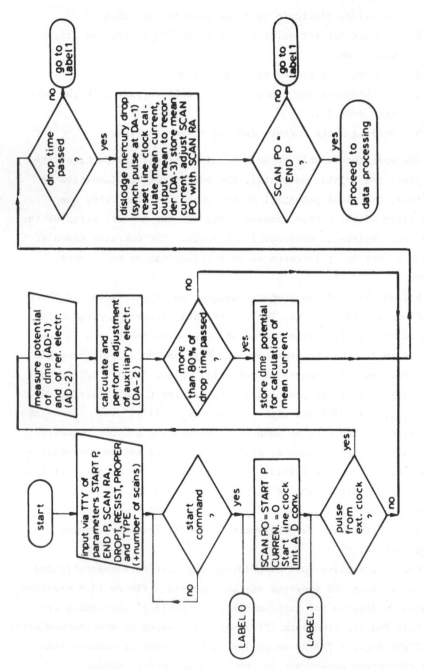

FIGURE 10.7 Flow diagram of a program for digital control of polarography. [See Table 10.1 for explanation of symbols.) [Reproduced from Anal. Chim. Acta 81, 21 (1976).]

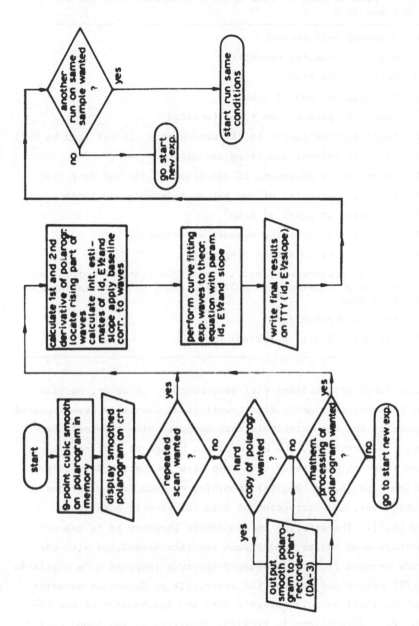

FIGURE 10.8 Flow diagram of program for processing current-sampled dc polarographic data. (See Table 10.1 for explanation of symbols.) [Reproduced from Anal. Chim. Acta *81*, 21 (1976).]

TABLE 10.1 Symbols Used in Flow Chart of Computer Programs in
Figs. 10.7 and 10.8

CURREN	Measured cell current
DA	Digital-to-analog converter
DROPT	Mercury drop time
ENDP	Potential of end of scan
NOPOL	Number of polarograms to be recorded
PROPOR	Amplification factor in iR compensation circuit (can be < 1)
RESIST	Value of current measuring resistor
SCANDRA	Increment or decrement of potential of dme per drop time
SCANPO	Value of potential of dme versus reference electrode
STARTP	Potential of start of scan
POTDME	Potential of dropping mercury electrode
POTREF	Potential of reference electrode
TYPE	Type of measurement, viz., P = single scan, V = single scan + data processing, S = multiple scan + data processing for each scan
$E_{1/2}$	Half-wave potential
i_d	Diffusion limiting current

resonance (nmr) and infrared (ir) spectroscopy. However, on-line
digital FFT applications in electroanalytical chemistry have appeared
only recently and are relatively rare compared with chemical spectroscopy. Nevertheless, the FFT approach brings to electrochemistry
the same inherent advantages that have been demonstrated frequently
in spectroscopy [5,6]. Digital smoothing of data is one obvious
application [6], and correction of data for kinetic nuances is
another [6,7]. The strategy in the latter instance is to measure
the heterogeneous and/or homogeneous kinetics associated with the
electrode process from the frequency-spectrum response made available
by the FFT method and compute the reversible or Nernstian behavior
so that the final ac polarographic data are independent of the kinetics. From discussions in previous chapters, it was found that
the reversible response is usually favored in the analytical use of

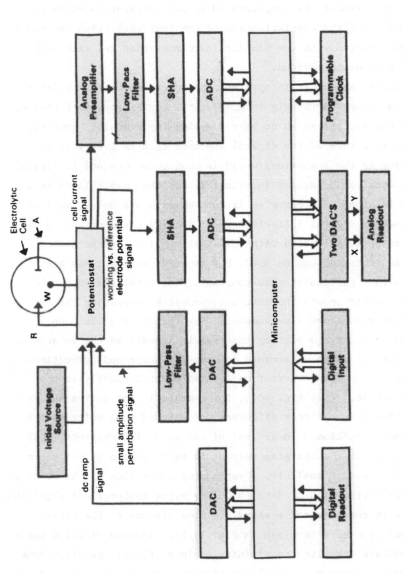

FIGURE 10.9 Schematic diagram of instrumentation for fast Fourier transform ac polaro-graphic and related small amplitude perturbation measurements: ⇒ = digital data lines; → = analog data or logic pulse lines; DAC = digital-to-analog converter; ADC = analog-to-digital converter; SHA = sample-and-hold amplifier; R = reference electrode; A = auxiliary electrode; W = working electrode. [Reproduced from Anal. Chem. 48, 221A (1976). American Chemical Society.]

modern polarographic techniques, and considerable advantage is gained
in this way. A flow diagram of a program for making electrochemical
kinetic measurements by computerized ac polarography is shown in
Fig. 10.10, and the sophistication of the instrumentation and capabilities compared with the simpler instrumentation and tasks considered earlier is obvious.

From the above, it is apparent that the computerized polarographs can range enormously in complexity, capability, performance,
and, of course, price, as do purely analog instruments. However,
the unique feature of the digital approach that is likely to be
attractive in the analytical world is that once obtained in digital
format, data can be stored indefinitely and then manipulated in any
desired fashion. For example, if the raw data are noisy, various
approaches of digital filtering can be applied consecutively until
the operator is satisfied with the final result. In the analog
world, RC filtering can be used, but once the data set has been obtained with a particular time constant, then nothing else can be
done to further smooth the data set, except by repeating the entire
experiment with a new time constant. By recording a blank on the
electrolyte alone and storing this result as well as the polarogram
of the species to be determined in memory, background corrections to
minimize the charging current may be implemented readily in the
digital format. (See Fig. 10.3, for example.) The equivalent analog situation is entirely different, as matched dual cells must be
used, and as mentioned in several of the preceding chapters, analog
compensation of the charging current in this fashion is not recommended for routine analysis. Overlapping waves also may be resolved
by simple subtraction of data [8] or by using mathematical algorithms
[9,10] with computerized systems. A flow diagram of the first-
mentioned procedure is given in Fig. 10.11. Instead of using positive feedback circuitry to minimize iR drop effects, as often done
with analog equipment, the alternative approach in the computerized
system is to compute the iR drop and subtract it from the raw data.
Computer-assisted optimization of modern polarographic techniques

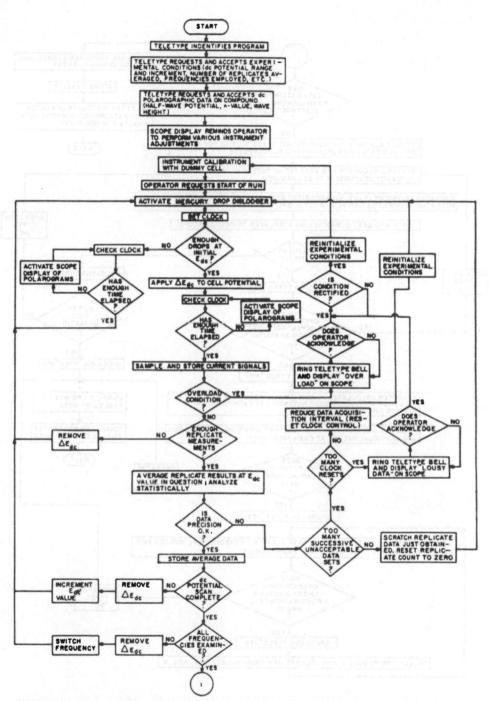

FIGURE 10.10a Flow diagram for electrochemical kinetic measurements
by computerized ac polarography: experimental initialization and
data acquisition stages. [Reproduced from Computers in Chemistry
and Instrumentation 2, 369 (1972).]

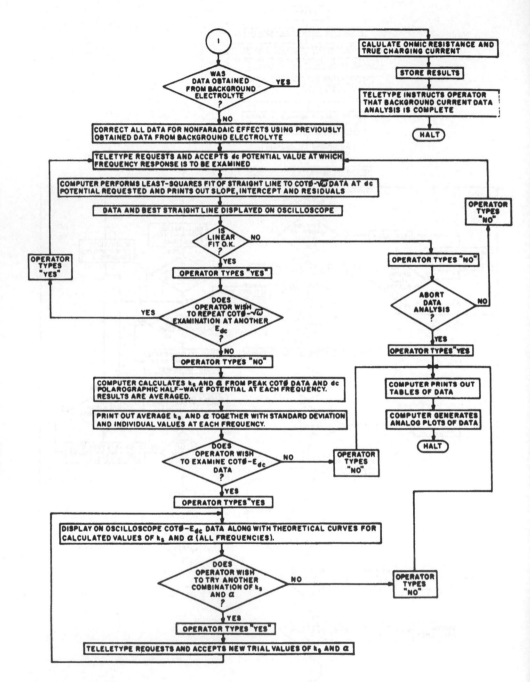

FIGURE 10.10b Flow diagram for electrochemical kinetic measurements
by computerized ac polarography: data analysis stage. [Reproduced
from Computers in Chemistry and Instrumentation 2, 369 (1972).]

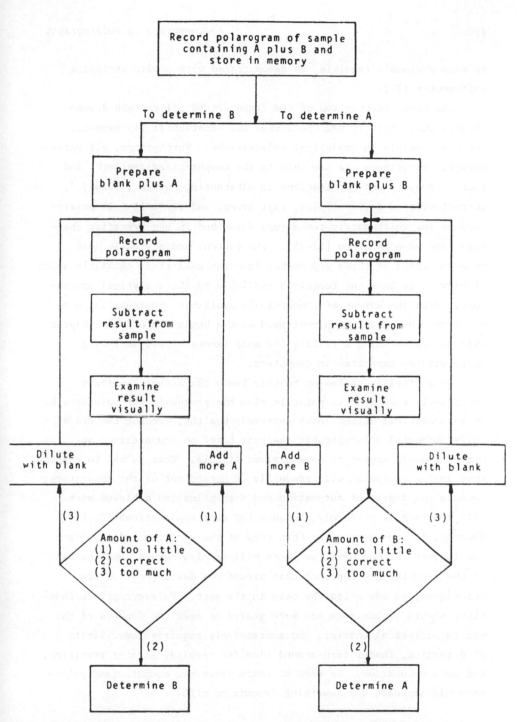

FIGURE 10.11   Flow diagram of a method for determining two species giving
rise to overlapping waves with a computerized polarograph.   [Reproduced
from Anal. Chem. *48*, 1624 (1976). © American Chemical Society.]

is also obviously feasible, as demonstrated with anodic stripping
voltammetry [11].

The above description of the computerized polarograph demonstrates that entirely new approaches and substantial improvement,
are now feasible in analytical polarography. Furthermore, all polarographic techniques are amenable to the computerized approach, and
indeed computerized applications in alternating current, pulse,
current-sampled dc, stripping, fast sweep, and virtually all polarographic and voltammetric techniques described in the preceding chapters have been reported [12-25]. The current and obviously most
relevant stage of these approaches from the analytical chemist's point
of view is to have the technique available in the analytical laboratory. With the advent of commercially available, inexpensive, computerized instrumentation mentioned at the beginning of this chapter,
this is now becoming a reality for many laboratories, including
those with no expertise in computers.

As a fitting conclusion to this book, the author can state
confidently that at this point in time the prognosis for polarography
as an analytical method looks extremely healthy, because the era of
fully automated polarographic analysis based on computerized approaches would appear to have arrived finally. This is not to suggest
that analog equipment will gradually be phased out of the laboratory,
because the degree of automation and sophistication achieved with
this approach is certainly adequate for many applications [26,27].
However, it does seem clear that many of the exciting and important
developments of the next few years will revolve more around the use
of the computerized approach than around the development of new
techniques, as was often the case in the past. Polarographic methodology should become more and more geared to meet the demands of the
modern analytical chemist, who increasingly requires lower limits
of detection, faster turn-around time for results, greater precision,
and more automation. In each of these areas the computerized polarographic approach has something dynamic to offer.

NOTES

1. R. E. Dessy, P. J. Van Vuuren, and J. A. Titus, Anal. Chem. *46*, 917A (1974).

2. R. E. Dessy, P. J. Van Vuuren, and J. A. Titus, Anal. Chem. *46*, 1055A (1974).

3. R. E. Dessy, Science *192*, 511 (1976).

4. M. Bos, Anal. Chim. Acta *81*, 21 (1976).

5. D. E. Smith, Anal. Chem. *48*, 221A (1976).

6. D. E. Smith, Anal. Chem. *48*, 517A (1976).

7. R. W. Schwall, A. M. Bond, and D. E. Smith, Anal. Chem. *49*, 1805 (1977).

8. A. M. Bond and B. S. Grabarić, Anal. Chem. *48*, 1624 (1976).

9. S. P. Perone and S. L. Sybrandt, Anal. Chem. *43*, 382 (1971).

10. S. P. Perone and W. F. Gutnecht, Anal. Chem. *42*, 906 (1970).

11. Q. V. Thomas, L. Kryger, and S. P. Perone, Anal. Chem. *48*, 761 (1976).

12. D. E. Smith, Computers in Chemistry and Instrumentation, *2*, 369 (1972).

13. R. A. Osteryoung, Computers in Chemistry and Instrumentation *2*, 353 (1972).

14. M. W. Overton, L. L. Alber, and D. E. Smith, Anal. Chem. *47*, 363A (1975).

15. L. Ramaley, Chem. Instrum. *2*, 415 (1970); *6*, 119 (1975).

16. J. Ogle, P. Tomsons, I. Tutani, and D. Stradins, Zavod. Lab. *36*, 1180 (1970).

17. L. Kryger, D. Jagner, and H. J. Skov, Anal. Chim. Acta *78*, 241 (1975).

18. L. Kryger and D. Jagner, Anal. Chim. Acta *78*, 251 (1975).

19. C. L. Pomernachi and J. E. Harrar, Anal. Chem. *47*, 1895 (1975).

20. G. Lauer and R. A. Osteryoung, Anal. Chem. *40*(10), 30A (1968).

21. T. Kugo, Y. Umezawa, and S. Fujiwara, Chem. Instrum. *2*, 189 (1969).

22. W. W. Goldsworthy and R. G. Clem, Anal. Chem. *43*, 918 (1971); *43*, 1718 (1971); *44*, 1360 (1972).

23. W. R. White, Anal. Lett. *5*, 875 (1972).

24. P. F. Seelig and H. N. Blount, Anal. Chem. *48*, 252 (1976).

25. H. E. Keller and R. A. Osteryoung, Anal. Chem. *43*, 342 (1971).

26. L. F. Cullen, M. P. Brindle, and G. J. Papariello, J. Pharm.
     Sci. *62*, 1708 (1973).

27. W. Lund and L. N. Opheim, Anal. Chim. Acta *79*, 35 (1975); *82*,
     245 (1976).

Author Index

Abdulah, M. I., 283(69), 286
Abel, R. H., 247(23), 285,
    425(107), 426(107),
    427(107), 428(107), 434
Adams, R. N., 41(29), 42(30),
    52(49), 80, 81, 172(1),
    176(1,22), 227(1), 228(1),
    231, 274(45,46), 283(75),
    285, 286, 287, 416(74),
    432, 443(26), 471
Adler, J. F., 176(24), 231
Agar, J. N., 227(102), 234
Agarwal, H. P., 362(68), 388
Airey, L., 157(71), 168
Alber, L. L., 491(14), 492
Alberg, W. J., 227(106,115),
    228(106,115), 235
Alden, J. R., 52(49), 81
Alexander, P. W., 111(37),
    112(37), 121
Allan, D. S., 66(87), 82
Anderson, D. P., 18(3), 19(3),
    64(3), 79, 256(30), 257(30),
    282(30), 285
Anderson, L. B., 414(65), 432
Anderson, R. E., 214(66), 233
Andrews, R. W., 470)68), 473
Andrieux, C. P., 216(70), 233
Anson, F. C., 192(37), 232
Anstine, W., 44(33), 80
Anstruc, M., 425(99,100,101,102,
    103,104,105), 427(105), 433
Athavale, V. T., 66(85,86), 82

Auerbach, C., 150(66,67), 168
    173(3), 231
Axt, A., 192(42), 232
Ayabe, Y., 11(12), 79, 105(27),
    191(27), 232

Bacon, J. R., 395(11), 430
Badinand, A., 376(88), 389
Baier, R. W., 459(61), 472
Bard, A. J., 227(114), 228(114),
    229(120), 235, 448(40,41),
    458(40,41), 472
Barendrecht, E., 438(9), 440(9),
    441(1), 470
Barker, G. C., 163(74), 168,
    173(2), 209(55), 231, 233,
    236(1,2), 243(4), 245(18),
    250(4), 280(4), 284, 290(2,
    3,6), 340(37), 341(37),
    362(67), 386, 387, 388,
    391(1,2,5), 393(5), 396(22,
    23,24,25,26), 397(26),
    399(5,31,32), 400(23,26),
    401(23), 403(31), 404(26,31),
    405(26), 424(31), 429, 430
Batley, G. E., 454(51), 459(51),
    460(51), 461(51), 465(51),
    466(51), 467(51), 469(51),
    470(51,72), 472, 473
Batrakov, V. V., 44(38), 80
Bauer, H. H., 20(9), 52(56),
    68(88,89), 69(88), 79, 81,

[Bauer, H. H.]
    82, 290(7), 296(7),
    306(17), 363(7), 367(78),
    364(80), 375(7,84), 386
    389(73), 389, 391(3), 429
Baxter, R. A., 161(73), 168
Belew, W. L., 52(65), 81,
    128(23), 143(53,54),
    144(53), 147(54), 148(54,
    56), 149(53,54,65),
    150(53), 151(53), 157(54,
    56), 166, 167, 168
Bellavance, M. I., 227(112),
    228(112), 235
Bennekom, W. P., 282(54),
    283(54), 286
Bentata, J. L., 425(100), 433
Beran, P., 227(116), 235
Berman, D. A., 124(10), 166
Bersier, P., 395(16), 430
Berzins, T., 414(56,57), 432
Besle, A., 140(39,42), 143(42),
    167
Bewick, A., 60(75), 82
Beyer, M. E., 353(56), 354(56),
    388
Beyerlein, F. H., 451(43), 472
Bezman, R., 52(62), 60(78), 81,
    82
Bichon, C., 376(88), 389
Bieler, R., 35(20), 80
Bird, B., 227(103), 234
Birk, J. R., 435(3), 441(20),
    470, 471
Biskupsy, V. S., 353(55),
    356(55), 388
Blanchard, R. J., 275(47), 286
Blažek, A., 40(25), 80
Blount, H. N., 490(24), 491
Blutstein, H., 270(40), 285
    341(39,44), 342(44),
    343(44), 347(44), 348(44),
    357(58), 362(64), 387, 388
    470(77,78), 473
Bockris, J. O'M., 20(8), 79
Bodya, V. I., 403(43), 431
Boehme, G. H., 225(99),
    226(99), 234
Boháčková, V., 20(7), 79
Bolzan, J. A., 245(18), 284
Bonastre, J., 425(99,100,101,
    102,103,104,105), 427(105),
    433

Bond, A. M., 37(22), 44(31),
    56(71), 80, 82, 101(13,14),
    103(13), 120, 123(9), 124(8,
    9,14,16), 125(16), 127(14),
    128(16), 129(14), 130(9,24),
    132(46), 134(16), 135(16),
    136(9,26), 137(9,26,27,28)
    140(25,45,46), 141(45),
    142(46), 152(46), 153(46),
    157(46), 165, 166, 167,
    261(32), 268(35), 270(38,40),
    273(43), 281(32), 285, 298(13),
    302(13), 304(13), 307(13,18,
    21), 308(13), 309(13),
    311(13), 312(13), 317(13),
    318(13), 319(13), 320(23),
    322(13), 323(13), 325(13),
    326(13), 327(13), 328(13),
    329(13), 330(23), 331(13),
    332(24,25,26,27), 333(29),
    337(29), 338(29), 339(13),
    341(39,44), 342(44), 343(44),
    347(44), 348(44,45,56),
    349(45,50), 351(52), 352(53,
    53), 353(29,53,54,55,56),
    354(54,56), 356(54,55),
    357(58,59), 362(64),
    375(29), 386, 387, 388
    396(27), 430, 451(44),
    454(50), 455(52), 459(52),
    460(52), 462(52), 463(52),
    468(62), 470(77,78), 472
    473, 487(8), 490(8), 492
Booman, G. L., 38(23), 39(23),
    52(51,52,60,61,68), 56(51,
    52), 59(51,60), 80, 81, 82
Booth, M. D., 367(79), 368(79),
    370(79), 371(79), 372(79),
    373(79), 389, 438(5), 470
Bordi, S., 173(7), 231
Bos, P., 409(53), 411(53), 431,
    481(4), 483(4), 484(4), 492
Bowning, G., Jr., 283(70), 287
Brabec, V., 283(74), 287
Brainina, Kh. Z., 438(14,15),
    455(14,15,54), 471, 472
Branica, M., 470(63), 472
Breukel, J. S. M. C., 290(4), 386
Breyer, B., 290(7), 296(7),
    363(7), 375(7), 386,
    391(3), 429
Brezina, M., 5(11), 7, 112(35),
    121

Bridicka, R., 35(19), 79, 107(19, 20,21,23), 109(19), 120
Brilmyer, G. H., 216(83), 217 (83), 234
Brindle, M. P., 491(26), 493
Brinkman, A. A. A. M., 244(7,8), 245(11,12), 246(8), 284
Britz, D., 374(80), 375(84), 389
Brocke, W. A., 362(66,69), 363 (70), 388
Brooks, M. A., 283(60), 286
Brown, E. R., 52(50,60,61), 56(50), 59(60), 81, 338(36), 340(36), 387
Bruckenstein, S., 227(106,110, 111,112), 228(106, 110, 111,112), 235, 448(38), 471
Bruk, B. S., 402(39), 403(39), 431
Bryskin, I. E., 402(40), 403(40), 431
Bubic, S., 470(63), 472
Buchanan, A. S., 44(31), 80, 124(16), 125(16), 128(16), 134(16), 135(16), 166
Buchanan, E. G., Jr., 395(11, 20), 430
Buckley, F., 126(19), 166
Budarin, L. I., 109(33), 120
Bull, G. C., 225(98), 234
Bull, R. H., 225(98), 234
Bullerwell, R. A. F., 111(37), 112(37), 121
Burangey, S. V., 65(85), 82
Burge, D. E., 240(42), 273(42), 285
Butler, J. A. V., 19(4), 20(4), 79

Cadle, S. H., 227(111), 228 (111), 235
Campbell, H. R., 367(78), 389
Canterford, D. R., 44(31), 80, 124(16), 125(16), 127(16), 128(16), 130(24), 134(16), 135(10), 140(25), 166, 247(23), 285, 320(23), 330(23), 332(27), 337(35), 387
Carr, P. W., 124(12), 166
Carritt, D. E., 442(24), 443 (24), 471

Casey, A. T., 101(13,14), 103(13), 120
Caton, R. D., Jr., 62(84), 82
Cermak, V., 176(15), 231
Cheng, K. H., 470(66), 472
Chernega, L. P., 403(43), 431
Chey, W. M., 283(75), 287
Chidester, D. H., 209(60), 210 (60), 212(60), 233
Christian, G. D., 441(21), 471
Christie, J. H., 212(58), 233, 237(3), 238(3), 245(20), 246(20), 251(25), 252(3), 258(3), 261(3), 262(3), 265(3), 266(3), 277(49,50), 278(49), 280(51), 281(51), 284, 285, 286, 446(28), 447(28), 453(28,47), 454(28, 47,48), 460(48), 471, 472
Chu-Xuan-Anh, 395(17), 430
Claeys, A., 439(16), 440(16), 471
Clem, R. G., 470(67), 473, 491(22), 492
Coetzec, J. F., 283(73), 287
Colovos, G., 438(6), 470
Connery, J. C., 123(7), 124(5,6,7, 11), 133(5,6), 165, 166
Connor, G. I., 225(99), 226(99), 234
Conway, B. E., 20(5), 79
Cooke, W. D., 176(20), 192(38), 231, 232
Copeland, T. R., 438(11), 440(11), 447(28), 453(28,47), 454(28, 47), 455(11), 471, 472
Costa, M., 173(7), 231
Courbusier, P., 173(10), 231
Cover, R. E., 123(7), 124(5,6,7, 11,13,15), 133(5,6,13,15), 134(13), 137(13,29), 138(13, 31,32), 139(31,32), 165, 166
Covington, J. R., 443(27), 471
Cox, J. A., 470(66), 472
Cox, P. L., 283(58), 286
Crow, D. R., 70(91), 82, 107(29), 120
Cullen, L. F., 491(26), 493

Damaskin, B. B., 44(38), 80
D'Arcoute, L. M., 283(60), 286
Date, Y., 101(10,11,12), 105(12), 120

Daum, P. H., 425(108), 429(108), 434

Davenport, R. J., 229(118), 235

Davis, D. G., 225(95), 234, 409 (52), 414(52), 431

Davis, H. M., 223(91,92), 234

Davis, J. A., 31(16), 32(16), 79

Davis, P. H., 470(73), 473

Dean, J. A., 149(65), 168

Deangelis, T. P., 470(80), 473

Deanhardt, M. L., 414(71), 432

Deford, D. D., 52(50), 56(50), 81

de Galan, L., 283(64), 286

Delahay, P., 24(10), 73(92), 79, 82, 107(27,28), 109(30), 120, 243(5), 284, 362(15), 388, 409(47,48), 414(56,57), 424(89,90,91,92,94,95,96), 431, 432, 433

de Levie, R., 375(83), 389

de Mars, R. D., 435(1), 470

del Rey, F., 425(103,104), 433

den Boef, G., 385(92), 389

deSilva, J. A. F., 283(60,61), 286

Dessy, R. E., 475(1,2,3), 492

Dévay, J., 60(76), 82, 357(60), 361(60), 388

Devries, W. T., 441(23), 442(45), 453(45), 471, 472

Dewolfs, R., 199(49), 201(49), 232

Dhaneshwar, M. R., 66(86), 82

Dhaneshwar, R. T., 65(85,86), 82

Diefenderfer, A. J., 375(81), 389

Dillard, J. W., 252(27), 285

Doskosil, J., 283(63), 286

Dracka, O., 414(58,59,60,62), 432

Drake, K. F., 270(38), 285

Dreiling, R., 274(46), 286

Duic, L., 470(74), 473

Dvořak, J., 20(7), 79

Eisner, V., 435(4), 447(4), 454(49), 460(49), 467(49), 470, 472

Elbel, A. W., 140(40,41), 143 (41), 167

Ellis, W. D., 438(13), 440(13) 444(13), 445(13), 446(13), 447(13), 471

Elving, P. J., 205(53), 233, 306(17), 386, 389(73), 389

Enke, C. G., 161(73), 168, 173(14), 174(14), 175(14), 231, 425 (107), 426(107), 427(107), 428(107), 434

Erkelem, C., 283(64), 286

Evans, D. H., 216(76,77), 218(76, 77), 233, 269(37), 270(37,39), 285

Everett, G. W., 409(49), 431

Evilia, R. F., 375(81), 389

Ewing, G. W., 46(39), 80

Faircloth, R. L., 290(2), 340(37), 341(37), 386, 399(32), 430

Fairless, C., 448(40,41), 458(40, 41), 472

Farrar, Y. J., 470(72), 473

Faulkener, L. R., 247(21), 285

Favero, P., 192(41), 232

Fekes, Zs., 229(117), 239(117), 235

Feldberg, S. W., 182(26), 183(26), 191(26), 231

Ferret, D. J., 393(8,9), 395(8,9), 430

Ferrier, D. R., 209(60), 210(60), 212(59,60), 225(98), 233, 234

Fick, A., 92(1), 119

Finston, H. L., 173(3), 231

Finston, H. O., 150(66,67), 168

Fischer, O., 414(58,59,61), 432

Fischer, R. B., 6(13), 7

Fisher, D. J., 52(65,69), 81, 82, 128(23), 143(49,50,51,52,53, 54), 147(54), 148(54,56), 149(49,50,51,52,53,54,56,65), 150(53), 151(53), 157(56), 166, 167, 168

Fisher, O., 414(58,59,61), 432

Fisher, R. B., 6(13), 7

Fleet, B., 176(24), 231, 288(1), 341(43), 344(43), 367(79), 368(79), 370(79), 371(79), 372(79), 373(79), 438(5), 470

Flemming, J., 245(19), 285
Florence, T. M., 2(1), 7, 224
  (94), 234, 442(25), 443
  (25), 445(25), 454(51),
  459(51), 460(51), 461(51),
  465(51), 466(51), 467(51),
  469(51), 470(51,71,72),
  471, 472, 473
Folliard, J. T., 124(15), 133
  (15), 137(29), 166
Fonds, A. W., 244(8), 245(13,
  14), 246(8), 284
Forsberg, G. A., 470(79), 473
Freese, F., 385(92), 389
Friedman, Y., 56(73), 60(73), 82
Frumkin, A. N., 205(52), 227(108,
  109), 228(108,109), 232, 235
Fujii, Y., 438(7), 470
Fujinaga, T., 416(75,76,77),
  417(76), 418(76,77), 432
Fujiwara, S., 491(21), 492
Fukunago, K., 275(47), 286
Furman, N. H., 416(74), 432
Furutani, S., 403(41,42,44), 431

Galvez, J., 244(9), 245(15,16),
  284
Garai, T., 357(60), 361(60),,
  388
Garber, R. W., 283(65), 286
Gardner, A. W., 236(2), 243(4),
  250(4), 280(4), 284, 396(23,
  25), 399(32), 400(23),
  401(23), 430
Gavan, M., 470(70), 473
Geissler, M., 395(10), 430
Gerischer, H., 52(58), 81
Gierst, L., 173(10), 231
Glickstein, J., 150(66,67), 168
  173(3), 231
Glover, D. E., 62(80), 82,
  375(85), 389
Goldner, H. J., 283(59), 286
Goldsworthy, W. W., 60(77), 82,
  491(22), 492
Goosey, M. H., 173(4), 231
Goto, M., 216(78,79,85), 219(88),
  221(88), 233, 234
Grabarić, B. S., 261(32), 281
  (32), 285, 352(53), 353(53),
  388, 487(8), 490(8), 492

Graham, D. C., 205(51), 232
Graneli, A., 457(60), 458(60), 472
Graud, R., 173(13), 231
Grenness, M. P., 216(73), 233
Griffiths, V. S., 176(21), 231
Grypa, R. D., 216(82), 217(82),
  234
Guidelli, R., 176(19), 231
Gupta, S. L., 334(32), 387
Gürtler, O., 395(17), 430
Gutnecht, W. F., 214(68), 233,
  487(10), 492

Hackman, M. R., 283(61), 286
Hahn, B. K., 173(14), 174(14),
  175(14), 231
Hamus, V., 107(19), 109(19), 120
Hanck, K. W., 252(27), 285,
  414(71), 432
Hans, W., 107(24), 120
Harrar, J. E., 52(42), 65(42), 80
  213(65), 214(65,66), 233
Harrison, M., 402(38), 431
Hart, J. B., 274(45), 285
Hasebe, K., 283(72), 287, 401(35),
  402(35), 403(35,45), 431
Hashinaga, F., 403(41), 431
Haul, R., 104(17), 120
Havas, J., 391(6), 394(6), 395(6),
  429
Hawkridge, F. M., 68(88,89),
  69(88), 82
Hayes, J. W., 52(53,56), 56(53),
  59(53), 81, 173(5), 231
Heath, G. A., 136(26), 137(26),
  166
Heijne, G. J. M., 252(26), 285
Heineman, W. R., 470(80), 473
Heotis, J. P., 283(58), 286
Herman, H. B., 389(73), 389,
  414(68,70), 432
Herrmann, C. C., 56(72), 60(72),
  82
Hetman, J. S., 385(91), 389
Heyrovský, J., 5(12), 7, 26(14),
  30(14), 31(14), 35(18),
  37(21), 44(33), 73(14),
  75(94), 79, 80, 83, 93(2),
  94(2), 95(2), 97(7), 107(2),
  119, 124(1), 126(20), 145(59,
  60), 165, 166, 168, 176(23)

[Heyrovský, J.]
231, 414(66), 415(66),
419(81), 424(86,87), 432,
433
Hickling, A., 46(40), 80
Hitchman, M. L., 227(115),
228(115), 235
Hoff, H. K., 385(96), 389
Hoffman, H., 414(64), 432
Holbrook, W. B., 52(51,52),
56(51,52), 59(51), 81
Horn, E., 389(75), 389, 399(29),
430
Hubbard, A. T., 271(44), 273(44),
274(44), 275(44,47), 276
(44), 285, 286
Huebert, B. J., 315(22), 387
Huff, J. R., 52(54), 56(54), 81
Hughey, J. L., 409(54), 431
Hung, H. L., 52(61), 81
Huntington, J. L., 225(95), 234
Husovsky, A. A., 375(83), 389

Ilkovic, D., 93(5,6), 97(7),
119, 391(7), 430
Imbeaux, J. C., 189(29), 216(71),
232, 233
Impens, R., 283(62), 286
Ishibashi, M., 416(75,76),
417(76), 418(76), 432
Ishii, I., 385(94,95), 386(94,
95), 389
Ishii, K., 219(88), 221(88),
234
Israel, Y., 419(80), 433
Ivanov, Yu. B., 227(107), 228
(107), 235
Ives, D. J. G., 62(83), 82

Jacobs, E. S., 435(2), 447(2,
32), 470, 471
Jacobsen, E., 2(3), 7, 383(90),
384(90), 385(96), 389
Jackson, J. S., 374(80), 389
Jackson, L. L., 278(49), 280(51),
281(51), 286
Jaenicke, W., 414(64), 432
Jagner, D., 457(60), 458(60),
491(17,18), 492
Jamieson, R. A., 197(44), 232

Janz, G. J., 62(83), 82
Jee, R. D., 288(1), 341(42,43),
344(43), 348(47), 386, 387
Jehring, H., 398(28), 399(28,29,
30), 430
Jelen, F., 283(74), 287
Jenkins, I. L., 173(2), 231,
290(3), 386
Jenring, H. J., 389(74,75,76,77),
389
Johansson, E., 173(6), 231
Johns, R. H., 409(49), 431
Johnson, C. L., 225(99), 227(99),
234
Johnson, D. C., 229(118), 235,
470(68), 473
Johnson, T. H., 283(57), 286
Jones, H. C., 52(65,69), 81, 82,
143(51,52,54), 147(54,56),
148(54,56), 149(51,52,54,56),
157(54,56), 167, 173(12), 231
Jones, R., 448(37), 471
Jongerius, C., 283(64), 286
Joyce, R. J., 52(43), 54(43),
67(43), 81
Juhasz, B., 391(6), 394(6),
395(6), 429
Juliard, A. L., 349(48,49), 388

Kadish, K. M., 283(71), 287
Kalvoda, R., 44(33), 80, 282(53),
286, 424(82), 433
Kambara, T., 229(119), 235,
401(35), 402(35), 403(35,45),
431, 433
Kane, P. O., 140(43), 167
Kanevskii, 157(69), 168
Katzenberger, J. M., 425(108),
429(108), 434
Kawakado, R., 101(10), 120
Kazi, G. H., 283(73), 287
Keller, H. E., 269(36), 270(36),
285, 491(25), 492
Kelley, M. T., 52(69), 82,
128(23), 143(49,50,51,52,53),
144(53), 147(56), 148(56),
149(49,50,51,52,53,56,65),
150(53), 151(53), 157(72),
162(53), 166, 167, 168
Kemula, W., 176(17), 192(42),
231, 232, 438(12), 471

Kenkel, J. V., 229(120), 235
Kho, B. T., 283(68), 286
Kiba, N., 385(93), 386(93), 389
Kies, H. L., 418(78,79), 433
Kinard, J. T., 470(64), 472
King, R. M., 414(63), 432
Kissel, G., 173(3), 231
Kissinger, P. T., 274(45), 285
Klatt, L., 448(37), 471
Klein, N., 56(74), 60(74), 82,
    282(53), 286
Kogan, N. B., 402(37), 431
Kolthoff, I. M., 5(6), 7, 44(32,
    35), 78(95), 80, 83, 93(4),
    94(4), 95(4), 104(16),
    109(33), 118(4), 119, 120,
    127(22), 138(33,34,35,36),
    166, 167, 334(32), 387
Kono, T., 395(18), 430
Koryta, J., 20(7), 40(26),
    41(24), 79, 80, 100(9),
    107(25), 119, 120
Koszegi, B., 391(6), 394(6),
    395(6), 429
Koutecký, J., 107(19,21,22,25),
    109(31), 120
Krause, M. S., Jr., 441(19), 471
Kronenberger, K., 140(40,47),
    142(47), 167
Kryger, L., 491(11,17,18), 492
Kublik, Z., 176(17), 231
Kudirka, J. M., 425(107),
    426(107), 427(107), 428
    (107), 434
Kugo, T., 491(21), 492
Kurosaki, Y., 101(11), 120
Kurosu, T., 197(45), 232
Kůta, J., 5(12), 7, 26(14),
    30(14), 31(14), 35(18),
    37(21), 44(36), 73(14),
    75(94), 79, 80, 83, 93(2),
    94(2), 95(2), 99(8),
    107(2), 109(32), 113(38),
    115(39), 119, 121, 124(1),
    126(20), 165, 166, 176(23),
    231, 414(66), 415(66), 432

Lacoste, R. J., 443(27), 471
Ladanyi, E., 470(70), 473
Lagrou, A., 283(56), 286

Laitinen, H. A., 438(8), 470,
    470(65), 472
Lamb, B., 66(87), 82
Lamy, C., 56(72), 60(72), 82,
    216(82,83), 217(82,83), 234
Lane, R. F., 271(44), 273(44),
    274(44), 275(44,47), 276(44),
    285, 286
Larson, B. V., 458(55,56,57), 472
Latimer, J. W., 199(50), 232
Laver, G., 52(55), 81, 192(37),
    232
Lehghel, B., 60(76), 82
Leleu, J., 425(99), 433
Leveque, M. P., 145(64), 168
Leveridge, B. A., 126(18), 166
Levich, B., 227(109), 228(109), 235
Levich, V. G., 227(101,107),
    228(107), 234, 235
Levit, D. I., 470(75), 473
Leydon, D. E., 173(5), 231
Li, C., 225(98), 234
Lightfoot, E. N., 227(103), 234
Lingane, J. J., 5(6), 7, 16(2),
    78(95), 79, 83, 93(4), 94(4),
    95(4), 118(4), 119, 126(18),
    127(22), 145(63), 166, 168,
    197(43), 212(58), 232, 233,
    409(50), 431
Lingane, P. J., 415(72), 432
Little, C. J., 341(43), 344(43),
    387
Los, J. M., 244(7,8), 245(11,12,
    13,17), 246(8), 284
Loveland, J. W., 205(53), 233
Lukashenkova, H. V., 402(36), 431
Lund, W., 458(55,56,57), 472,
    491(27), 493
Luong, L., 447(35), 448(35),
    458(58,59), 472

Maas, J., 126(17), 166
Macero, D. J., 414(65), 432
Maertens, W., 283(64), 286
Maienthall, E. J., 4(5), 7,
    197(47), 199(47), 209(54),
    224(47,54,93), 232, 233
Mairanovskii, S. G., 112(36),
    121, 124(4), 165
Malatesta, F., 447(34), 471

Maloy, J. T., 216(82,83), 217(82, 83), 234

Mamantov, C., 409(47), 431

Mann, C. K., 212(56), 233

Mann, S., 145(57), 167

Marin, R. L., 136(26), 137(26), 166

Mark, H. B., 140(44), 167, 435(4), 447(4), 470

Marshall, J. C., 334(32), 387

Matson, W. R., 442(24), 443(24), 471

Matsuda, H., 11(12), 79, 185(27), 191(27), 227(104,105), 232, 235, 422(85), 433

Matsui, Y., 101)10,11,12), 105(12), 120

Matsumoto, K., 403(41), 431

Mattax, C. C., 414(57), 432

McCord, T. G., 52(50,61), 56(50), 81, 338(36), 340(36), 387

McCreery, R. L., 274(46), 286

McKinney, P. S., 52(62,66), 56(66), 81, 334(33), 380(33), 387

Meites, L., 5(8), 7, 45(1), 55(1), 73(1), 78(1), 79, 93(3), 94(3), 95(3), 107(3), 118(3), 119, 127(21), 166

Melabs, 50(41), 55(41), 80

Meszáros, L., 357(60), 361(60), 388

Meyers, R. L., 225(96), 234

Mezarus, L., 60(76), 82

Michielli, R., 283(70), 287

Micka, K., 422(83), 424(87), 433

Miller, B., 227(112), 228(112), 235

Miller, C. S., 44(32), 80, 104(16), 120

Miller, H. H., 157(72), 168

Milner, G. W. C., 5(7), 7, 393 (8,9), 395(8,9), 396(22), 430

Miri, A. M., 192(41), 232

Miwa, T., 395(13), 430, 438(7), 470

Mizuike, A., 395(13,18), 430, 438(7), 470

Mogers, J. L., 438(6), 470

Moorhead, E. D., 470(73,79), 473

Morring, C. I., 283(64), 286, 341(40), 387

Morris, J. L., Jr., 247(21), 285

Mortko, H. J., 138(31,32), 139(31,32), 166

Moussa, A. A., 307(20), 309(20), 387

Mueller, T. R., 52(65), 81, 213(63), 214(67), 224(63), 233

Mukoyama, T., 385(93), 386(93), 389

Muller, O. H., 5(9), 7

Murray, R. W., 414(67), 432

Musha, S., 385(94,95), 386(94,95), 389

M'vunzu, Z., 283(62), 286

Myers, D. J., 262(31), 264(31), 265(31), 272(41), 283(59), 285, 286

Nadjo, L., 191(35,36), 216(70), 232, 233

Nagy, G., 229(117), 230(117), 235

Nakashima, M., 403(41,44), 431

Namee, L., 52(70), 56(70), 82, 334(31), 387

Nangniot, P., 283(62), 286

Nebergall, W. H., 31(16), 32(16), 79

Neeb, R., 62(82), 82, 173(8), 231, 341(41), 362(62,63), 387, 388, 438(10), 470

Nekrasov, L. N., 227(108,109), 228(108,109), 235

Nemec, L., 143(55), 167

Newman, J., 227(100), 234

Nicholson, M. M., 455(53), 472

Nicholson, R. S., 182(25), 183 (25), 186(28), 187(28), 188(28), 191(25,28), 217(86), 231, 232, 234, 415(73), 432, 451(43), 472

Nickels, W., 140(47), 142(47), 167

Niedrach, L. W., 197(43), 232

Nigmatullin, R. S., 212(57), 233

Norris, A., 362(64), 388

Noshiro, M., 395(12), 430

Novak, J. V. A., 138(30), 166

Novosel, B., 395(20), 430
Nurnberg, H. W., 283(67), 286,
    362(66), 388, 403(46), 431
Nygard, B., 173(6), 231

O'Brien, G. E., 409(54), 431
O'Deen, W., 247(24), 285
O'Dom, G., 441(22), 443(22),
    444(22), 471
O'Donnell, T. A., 137(27), 166
    268(35), 285, 454(50), 472
Ogle, J., 491(16), 492
O'Halloran, R. J., 124(14),
    127(14), 129(14), 132(46),
    140(46), 142(46), 152(46),
    153(46), 157(46), 166, 167,
    273(43), 285, 348(45),
    349(45,50), 351(52),
    352(52), 387, 388, 451(44),
    472
Oki, S., 395(13), 430
Okinaka, Y., 44(35), 80,
    138(33,34,35,36), 167
Okochi, H., 395(15), 430
Oldham, K. B., 216(72,73,74,78,
    79,84,85), 219(87), 233,
    234, 245(10), 255(29),
    258(29), 267(33), 268(33),
    269(33), 283(66), 284,
    285, 286
Olofsson, J., 173(6), 231
Oostervink, R., 385(92), 389
Opekar, F., 227(116), 235
Opheim, L. N., 491(27), 493
Orth, G. L., 111(37), 112(37),
    121
Osajima, Y., 403(41,44), 431
O'Shea, T. A., 41(24), 80,
    109(33), 120
Osteryoung, J., 251(25),
    262(31), 264(31), 265(31),
    272(41), 277(50), 283(59,
    72), 285, 286, 287
Osteryoung, R. A., 52(55), 81,
    192(37), 232, 237(3),
    238(3), 245(20), 246(20),
    247(23,24), 251(25),
    252(3), 258(3), 261(3),
    262(3), 263(3), 265(3),
    266(3), 268(34), 269(36),
    270(36), 272(41), 277(48,
    49,50), 280(51), 281(51),
    284, 285, 286, 446(28,29),
    447(28), 453(28,29,47),

[Osteryoung, R. A.]
    454(28,47,48,49), 460(48,
    49), 467(49), 471, 472,
    491(13,20,25), 492
Overton, M. W., 491(14), 492

Pac, R. G., 395(14), 430
Palagyi-Fenyes, B., 357(60),
    361(60), 388
Palecek, E., 283(63,74), 286, 287
Papariello, G. J., 491(26), 493
Papeschi, G., 173(7), 231
Parker, G. A., 41(24), 80,
    109(33), 120
Parker, W. J., 176(21), 231
Parry, E. P., 18(3), 19(3),
    64(3), 79, 109(33), 120,
    243(6,10), 245(10,20),
    252(6), 253(6), 254(6),
    255(29), 256(30), 257(30),
    258(29), 267(33), 268(33,34),
    269(33), 283(57,65,66), 284,
    285, 286
Pearson, J. C., 52(54), 56(54), 81
Peover, M. E., 52(57), 81
Perone, S. P., 197(44), 212(61),
    213(63), 214(66,68,69),
    224(63), 232, 233, 435(3),
    441(20), 447(30,31,33),
    470, 471, 487(9), 487(10),
    491(11), 492
Petak, P., 448(36), 454(36), 471
Petrie, L. M., 459(61), 472
Petrii, O. A., 44(38), 80
Piccardi, G., 176(19), 231
Pilkington, E. S., 468(62), 472
Polin, D., 283(58), 286
Poojary, A., 163(75), 168
Pool, K. H., 225(99), 226(99), 234
Porthault, M., 470(69), 473
Pouli, D., 52(54), 56(54), 81
Powell, K. G., 157(70), 168
Powell, J. S., 52(57), 81
Prater, K. B., 227(114), 228(114),
    235
Probst, R. C., 173(4), 231,
    470(64), 472
Prue, D. G., 283(68), 286
Pungor, E., 229(117), 230(117),
    235

Raaen, H. P., 173(12), 231
Radulescu, U. D. N., 470(70), 473

Rafikov, R. K., 363(72), 388
Rajagopalan, S. R., 163(75,76),
    165(76), 168
Ramaley, L., 441(19), 471, 491
    (15), 492
Randles, J. E. B., 176(16),
    231, 307(19), 387
Rangarajan, S. K., 163(76),
    165(76), 168
Rankovitz, S., 150(66), 168
Raspi, G., 447(34), 471
Reddy, A. K. N., 20(8), 79
Reilley, C. N., 52(53), 56(53),
    59(53), 81, 140(44), 167,
    173(5), 219(89), 221(89),
    223(89), 231, 234, 409(49),
    414(63), 416(74), 431, 432
Rein, J. E., 38(23), 39(23), 80
Reinmuth, W. H., 35(17), 52(47),
    79, 81, 191(30), 232,
    361(61), 362(61), 388,
    399(33), 409(51), 414(69),
    424(97,98), 430, 431, 432,
    433, 451(42), 472
Reklat, A., 364(75), 389,
    399(29), 430
Retajczyk, T. F., 375(82), 389
Revenda, J., 104(15), 120
Reynolds, G. F., 157(70),
    168, 197(48), 199(48),
    200(48), 232
Riccoboni, L., 157(68), 168
Riddiford, A. C., 227(113),
    228(113), 235
Ridgeway, T. H., 219(89),
    221(89), 223(89), 234
Rifkin, S. C., 269(37), 270(37,
    39), 285
Riha, J., 145(61,62), 168,
    173(11), 176(11), 231
Riley, C. N., 44(34), 80
Roe, D. K., 375(82), 389
    442(24), 443(24), 471
Roeleveld, L. F., 245(17), 284
Roffia, S., 176(18), 231
Rojahn, T., 2(3), 7, 383(90),
    384(90), 389
Rooney, R. C., 212(62), 213(62),
    214(62), 215(62), 233
Rose, G. M., 283(58), 286
Roth, F., 145(64), 168
Roughton, C. L., 402(38), 431
Roux, J. P., 470(69), 473

Royle, L. G., 283(69), 286
Rüetschi, P., 107(26), 120
Ruzić, I., 252(28), 285, 290(5),
    305(16), 348(45), 349(45),
    386, 387, 395(21), 430,
    451(44), 472

Salikhdzhanova, R. M., 402(40),
    403(40), 431
Sammour, H. M., 307(20), 309(20),
    387
Sanders, P. R., 124(10), 166
Sauer, D., 362(63), 388
Saveant, J. M., 189(29), 191(31,
    32,33,35,36), 216(70,71,75,
    80,81), 232, 233, 234
Schaap, W. B., 31(16), 32(16),
    52(64,66), 56(66), 79, 81,
    334(33), 380(33), 387
Schmidt, H., 140(37), 167, 197(46),
    232, 391(4), 400(4), 419(4),
    424(4), 429
Scholz, E., 104(17), 120
Schroeder, R. R., 209(60),
    210(60), 212(59,60), 255(98),
    233, 234
Schroeder, T. D., 395(20), 430
Schute, J. B., 282(54), 283(54),
    286
Schwall, R. W., 485(7), 492
Seaborn, J. E., 223(91), 234
Seelig, P. F., 491(24), 492
Seitz, W. R., 448(37), 471
Semerano, G., 157(68), 168,
    391(7), 430
Serna, A., 244(9), 245(15,16), 284
Shain, I., 52(48), 81, 182(25),
    183(25), 191(25), 192(39,40),
    193(39), 194(39), 225(96),
    231, 234, 341(38), 345(38),
    387, 435(1), 440(17), 441(18),
    453(18), 455(18), 470, 471
Shalgosky, H. I., 223(92), 234,
    396(22), 430
Shallal, A. K., 367(78), 389
Shinagawa, M., 197(45), 232
Siegerman, H., 441(22), 443(22),
    444(22), 471
Sioda, P. E., 229(119), 235
Skogerboe, R. K., 438(11),
    440(11), 453(47), 454(47),
    455(11), 471, 472

Skov, H. J., 491(17), 492
Sluyters, J. H., 290(4), 291(10),
    292(10), 298(14), 313(14),
    386
Sluyters-Rehbach, M., 290(4),
    291(10), 292(10), 298(14),
    313(14), 386
Smales, A. A., 157(71), 168
Smith, D. E., 3(4), 7, 52(44,45,
    46,50,60,61,67), 56(44,45,
    46,50), 59(60), 62(80),
    81, 82, 219(90), 222(90),
    223(90), 234, 291(9,11),
    292(9,11), 293(12), 294(9,
    12), 295(12), 296(11,12),
    298(9,11,15), 301(9),
    306(9), 209(9), 313(15),
    315(9,11,22), 332(11),
    335(9), 338(36), 340(36),
    348(45), 349(45), 357(9,
    11,59), 359(9), 360(9,11),
    362(9), 375(11,85), 377(89),
    381(89), 386, 387, 388, 389,
    399(34), 431, 485(5,6,7),
    486(5), 489(12), 491(14),
    492
Smith, G. S., 333(30), 375(30),
    387
Snow, M. R., 37(22), 80
Somerton, K. W., 307(19), 387
Sorensen, R. W., 425(106), 434
Sour, H., 40(27), 80
Spiehler, V. R., 283(71), 287
Springer, J. S., 225(97), 234
Spurgeon, J. C., 283(73), 287
Srinivasan, S., 470(74), 473
Stackelberg, M. Von, 140(37),
    167, 197(46), 232
Stapelfeldt, H. E., 447(30), 471
Staubach, K. E., 52(58), 81
Stelzner, R. M., 52(65), 81
Sternberg, B. M., 402(39),
    403(39), 431
Stevens, F. B., 213(65), 214(65,
    66), 233
Stewart, W. E., 227(103), 234
Stiehl, G. L., 109(30), 120
Stolle, W., 367(76), 389(75,
    76), 389, 399(29,30), 430
Stradins, D., 491(16), 492
Strehlow, H., 140(40), 167
Streuli, C. A., 176(20), 192(38),
    231, 232

Stulikcva, M., 448(36), 454(36),
    471
Stumm, W., 44(34), 80
Sturrock, P. E., 409(54,55), 431
Sudo, E., 395(15), 430
Sugisaki, M., 395(12), 430
Surfleet, B., 402(38), 431
Surprenant, H. L., 219(89),
    221(89), 223(89), 234
Syalikov, S., 403(43), 431
Sybrandt, L. B., 214(69), 233,
    487(9), 492
Sympson, R. F., 425(106), 434
Szechter, S., 470(74), 473

Tacussel, J. R., 52(63), 62(79),
    81, 82
Tanaka, M., 385(93), 386(93), 389
Tanaka, S., 401(35), 402(35),
    403(35,45), 431
Taylor, J. K., 126(19), 166,
    197(47), 199(47), 224(47),
    232
Taylor, R. J., 268(35), 285,
    307(21), 387, 454(50, 472
Teasdale, D., 66(87), 82
Temmerman, E., 280(55), 283(55),
    286
Terry, E. A., 197(48), 199(48),
    200(48), 232
Tessier, D., 216(75), 233
Testa, A. C., 35(17), 79, 414(69),
    432
Thackeray, J. R., 101(13,14),
    103(13), 120
Thomas, Q. V., 491(11), 492
Thomas, W. E., Jr., 52(64), 81
Tindall, G. W., 448(38), 471
Titus, J. A., 475(1,2), 492
Tomsons, P., 491(16), 492
Trojanek, A., 282(53), 286
Trumpler, G., 35(20), 80,
    107(26), 120
Tsuji, K., 44(37), 80
Turner, J. A., 247(23), 285,
    454(59), 460(49), 467(49),
    472
Tutani, I., 491(16), 492

Umezawa, Y., 491(21), 492
Underkofler, W. L., 52(48), 81,

[Underkofler, W. L.,]
    341(38), 345(38), 387,
    440(17), 441(18), 453(18),
    455(18), 471

Valenta, P., 52(59), 81
Vallon, J. J., 376(88), 389
van Dalen, E., 409(53), 411(53),
    431, 452(45), 453(45), 472
van den Born, H. W., 216(76,77),
    218(76,77), 233
van der Linden, W. E., 252(26),
    285
Vandreuil, B., 409(54,55), 431
van duyne, R. P., 270(38), 285
van vuuren, P. J., 475(1,2), 492
Varel, P., 173(13), 231
Vasileva, L. N., 402(36,37), 431
Vassos, B. H., 277(48), 286
Vavricka, S., 100(9), 119
Velghe, N., 439(16), 449(16),
    471
Verbeek, F., 199(49), 201(49),
    232, 280(55), 283(55,56),
    286
Verdier, E., 173(13), 231
Vesely, K., 35(19), 79
Vetter, K. J., 20(6), 79
Vianello, E., 176(18), 191(31,
    32,33), 231, 232
Vicente, V. A., 227(110),
    228(110), 235
Vittori, O., 470(69), 473
Vleck, A. A., 104(18), 120
Vogel, J., 52(59), 81, 145(61),
    168
von Sturm, F. Z., 395(16), 430
Vyaselev, M. R., 212(57), 233
Vydra, F., 447(35), 448(35,36),
    454(36), 458(58,59), 471,
    472

Wahlin, E., 140(38,39), 167
Walker, D. E., 52(49), 81
Wark, D. A., 353(55), 356(55),
    388
Warner, C. R., 283(68), 286

Watkins, N. H., 470(65), 472
Weeks, C. H., 468(62), 472
Wetsemar, B. S. C., 245(17), 284
White, W., 176(16), 231
White, W. R., 491(23), 492
Whitson, P. E., 216(76), 218(76),
    233
Wienhold, K., 40(27), 80
Wiesner, K., 107(20), 120
Willems, G. G., 62(82), 82,
    173(8), 231
Williams, M. J., 396(23,25),
    400(23), 401(23), 430
Williams, R., 145(63), 168
Wilson, C. E., 283(65), 286
Wilson, G. S., 438(6), 470
Wimmer, F. L., 37(22), 80
Winzler, R. J., 124(10), 166
Wolf, S., 124(2,3), 127(3), 165
Wolff, G., 283(67), 286, 403(46),
    431
Woodson, A. L., 377(89), 381(89),
    389
Wopschall, R. H., 192(39,40),
    193(39), 194(39), 232

Yamada, Y., 227(105), 235
Yamata, N., 395(19), 430
Yano, N., 197(45), 232
Yarnitzky, C., 57(73,74), 60(73,
    74), 82, 173(9), 231,
    282(52), 286
Yasumori, Y., 140(48), 167
Yatsimirzki, K. B., 109(33), 120
Yemane, T., 385(93), 386(93), 389
Yllo, M. S., 283(75), 287
Young, C. C., 438(8), 470

Zagorski, Z. P., 70(90), 82
Zheleztsov, A. V., 363(71,72),
    388
Zipper, J. J., 212(61), 233
Zublick, Z., 192(42), 232
Zuman, P., 5(10,11), 7, 11(12),
    29(12,13), 30(15), 79,
    112(35), 121

AC cyclic voltammetry, 349, 350,
  436, 437
AC polarography
  basic principles of, 292–298
  charging current in, 73–75,
    294–296, 317, 321–329,
    333–338
  with chemical reactions or
    adsorption, 314–316
  comparison of differential
    pulse and, 355
  comparison of Faradaic im-
    pedance measurements
    and, 290,291
  with computerized instru-
    mentation, 64, 485–489
  current sampled, 339
  definition of, 288, 290–292
  and disproportionation, 39
  and EC reactions, 38
  examples of the use of,
    374–386
  external voltage sources
    in, 52
  Faradaic current in, 317–329
  fourth harmonic, 296, 297
  fundamental harmonic, 6, 293,
    298–316
  influence of resistance in,
    59, 294, 328, 333–338
  instrumental aspects of,
    374, 375
  intermodulation effects in,
    296, 298, 361, 362
  irreversible electrode proc-
    esses in, 313, 314
  and migration current, 78
  optimum conditions for use
    of, 329–332
  phase-angle measurements in,
    295, 315, 325
  phase selective, 4, 6, 294–298,
    316–332
  quasi-reversible electrode
    processes in, 305–313,
    318–321
  reversible electrode processes
    in, 24, 300–305, 317, 318
  second harmonic, 6, 15, 293,
    296–298, 356–363
  short controlled drop time,
    332–339
  in the subtractive mode, 340
  systematic use of, 298, 299
  tensammetric polarographic
    responses with, 363–374
  third harmonic, 296, 297, 358
  time domain of, 24
  use of high frequencies in,
    353–356
  use of pulsed dc potentials
    in, 349, 351–353
  validation of data with, 314,
    316
  variation of potential with
    time in, 74
  waveforms used in, 288–290
  zero-order terms in, 296, 298

AC stripping voltammetry, 441,
    448, 453, 455, 460–465
AC techniques for charging
    current compensation,
    163–165
AC voltammetry, 341–349
  analytical aspects of, 347,
    348
  charging current in, 345–347
  faradaic current in, 342, 345
  fast sweep techniques at dme
    in, 341–348
  phase-selective, 342, 343,
    348
  reproducibility of, 347
  second harmonic, 342, 343,
    348
N-acetylisoniazide, determina-
    tion by ac polarography,
    376, 377
Adrenaline, determination by
    linear sweep voltam-
    metry, 203
Adsorption, 43, 44, 71, 75,
    76, 112–117
  controlled currents, 10, 116
  dc polarographic theory of,
    113–117
  influence on drop time of, 75
  isotherms, 113
  and rapid dc polarography,
    124, 127, 128, 133
  tensammetric waves arising
    from, 363–374
Alternate drop differential
    pulse polarography,
    277–279
Alternate drop normal pulse
    polarography, 278
Amperometric titrations, 265
Anodic current, 9, 16, 20
Anodic stripping voltammetry,
    436, 437 (see also
    Stripping voltammetry)
Antimony(III)
  ac polarographic detection
    limit for determina-
    tion of, 304
  determination by stripping
    voltammetry, 468, 469

Antimony(V), determination by
    stripping voltammetry,
    469
Aromatic hydrocarbons, reduction
    of, 41
Arsenic(III), determination
    by differential pulse
    polarography, 264, 477
Automated polarographic anal-
    ysis, 90, 491
Auxiliary electrodes, 47–52, 62
  and controlled potential
    electrolysis, 65
  positioning of, 66
  use of salt bridges with, 65

Background current correction,
    477, 482, 487 (see also
    Subtractive polarography)
Bismuth
  ac polarography of, 304,
    306, 307, 309, 311
  ac voltammetry of, 343
  determination by anodic
    stripping voltammetry,
    440
Boron, determination by linear
    sweep voltammetry,
    198–200
Boundary-value problems, 92, 93
Bromide
  determination by cathodic
    stripping voltammetry,
    438
  normal pulse polarograms
    in fused salt media, 248

Cadmium
  ac and dc cyclic voltammetry
    for reduction of, 350,
    437
  ac differential pulse polaro-
    grams of, 352
  ac polarographic reduction
    of, 304, 318, 323, 327,
    329, 339, 354
  ac voltammetry of, 342, 343
  charge step polarography of,
    427, 428

[Cadmium]
current-averaged polarograms
of, 144
cyclic, semi-integral and
semi-differential voltam-
mograms of, 220, 221
dc voltammetry of, 342, 343
derivative polarograms of,
146, 149, 151
determination
by dc polarography
with controlled current,
416, 417
with charging current
correction, 162, 165
of in presence of more
positively reduced cop-
per, 88, 141
by stripping voltammetry,
440, 447, 456, 457,
460, 461, 462, 465,
466, 468
using current sampled dc
polarography, 140
using subtractive linear
sweep voltammetry, 224
differential pulse polarograms of, 254, 354
linear sweep voltammogram of,
179
normal and derivative linear
sweep voltammograms of,
213-216
polarograms in the presence
of tribenzylamine, 126
radio frequency polarogram
of, 404-406
reduction of, 12, 13, 62
short drop time polarography
of, 134
square wave intermodular
voltammogram of, 396
square wave polarograms of,
394, 395, 396
Caffeine, determination by
linear sweep voltammetry,
202
Carbaryl (Sevin), determination
of nitrosated compound
by differential pulse
polarography, 283, 284
Catalytic mechanisms, 38-41,
108-112

Catecholamine, determination
by differential double-
pulse voltammetry, 274
Cathode-ray polarography, 170
(see Linear sweep voltam-
metry)
Cathodic current, 9, 16, 20
Cathodic stripping voltammetry,
438 (see also Stripping
voltammetry)
CE mechanism, 35, 36, 106-108
Charge-step polarography, 424-429
in high resistance media, 426
instrumental aspects of,
426-429
resolution in, 426
sensitivity of, 426
theory of, 425, 426
use of computerized instrumentation in, 426
Charge transfer coefficient, 21
Charging current, 48, 72-76,
88-90, 160-165 (see also
underheadings for dif-
ferent techniques)
compensation with superimposed
ac signal, 163-165
correction for, 89, 160-165
in nonaqueous media, 89
Chemically modified platinum
electrodes, 275, 276
Chloramphenicol, differential
pulse and dc polaro-
grams of, 242
Chloride
determination by cathodic
stripping voltammetry,
438
oxidation of mercury in the
presence of, 104-106
Chronopotentiometry, 66, 67,
407-415
charging current in, 410, 411
with controlled alternating
current, 419-424
charging current in, 420,
421, 423
different types of readout
form in, 422
effect of adsorption in, 423
theory of, 422, 423
time lags in, 421
cyclic, 414, 415, 420

[Chronopotentiometry]
  disadvantages of, 409, 410
  theory of, 412–415
Computerized instrumentation,
      474–491
  in charge-step polarography,
      426
  in current sampled dc polar-
      ography, 140
  data manipulation procedures
      using, 476–478, 482,
      485, 487, 490
  in linear sweep voltammetry,
      224
  in pulse polarography, 281,
      282
Concentration gradient at
      electrode surface, 92,
      94
Constant potential pulse polar-
      ography, 280, 281
Continuous flow measurement,
      229, 230
Convection, 76, 77, 226
Convolution techniques in
      voltammetry, 216–223
Copper
  ac polarography of, 311, 312,
      318–322, 330, 331, 354
  dc polarography of, 68, 69,
      90, 140, 141
  determination by stripping
      voltammetry, 440, 456,
      457, 468
  differential pulse polarogra-
      phy of, 241, 354
  normal pulse polarograms
      of, 240, 241
  radio frequency polarograms
      of, 404, 405
Cottrell equation, 244, 245,
      249, 250
Coulostatic polarography (see
      Charge-step polarog-
      raphy)
Counter electrodes (see
      Auxiliary electrodes)
Current-averaged dc polarogra-
      phy, 142–145
  instrumentation in, 143, 145

Current-sampled dc polarography,
      139–142
  with computerized instrumen-
      tation, 140, 481–485
  improved Faradaic-to-charging
      current ratio in, 139
  method of current measurement
      in, 139
  readout improvement gained
      in, 140
  use of sample-and-hold
      circuitry in, 139, 142
Current-time curves of sulfide,
      135
Cyclic voltammetry, 172, 436,
      437, 451
  diagnostic criteria for assign-
      ing mechanisms in, 194–196
  effect of uncompensated resist-
      ance in, 188–190
  influence of adsorption in,
      192–194, 205
  influence of coupled chemical
      reactions in, 190–192
  theory of, 180–196
  use of convolution of semi-
      integral techniques in,
      216–219
Cyclohexanol, tensammetric
      polarographic waves for,
      365, 366

DC limiting currents, 9, 84, 85
  adsorption controlled, 112–117
  and catalytic mechanisms, 40, 109
  and CE mechanism, 106–108
  dependence on concentration
      of, 116
  dependence on mercury column
      height of, 117
  and EC mechanism, 108
  influence of supporting
      electrolyte concentration
      on, 78
  in the presence of dispropor-
      tionation, 38
DC polarography
  as an absolute method of
      analysis, 118, 119

[DC polarography]
and adsorption controlled
processes, 112-117
and catalytic mechanisms, 108-
112
and CE mechanisms, 106-108
and charging current, 73-75,
84, 88-90, 117
comparison of linear sweep
voltammetry and, 207-209
with controlled current, 416-419
theory of, 418
use of controlled current
density in, 419
conventional forms of, 1-6,
15, 87-91
current-averaged, 142-145
current-sampled, 139-142
derivative, 85, 145-157
detection limits in, 88
difficulties encountered in,
68, 69, 84-91
and disproportionation mechanism, 109
and EC mechanisms, 37, 108
and film formation, 112
and formation of insoluble
products, 112
equations for diffusion controlled processes in,
96-99
history of, 1-5
irreversible electrode processes in, 24
and kinetically controlled
currents, 106-108
maxima and, 44, 68, 69, 85, 86
position of waves with respect
to E° in, 25
quasi-reversible electrode
processes in, 24
rapid, 122-137
reversible electrode processes
in, 24, 96-99
scan rates of potential in,
122, 123, 129
simple concepts of, 8-11
subtractive, 157-160
systematic use of theory in,
91-117
DC stripping voltammetry, 446-
448, 454, 456, 457, 462,
463, 465

DC voltammetry (see Linear Sweep
voltammetry)
Deconvolution techniques in
voltammetry, 219-223
Derivative dc polarography, 85,
145-157
charging current contribution
in, 149
circuits for, 147-149
current-sampled, 147, 150,
152, 153
first and second derivatives
in, 147-151
limits of detection in, 157
preparation of calibration
curves with, 153
theory of, 154-157
time-averaged, 147-149
use of RC circuits in, 145-147
Derivative linear sweep voltammetry, 212-216
accuracy and reproducibility
of, 212, 213
resolution of, 213, 214
sensitivity of, 213, 214
theory of, 213
use of second derivatives in,
213, 214
Derivative pulse polarography,
236 (see also Pseudo-
derivative pulse polar-
ography)
Derivative stripping voltammetry,
441, 463
Desorption, 75
Detection limits in polarography,
4
Dialkyl phenacyl sulfonium ions,
reduction of, 30
3-Diazocamphor, reduction of, 87
Differential capacity, 75
Differential double-pulse voltammetry, 273-276
minimization of film formation problems by, 273,
274
waveform used in, 274
Differential pulse polarography
alternate drop, 277-279
amperometric titrations with,
265
assessment of reversibility
in, 251

[Differential pulse polarog-
raphy]
  charging current in, 253, 258,
      261, 265
  comparison of ac and, 355
  current measurement in, 240
  dc effects in, 258, 262
  external voltage sources in,
      52
  half-widths in, 250, 251
  influence of resistance in,
      253
  influence of surfacants on
      background in, 264, 265
  instrumental artifacts in, 251
  with low supporting electro-
      lyte concentrations, 253,
      254, 261, 262
  with microprocessor controlled
      instrumentation, 476–478
  resolution in, 250, 251
  sensitivity in, 262–264
  theory of, 249
  waveform used in, 239, 241
Differential pulse stripping
      voltammetry, 441, 444,
      446–448, 457, 460, 465
Differential pulse voltammetry,
      269–271
  at a dropping mercury elec-
      trode, 270, 271
  problems with, 275, 276
  at rotated electrodes, 272
Differentiation-integrator
      method for charging cur-
      rent correction, 164
Diffusion, 76, 77
  coefficients, 92, 97
  controlled currents, 10, 92
  current constant, 118
  cylindrical, 92
  at a dme, 93
  electrode processes controlled
      by rate of, 16
  Fick's law of, 92
  linear, 92, 93
  spherical, 92, 93
Dithiocarbamates, dc polarog-
      raphy of, 128
Dopamine, differential pulse and
      differential double-pulse
      voltammetry of, 275, 276

Double layer, 73
  capacity, 74
  differential capacity of, 75
Drop knockers, 60, 61
Dropping mercury electrodes, 5,
      6, 45, 46, 61
  capillary response phenomena
      and, 280
  controlled drop time, 122, 123
  conventional drop time of,
      122, 123
  and film formation, 112
  mechanical control of drop
      time with, 60
  spinning, 138, 139
  vibrating, 124
Drop time
  and adsorption, 75
  gravity control of, 60
  mechanical control of, 60, 91
  potential dependence of, 91
Dual cell polarography (see
      Subtractive polarography)
Dual-drop differential pulse
      polarography (see Alter-
      nate drop differential
      pulse polarography)
Dual-drop normal pulse polarog-
      raphy (see Alternate
      drop normal pulse polar-
      ography)

$E_{1/4} - E_{3/4}$ values, 98, 99, 102,
      104, 118
EC mechanism, 36–38, 108
Electrocapillary curve in
      acetone, 136
Electrode processes
  adsorption controlled, 16
  and absolute rate theory, 14,
      19
  catalytic, 38–41
  CE, 35, 36
  classification of, 15, 35–43
  and coupled chemical reactions,
      43
  and dependence on stoichiom-
      etry, 101
  diffusion controlled, 16
  distinction between reversible
      and irreversible, 99

[Electrode processes]
  EC, 36–38
  ECE, 41–43
  influence of disproportiona-
    tion on, 38–39
  irreversible, 15, 24, 26, 43,
    99, 100
  kinetically controlled, 10, 16
    89, 107, 108
  kinetic description of, 14, 19
  quasi-reversible, 24, 100
  and rate determining steps,
    16, 43
  rates of, 23
  regenerative, 38–41
  reversible, 15, 16, 23–27,
    96–99, 104
  and sensitivity, 36
  theory of, 20–35
Electrodes
  boron carbide, 176
  capacitance of, 73
  carbon paste, 176, 443
  charge on, 73
  chemically modified, 275, 276
  conical, 226, 228
  disc, 448
  glassy carbon, 123, 176, 443,
    445, 446, 447, 454, 461
  gold, 176
  graphite, 446
  hanging mercury drop, 123, 176,
    436, 440, 441, 451–453,
    460, 462, 464–470
  hanging mercury drop suspended
    from platinum wire, 445
  hydrodynamic, 226, 227
  iodide treated platinum, 276
  iridium, 176
  mercury pool, 176
  mercury thin film, 441–443,
    451–453, 460–462, 464–470
  plate, 227, 228
  platinum, 123, 176
  pyrollytic graphite, 202, 269,
    443, 447
  rhodium, 176
  rotating disc, 226, 228, 229
  rotating glassy carbon, 445,
    446
  rotation of, 76

[Electrodes]
  silicone rubber-based graphite,
    203
  silver, 176
  slowly growing mercury drop,
    439, 440
  static mercury drop, 165
  streaming mercury, 226
  tubular mercury-covered
    graphite, 448
  used in stripping voltammetry,
    439–448
  wax-impregnated graphite, 176,
    443, 447
Electron transfer, 12, 13, 15,
  19
  rate of, 17, 20, 71
Exchange current, 23
Explosives, ac polarographic
    determination of, 385

Faradaic currents, 71, 72, 117
Faradaic impedance measurements,
    290, 291
Faradaic rectification polarog-
    raphy, 296, 298, 361–363
Faraday's law, 115, 450
Fast Fourier transform in elec-
    trochemistry, 482, 485–487
Fast sweep polarography, 170 (see
    Linear sweep voltammetry)
Fick's law of diffusion, 92
Film formation, 43, 44 (see also
    under various techniques)
Flameless atomic absorption
    spectrometry and electro-
    chemistry, 448
Flowing solutions (see Stirred
    solutions)
Formaldehyde
  reduction of, 35
  short drop time polarography
    of, 134
Free energy, 114
Fundamental harmonic ac polarog-
    raphy (see AC polarography,
    fundamental harmonic)

Galvanostatic control, 66, 67, 439

Gold
  determination by stripping
    voltammetry, 446
  normal pulse voltammogram for
    reduction of, 269

Half-wave potential
  for adsorption controlled
    processes, 114
  calculation of, 97
  definition of, 9
  dependence on concentration
    of, 10, 11, 97, 102,
    104-106
  dependence on nature of elec-
    troactive species on, 10,
    11
  dependence on solution compo-
    sition of, 10, 11
  influence of chemical reactions
    on, 29
  influence of complex formation
    on, 31-34
  influence of co-ordination
    number on, 33
  influence of follow-up reac-
    tion on, 37
  influence of pH on, 29, 30
  irreversible, 25
  problems with calculation
    of, 84, 85
  and quasi-reversible electrode
    processes, 100
  relationship to $E°$, 97, 98
  reversible, 25
  for reversible electrode
    processes, 97-99
Heterogeneous charge transfer
    rate constants
  definition of, 22
  table of, 26
Heterogeneous kinetics, 13, 15,
    20
Hexachlorophene, determination
    by ac polarography,
    382-385
Heyrovský-Ilkovic equation, 97
Homogeneous kinetics, 13, 15,
    20, 21, 27, 28
Hydrodynamic electrodes, 226,
    227

Hydrodynamic voltammetry, 227,
    229
Hydrogen reduction waves, cata-
    lytic, 109, 111, 112
2-hydroxyphenazine, determination
    of, 145

Ilkovic equation, 94, 95, 105,
    118, 119
Indium
  ac polarographic data for re-
    duction of, 304
  dc polarogram of, 86
  derivative linear sweep voltam-
    mogram of, 215
  determination by anodic strip-
    pint voltammetry, 440
  determination by derivative dc
    polarography, 151
  square wave polarogram of, 394
Insoluble product formation, 112,
    125
Instrumentation, 44-70
  and artifacts in polarography,
    68
  automation of, 55, 90, 491
  computerized, 3, 54, 55, 58,
    62, 64, 474-491
  general recommendations in,
    67-70
  history of, 1-4
  multi functional, 62
  readout forms associated with,
    52, 54-56, 62, 70, 480
  use of Faraday cage to achieve
    electrical shielding in,
    61
Integral pulse polarography, 236,
    238 (see Normal pulse
    polarography)
Intermodulation polarography, 296,
    298, 361-363 (see also AC
    polarography)
Interrupted voltage ramps, use of
    in linear sweep voltam-
    metry, 225-226
Inverse voltammetry (see Stripping
    voltammetry)
Iodide, dc polarography of, 127

iR drop, 46, 48, 49, 52, 55,
    56, 59, 60, 67, 72,
    76, 77 (see also Resist-
    ance, and sections
    associated with individ-
    ual techniques)
Iron(III)
  ac polarography for reduction
    of, 340
  determination by derivative
    dc polarography, 153
  determination by linear sweep
    voltammetry, 199
Iron oxalate complexes, reduction
    and oxidation of, 16-19
Isoniazide, determination by
    ac polarography, 376,
    377
Isonictonic acid, determination
    by ac polarography, 376,
    377

Kinetics, corrections for effects
    of, 485, 486

Langmuir isotherm, 115
Lead
  ac polarography of, 354
  ac voltammetry of, 343
  alternate drop and ordinary
    differential pulse
    polarograms of, 279
  constant potential pulse
    polarogram of, 281
  determination
    with charging current com-
      pensation, 165
    by dc polarography with con-
      trolled current, 416, 417
    by derivative dc polarog-
      raphy, 150
    by lubricating oil by ac
      polarography, 386
    by stripping analysis, 440,
      445, 446, 456, 457,
      459-562, 465, 466
  differential pulse polarog-
    raphy of, 354

[Lead]
  linear and derivative linear
    sweep voltammograms of,
    214
  radio frequency polarogram of,
    405
  reduction as a function of
    supporting electro-
    lyte concentration, 78
  simulated differential pulse
    polarograms of, 261-263
  simulated normal pulse polar-
    ograms of, 266
  square wave intermodular vol-
    tammogram of, 396
  square wave voltammogram of,
    396
  staircase voltammograms of, 210
Levich equation, 449
Librium, determination of, 476
Limiting currents
  adsorption controlled, 114-117
  dependence
    on concentration, 116
    on mercury column height, 117
  difficulties in measurement
    of, 85-87
  diffusion controlled, 88
  dependence
    on capillary characteristic
      of, 95
    on concentration of, 95
    on mercury column heights of,
      95, 96
    on temperature of, 95
  drop time dependence of, 91
  kinetically controlled, 106-112
  potential dependence of, 91
  in rapid dc polarography, 130
  theory of, 92-96
Linear sweep voltammetry, 3, 4,
    24, 342-362, 348 (see
    also Derivative linear
    sweep voltammetry)
  adsorption in, 177, 192-194, 205
  analytical applications of,
    197-203
  and charging current, 204-205
  comparison of dc polarography
    and, 178-180, 207-209

[Linear sweep voltammetry]
 derivative, 212–216
 diagnostic criteria for assigning mechanisms in, 194–
   196
 at a dme, 6, 169, 172–176
 effect of uncompensated resistance in, 188–190
 influence of coupled chemical
   reactions in, 190–192
 influence of forced convection
   in, 177, 178
 influence of scan rate in,
   178–180
 limits of detection in, 206
 and non-reversible electrode
   processes, 185–188
 reproducibility of, 207
 resolution in, 207–208
 and reversible electrode
   processes, 182–185
 shape of curves in, 178–180
 subtraction of background in,
   223–225
 theory of, 177–196
 use of interrupted voltage
   ramps in, 225–226
 waveforms used in, 171, 175

Mass transfer, mechanisms of, 76,
   77
Maxima, 86, 112
 two electrode and three electrode instrumentation
   and, 68, 69
Mercury
 determination by stripping
   voltammetry, 446
 oxidation, in presence of
   halides and xanthates,
   101–105
Mesityl oxide, ac polarography
   and voltammetry of, 344
Meta-dinitrobenzene, convoluted
   and linear sweep voltammograms of, 217
Methylcarbamates, tensammetric
   polarographic waves of,
   367–374
Methylene blue, cyclic voltammograms of, 193

Microprocessors, 474–479
Migration current, 76–78, 91
Minicomputers, 90, 474, 476,
   481–491
Minima, 86, 112
Molybdenum, metal as a reference
   electrode, 65, 66
Molybdenum carbonyl complexes,
   oxidation of isomers of,
   37

Nernst equation, 14, 18, 23,
   94, 99, 101, 102
Nickel
 charge step polarography of,
   428
 square wave intermodular
   voltammogram of, 397
 square wave voltammogram of,
   397
Nitrilotriacetic acid, determination by ac polarography, 385
P-nitrophenol, reduction of, 42
Nonaqueous solvents, charging
   current in, 89
Non-faradaic processes, 71, 72
Non-Nernstian electrode processes (see Irreversible
   electrode processes)
Normal pulse polarography
 alternate drop, 278
 characterization of reversibility in, 254, 257
 charging current in, 238, 247,
   266
 current measurement in, 238
 dc effects in, 237, 238, 258,
   265, 266, 277
 determining species in different oxidation states
   by, 256, 257
 influence of adsorption in,
   245, 257
 influence of coupled chemical
   reactions in, 245, 247
 instrument types in, 238
 pseudo-derivative, 273
 and sensitivity relative to
   dc polarography, 244

[Normal pulse polarography]
  setting of initial potential
      in, 237, 243, 247
  theory of, 243-248
  waveform used in, 238, 239
Normal pulse stripping voltam-
      metry, 441, 453, 454,
      462, 463, 465
Normal pulse voltammetry, 267-269
  depletion effects in, 268
  at a dropping mercury elec-
      trode, 270
  importance of initial poten-
      tial in, 267, 268
  at rotated electrodes, 271, 272
  stirring effects in, 268, 269

Octyl alcohol, linear sweep
      voltammogram of, 205
Ohmic (iR) compensation, 482,
      487 (see also iR drop
      and resistance)
Organic reduction processes,
      influence of protonation
      reactions in, 36, 37
Oscillographic polarography, 170
      (see Linear sweep voltam-
      metry)
Overlapping waves, resolution
      of, 487, 490
Oxygen, determination in gasoline
      by ac polarography, 386

Peak polarography, 170 (see
      Linear sweep voltammetry)
Pharmaceuticals, use of ac polar-
      ography in nonaqueous
      solvents to determine,
      377-382
Phase-selective ac polarography
      (see AC polarography,
      phase-selective)
Phenazine-1-carboxylic acid,
      determination of, 145
Platinum, polarography of com-
      plexes of, 136, 137
Polarography
  analogy with absorbance spec-
      troscopy, 11

[Polarography]
  definition of, 6
  different time domains in, 15
  free energy considerations
      in, 34
  historical aspects of, 46
  influence of chemical reactions
      in, 28
  influence of different time
      scales in, 34
  kinetic aspects of, 3
  in nonaqueous solvents, 49, 62
  relationship of potentiometric
      titrations and, 30
  theoretical aspects of, 3
  use of buffers in, 30
Polyethylene, determination
      using square-wave polar-
      ography, 398
Polyethylene glycol, molecular
      weight determination
      by polarography, 367
Potassium chloride, dc charging
      current in, 161
Potassium nitrate, variation of
      dc limiting current with
      concentration of, 78
Potassium oxalate, charging
      current in, 132
Potentiometric stripping anal-
      ysis, 456-458
Potentiostats, 49-52, 46-60,
      439, 474, 479, 480
Programming of computerized
      instrumentation, 478, 479,
      482-485
Propylon, determination in flow-
      ing media, 230
Pseudo-derivative dc polarog-
      raphy (see Derivative
      dc polarography, cur-
      rent sampled)
Pseudo-derivative pulse polar-
      ography, 273
Pseudo-derivative stripping
      voltammetry, 462
Pulse polarography, 3, 4, 6
  analytical usefulness of, 282-
      284
  charging current in, 237, 238
  derivative, 6
  differential, 6

[Pulse polarography]
  and disproportionation, 39
  and migration current, 78
  miscellaneous techniques in,
      281, 282
  normal, 6, 18
  subtractive techniques in, 281,
      282
  and voltammetry (see under
      various methods)

Radio frequency polarography,
      399–406
  resistance effects in, 402, 403
Randles-Sevcik equation, 451
Rapid dc polarography, 122–137
  and adsorption, 124, 127, 128,
      132, 133
  advantages of, 124–128
  applications of, 137
  charging current in, 131–132
  calibration curves in, 128, 130,
      131
  dependence of limiting current
      on various parameters in,
      130–132
  derivative techniques in, 152
  determination of n in consec-
      utive electrode processes
      by, 135–137
  historical reasons for lack of
      use of, 125–127
  influence of kinetic processes
      in, 133
  limits of detection in, 130–132
  minimization of undesirable
      surface reactions in,
      124–127, 135
  recording of curves in, 129, 130
  reproducibility in, 130
  scan rate of potential in,
      122–124
  suppression of maxima in, 124
  theory of, 127–130
  used in stirred solutions of,
      138
Rate constants
  heterogeneous, 16, 18, 20
  homogeneous, 16, 20

Reference electrodes, 9, 47–50,
      62, 65, 66, 77
  commercially available, 62
  in nonaqueous solvents, 62
  positioning of, 48, 66, 67
  molybdenum, 65, 66
  saturated calomel, 62
  silver/silver chloride, 62
  and use of salt bridges, 62
Resistance (see also iR drop)
  effects in two-electrode
      polarographs, 47, 48
  minimization of effects of, 55
  uncompensated, 55, 56, 59, 60
  use of positive feedback cir-
      cuitry to minimize, 59,
      60
Resolution, 27
Reversible electrode processes,
      15, 16, 17
  practical definition of, 14,
      23
  thermodynamic definition of,
      14, 17, 23
Rhodium, catalytic hydrogen
      wave for reduction of,
      111

Sand's equation, 408
Scan reversal pulse polarography,
      255–258
Second harmonic ac polarography
      (see AC polarography,
      second harmonic)
Semidifferential (see Deconvo-
      lution)
Semi-integral (see Convolution)
Short controlled drop time
      polarography (see Rapid
      dc polarography)
Silver
  determination by stripping
      voltammetry, 435, 446
  determination of, 66
Sodium fluoride, charging cur-
      rent in, 74
Software (see Programming)
Solid electrodes, comparison of
      mercury electrodes and,
      176–178

Sphericity of electrodes, 437, 451

Spinning dropping mercury electrodes, 138, 139

Square-wave intermodulation polarography, 396-399

Square-wave polarography, 91, 391-399
  analytical use of, 395
  charging current in, 391-393
  theory of, 393
  waveform used in, 391, 392

Square-wave stripping voltammetry, 441

Staircase voltammetry, 209-212
  discrimination against charging current in, 210, 211
  theory of, 212
  waveform used in, 209

Staircase waveform in stripping voltammetry, 454

Standard rate constants, 22

Standard redox potential, 10, 22

Static mercury drop electrode, 165

Stationary electrodes in flowing solutions, 227-230

Stationary electrode polarography (see Linear sweep voltammetry)

Stationary electrode voltammetry (see Linear sweep voltammetry)

Stirred solutions
  continuous analysis in, 138
  limitations of conventional dc polarography in, 138
  use of short controlled drop time dme in, 138
  use of vibrating dme in, 138

Streaming mercury electrodes, 123, 124

Stripping voltammetry
  broadening of peaks in, 441
  and calibration procedures, 449, 455
  cell for, 70, 444
  and charging current, 435, 448, 454
  concentration step in, 436, 437
  comparison of methods in, 459-467

[Stripping voltammetry]
  controlled current methods in, 458
  definition of anodic, 436-438
  definition of cathodic, 438
  electrodes used in, 438-448, 455, 458, 460, 469
  and environmental investigations, 438, 466
  equilibration period in, 448
  examples of practical analysis by, 468-470
  and flameless atomic absorption spectrometry, 448
  in a flowing system, 448
  influence of surfacants in, 460
  interference in, 448
  intermetallic compound formation and, 441
  limits of detection in, 462, 464, 465
  oxide films and, 441
  practical aspects of, 438
  pre-electrolysis step in, 448, 449, 450
  resolution in, 463, 465, 467
  sensitivity of, 435, 439, 441, 464, 465
  and solution contamination, 459
  theory of, 448-454

Strobe polarography (see Current sampled dc polarography)

Subtractive polarography, 65, 157-160 (see also under various techniques)
  advantages of, 158
  disadvantages of, 157, 160
  use of computers in, 159, 160, 487
  utilization of twin dropping mercury electrodes in, 157-159

Sulfide
  current-time curves for reduction of, 135
  determination by cathodic stripping voltammetry, 438
  and formation of insoluble products at dme, 113
  oxidation of mercury in presence of, 112, 113

Supporting electrolytes, 76-78, 91

Tast polarography (see Current sampled dc polarography)
Tellurium
  dc polarography of, 197
  linear sweep voltammetry of, 197-199
Tensammetry, 75, 315, 363-374, 398
Thallium
  ac polarography of, 302, 304
  current-averaged polarograms of, 144
  determination by stripping voltammetry, 468
  differential pulse polarographic data for, 252
  radio frequency polarogram of, 404
Theo-bromine, determination by linear sweep voltammetry, 202
Third electrodes (see Auxiliary electrodes)
Tin, determination
  by ac polarography, 385, 386
  using subtractive linear sweep voltammetry, 224
Tin(II), reduction and oxidation of, 31-33, 307-309
Tin(IV), dc polarography of, 125, 131
Titanium(IV) oxalate, influence of hydroxylamine on reduction of, 40
Transfer coefficient, 21
Transference number, 77
Transition state theory, 21
Transition time in chronopotentiometry, 408
Tribenzylamine, influence on cadmium reduction, 126, 134
Tungsten, catalytic wave of, 40, 41, 109, 110

Uranium
  determination by high frequency ac polarography, 353, 354
  reduction of, 38, 39
    at a gold disc electrode, 228, 229
  short drop time polarography of, 134

Vibrating dropping mercury electrode (see Dropping mercury electrode, vibrating)
Voltammetry (see Linear sweep voltammetry)
  with forced convection, 226-230

Working electrodes, 47-52, 77
  positioning of, 48,66

Xanthate, oxidation of mercury in presence of, 101=104

Zinc
  ac polarography of, 309-312, 354
  charge step polarography of, 427, 428
  dc polarography of, 309, 310
  determination
    with charging current compensation, 165
    by differential pulse stripping voltammetry, 444
    by stripping analysis, 456, 457
  linear sweep voltammogram of, 204
  normal pulse polarograms of, 246
  quasi-reversible reduction of, 100
  radio frequency polarogram of, 405
  radio frequency voltammograms of, 401
  square-wave intermodular voltammogram of, 397
  square-wave voltammogram of, 397
Zinc concentrate, determination of copper, lead, cadmium, and zinc in, 354